Fine Lines

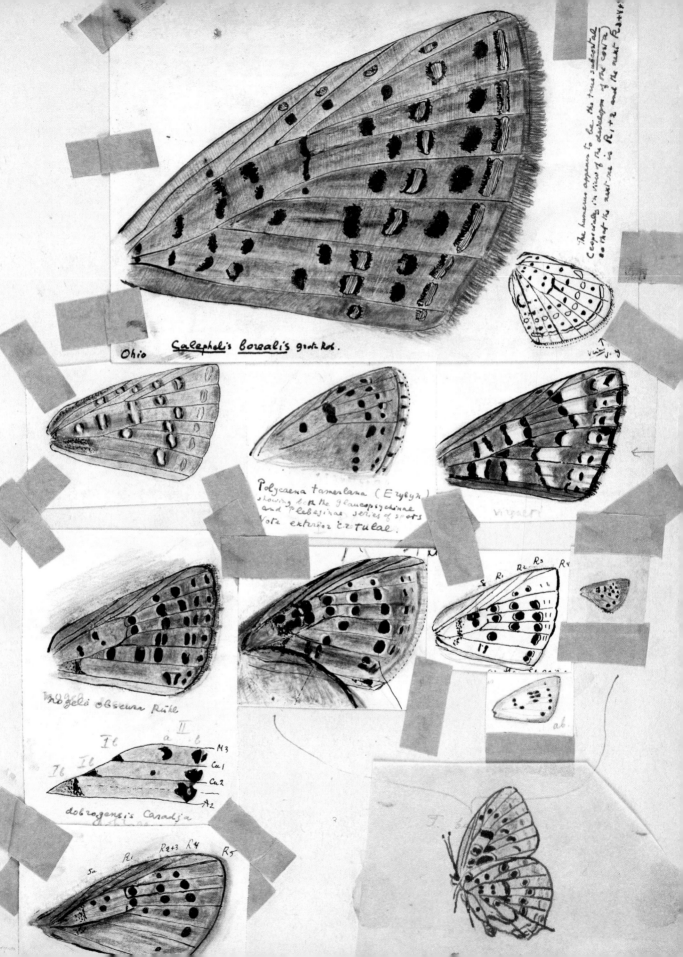

Ohio *Calephelis Borealis* grote-rob.

Polycaena tamerlana (Erybgh.)
showing both the glaucopsychinae
and plebesinae series of spots
Note exterior c. tulal.

virgaeri

mogeli obscura Rühl

dobrogensis Caradja

ab.

Fine Lines

Vladimir Nabokov's Scientific Art

Edited by Stephen H. Blackwell
& Kurt Johnson

Yale UNIVERSITY PRESS

New Haven and London

Published with assistance from the foundation established in memory of Philip Hamilton McMillan of the Class of 1894, Yale College.

Yale University Press books may be purchased in quantity for educational, business, or promotional use. For information, please e-mail sales.press@yale.edu (US office) or sales@yaleup.co.uk (UK office).

Designed and typeset by Chris Crochetière, BW&A Books, Inc.
Printed in China.
Frontispiece: Color Plate 21 (detail)

Library of Congress Control Number: 2015951601
ISBN 978-0-300-19455-5

A catalogue record for this book is available from the British Library.

This paper meets the requirements of ANSI/NISO z39.48–1992 (Permanence of Paper).

10 9 8 7 6 5 4 3 2 1

*To Dieter E. Zimmer, an inspiration to all
who chase after Nabokov's receding footsteps*

Minister: *No, you always draw such complicated things.*
And look, you've even added shading. It's revolting.

—Vladimir Nabokov, *The Waltz Invention*, 1938

Contents

Preface

This book represents a landmark moment in the understanding of Vladimir Nabokov as a scientist. After three decades of neglect and two decades of scientific confirmation and revision, it is finally possible to present a comprehensive review of his achievement. The collection of drawings gathered here also offers a rich window into his visual engagement with the world of microscopic structures found inside butterflies. Nabokov's visual passion resulted in well over a thousand drawings, one hundred and forty-eight of which are included here. These drawings, and the way Nabokov used them in his research, give important insight into how he perceived the world and attempted to know it, and these insights, in turn, shed light on his artistic perception and creativity.

Never before has any appraisal of Nabokov's research been based on a thorough review of his laboratory notes, which, combined with his published scientific papers and the recent DNA-based confirmations of two of his most controversial hypotheses, allow new clarity in understanding the significance of his work and locating it in the contexts of evolutionary biology and systematics. Jotted notes accompanying many drawings reveal details of his discovery process, showing how he sought to understand the evolutionary diversity of Blue butterflies by means of morphology. The notes clarify Nabokov's uncanny sense of the interplay of temporal, spatial, and developmental factors underlying the major evolutionary and taxonomic questions of his day. Viewed in extended sequence across many cards, the drawings allow an understanding of how he sought to trace the temporal—indeed the evolutionary—history of the species he studied. Aesthetically, too, they present an unusual temporal picture—one that brings to mind Nabokov's habit of writing his novels on similar note cards, working in nonlinear fashion, and thinking of novelistic time as a painted canvas or foldable magic carpet.

After an Introduction that provides a full historical and theoretical context for Nabokov's evolutionary work and its relation to his art, the captions to the drawings tell a fascinating story of how Nabokov conducted his research, how he chose his comparison groups, and how he identified new distinguishing features of butterfly genitalia. We place the drawings together, in two groups (ninety-two black-and-white, fifty-six in color) between the Introduction and the essays, so that the reader can view them all in close contiguity to one another, much as Nabokov might have reviewed large groups of drawings as he worked to discover relationships between butterfly species and construct hypotheses for their origins. Extended contemplation of the drawings opens the opportunity for a cumulative aesthetic impact, as the drawings' composition and strange contents work their way into the viewer's perceptual framework. We intend this organization to facilitate the research of historians of science as well as that of those who undertake to study the drawings as aesthetic artifacts.

The ten essays that follow the images, written by leading Nabokov specialists and by scientists studying the same groups of butterflies he did, are intended as a first collective effort to understand the significance of science in Nabokov's life and art. Rather than exhausting the topic of Nabokov as a scientist who also practiced art, or as an artist who happened to pursue scientific knowledge, these essays demonstrate the rich soil for further sustained attention and research. We hope that future writers will find inspiration in these essays.

Acknowledgments

We wish to thank all the participants for their generous contributions of time and critical acumen to this volume: more than many such collections, this one has involved a process of constant communication and feedback among the participants, beginning with an online symposium in November 2012. Almost every part of this book has been crafted in a collaborative spirit (not something Nabokov would have necessarily enjoyed!). Among the authors in this volume, we wish to single out Victoria N. Alexander for her extra attention to all the texts and her help resolving many early complexities, and Brian Boyd for his constant support and encouragement, as well as his contribution of much more time and material than he thought he would be able to provide. The Nabokov Estate and the Andrew Wylie Agency have been extremely helpful and supportive at every stage of this project. We also want to express our deepest gratitude to Dubi Benyamini and Zsolt Bálint, the first for his photographs of *Pseudolucia* butterflies, and both for their groundbreaking work that made this book possible, as well as for their help, patience, and indulgence throughout this project. We also thank Dieter Zimmer and Steve Coates for their pioneering work in the field of Nabokov's science.

Generous financial support for this project was provided by the University of Tennessee, Knoxville's Department of Modern Foreign Languages and Literatures; its Exhibition, Performance, and Publication Expense fund; and its Tennessee Humanities Center. Alan Rutenberg of the Office of Research helped us refine our proposals. We also thank the McGuire Center for Lepidoptera and Biodiversity, University of Florida at Gainesville, which houses Kurt Johnson's Vladimir Nabokov Lepidoptera Archives.

The staff of the New York Public Library, including Isaac Gewirtz and his colleagues at the Berg Collection, and Thomas Lisanti and his team at the Office of Permissions, were extraordinarily helpful and efficient. To say that we could not have done this book without them is a platitude that so vastly understates reality that one blushes to include it. Yet there it is. Pam Hughes and the office staff of the Department of Modern Foreign Languages and Literatures at the University of Tennessee were always gracious and helpful, especially when we needed it most.

Sarah Funke Butler of Glenn Horowitz Bookseller was most obliging, securing generous permission to use their reproductions of Nabokov's inscription drawings. Doug Canfield of the University of Tennessee Department of Modern Foreign Languages and Literatures assisted with technical features of the online symposium held November 5–9, 2012. We are also grateful to friends and loved ones who watched, tolerated, and encouraged us, as well as made suggestions about image choices and the book's title.

Jean Thomson Black and the staff at Yale University Press have been extremely helpful in their constant enthusiasm, patience, and flexibility as our project wended its way toward completion. We also thank our anonymous readers for their constructive and challenging observations on our proposal and the essay manuscripts.

This project was begun with the kind approval of Dmitri Nabokov, who, sadly, passed away too soon to see its completion. We certainly intend the book as something that would have delighted him, and his father, and these pages embody our gratitude.

Anatomical Diagrams

Structures of the genital anatomy most frequently illustrated by Nabokov, assembled from drawings on his laboratory cards.

Male: *aedeagus,* the penis; *dorsal terminal elements* (*uncus,* a dorso-terminal lobate element; *falx/humerulus,* a contiguous but bent element, *humerulus* at base, *falx* bent at *elbow* to make distinct *forearm*); *furca,* a U- or V-shaped structure supporting the penis; *genital ring* (or vinculum), a ringlike structure holding the other genital parts; *sagum,* a variously sized enfolded structure prominently surrounding the aedeagus in some Blues; the *valve* (plural *valvae*) (=male clasper[s]), the male grasping organ, often with very specific terminal structures.

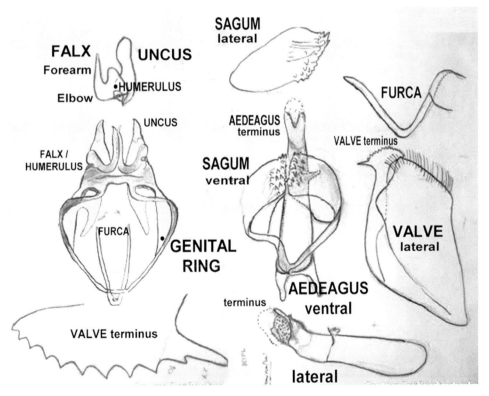

Diagram 1: Structures of Male Genitalia

Female: *lamella* (plural *lamellae*): terminal liplike elements surrounding the female genital opening; *corpus bursa* and *ductus bursa*, contiguous caudal and cephalic elements of the female genital duct; *papillae anales*, paired terminal flaplike structures at the female genital terminus.

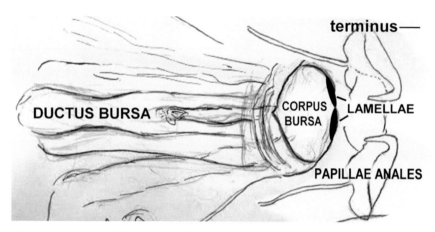

Diagram 2: Structures of Female Genitalia

Fine Lines

Introduction

Stephen H. Blackwell and Kurt Johnson

In a 1959 interview for *Sports Illustrated*, Vladimir Nabokov said to Robert H. Boyle, "I cannot separate the aesthetic pleasure of seeing a butterfly and the scientific pleasure of knowing what it is."[1] Nabokov linked the driving force behind his two passions, artistic literature and lepidoptery, in many ways. One especially apt method manifests itself in his chosen research and writing medium: four-by-six-inch index cards. Nabokov saw his scientific work as filling in gaps in knowledge about specific organisms, their interrelationships, and their evolutionary course. In his artistic work, analogously recorded on note cards, he would "just fill in the gaps of the picture, of this jigsaw puzzle which is quite clear in my mind, picking out a piece here and a piece there." Later on, he would qualify the phrase "quite clear" with the more specific "utmost truthfulness and perception" of his effort to copy on cards the "conceived picture . . . as faithfully as physically possible—a few vacant lots always remain, alas."[2] In interviews, lectures, essays, and letters, Nabokov expresses his sense that artistic creation is a matter of careful perception and also one of necessarily incomplete understanding—just as science is also always incomplete. Knowledge and artworks are continually in progress—and even the artist's perception of inspiration and the created artwork's significance are necessarily gappy. As he wrote to the scholar Carl Proffer on this very topic: "Many of the delightful combinations and clues [in Proffer's *Keys to "Lolita"*], though quite acceptable, never entered my head or are the result of an author's intuition and inspiration, not calculation and craft. Otherwise why bother at all—in your case as well as mine."[3] Just as the human quest for knowledge, of even the smallest corner of the natural world, is "infinite . . . unquenchable, unattainable," so also the artist—and the reader or viewer of art—faces an infinite task of expanding depth.[4] This affinity between Nabokov's views of artistic and scientific knowledge should give pause to those who feel that he intended, desired, or ever believed it possible to control every facet of his works' form, meaning, and interpretive history. His views also suggest that—to the extent that he loved nature, its complexity and its elusiveness—he would wish his novels to participate in and stimulate similarly unquenchable adventures.[5]

This book's hybridity springs from two impulses: an intangible sense, on the one hand, that everyone should have the opportunity to see these strange, often beautiful drawings; and a recognition, on the other, that Nabokov ended up making a highly significant contribution to the sciences of evolutionary biology and biogeography—one that these fields were slow to recognize and appreciate. He did much more than name a few species and genera, and the time is ripe to assess his complete scientific legacy. To that end, this volume presents ten essays by scholars: scientists who have continued or built on Nabokov's lepidoptery and humanists who probe links between his scientific work and his artistic explorations.

This volume offers a rare glimpse into the workshop of an explorer of inner space. For the Nabokov enthusiast, these 148 illustrations offer a chance to accompany him as he journeyed in the microscopic world of butterfly anatomy, seeking out structures and patterns no one else had ever seen. To the historian of science, the drawings represent empirical evidence of Nabokov's research method, an approach that allowed him to erect ten genera and name eleven species and subspecies.[6] Though some viewed him as an overzealous taxonomic splitter—someone who sliced up groups into unnecessarily fine subdivisions—time and subsequent research have shown that Nabokov's proposals were predominantly accurate and were, in fact, ahead of their peers in method and scope. These drawings and the accompanying essays should help historians better delineate the precise benefits of morphological research at a time when it was vying with other ascendant methodologies. To all, they offer the vicarious thrill of watching an explorer at work, studying minute distinctions in the natural world with extraordinary care—and discovering form, function, variety, and beauty.

A handful of these drawings appeared in the compendium edited by Brian Boyd and Robert Michael Pyle, *Nabokov's Butterflies*, but due to the necessarily small selection, reductions in size, and the limited number of high-resolution plates, that printing failed to capture the depth of Nabokov's visual passion for his object of study. Gennady Barabtarlo has written that Nabokov exhibits "the 'eye-thirsty' love for the created world, in all its micro- and macro-forms, for things small and large, unnoticed or unworded before and thus begging to be brought to life by precise and fresh description."[7] In his published works and comments, Nabokov never specifically endorsed a "creator" other than "Nature"; he told his sister that such work "enraptured" him and was "so enticing that [he] cannot describe it."[8] Whether love, obsession, or passion, the desire to see the never-before-seen drove Nabokov in his effort to comprehend as much of the world around him as his physical body would allow.

If we imagine these drawings, not just as artifacts, but as the result of hours of effort—if we envision the careful preparation of the organs, the time-consuming process of fixing, projecting, tracing, and refining each image—perhaps we get some sense of Nabokov's deep connection to each natural form. Subsequently, having considered each image carefully, we can think of them all in series and contemplate the similarities, idiosyncrasies, and subtle distinctions that hint at the flux of natural evolution. This alternation between frozen moments of time in evolution—"stopping places"—and an almost cinematographic contemplation of time's shadow, manifest in synchronic forms—rapid comparisons of many geographical variations—marks the distinctive component of Nabokov's scientific methodology.

NABOKOV ONCE SAID THAT HE WAS "born a painter."[9] As a boy, he took drawing lessons from Mstislav Dobuzhinsky, the renowned artist (to whom he wrote with a boy's pride at the age of forty-five of the five hundred drawings of butterflies he had made).[10] Although the drawings collected here were mostly produced with the aid of a camera lucida (a projection device attached to a microscope), when we look at them we sense a rich exploration of compositional form in nature. In many, the precision, shading, ciliation, and asymmetry appear to serve aesthetic ends in addition to real and anticipated scientific ones. The careful shading and the representation of cilia embody a heightened and extended attention to these structures: it is as though Nabokov wished to stare at these objects for as long as possible, to extract from them every iota of information and knowledge and aesthetic pleasure that his mind and visual acuity could discover. He did this for up to fourteen hours a day, even to the point of severe eye strain and the possibility of damage to his vision.[11]

Although early on, from childhood into early middle age, Nabokov was keen to discover an unnamed butterfly—to be a "first describer" and have his name attached to a species name—he adopted a new perspective once he became engrossed in the study of microscopic butterfly organs and their morphology in the laboratory. In this world of invisible structures—things that "no one before [him] has seen"—he found a terra incognita to explore, through the microscope, just as his real-life heroes, explorer-naturalists Grigory Grumm-Grzhimaylo and Nikolay Przhevalsky, had marched off into Eurasia to describe previously unknown parts of the natural world—"named the nameless . . . at every step," as Fyodor puts it in *The Gift*. In the genitalia of butterflies, Nabokov discovered an inner space and forms of strange symmetry and asymmetry: beautiful hooks and hoods, forearms, pugilists, and clasps and combs, spurs and brushes and elbows, even hints of Klansmen (Figures 1 and 20; Color Plate 35) and tiny caterpillars (Figure 14) or elephants (Figure 69)! He valued careful depiction, greatly admiring Nikolai Kuznetsov's study of butterfly morphology, which he called "unsurpassed."[12]

We follow Nabokov into this visual world both to experience his sense of wonder and enthrallment before the amazing devices he found and to grasp, tangibly, how extensively and diligently he labored, what extremes of careful delineation and distinction he recorded between forms that might, to most other observers, appear essentially the same. All this he did in the hope "by luck [to] hit upon some scrap of knowledge . . . that has not yet become common knowledge," scraps that lead to "inestimable happiness."[13] The drawings probe nature's extraordinary engineering as well as the aesthetic qualities inherent in such perfection. In those with extra shading and colors and three-dimensional accuracy, he vividly, irrepressibly, performs the kind of loving attention to these structures that he held as the ideal form of knowledge quest. "Lovingly finger the links of the many chains that connect [a] subject to the past and the future," he exhorted his students at Wellesley College, after a few years of his most productive research.[14] These chains are characteristic, for Nabokov always wanted to understand—even to experience directly—the connections between objects in the present and their journey from the past to today; his attitude toward the tangibility of the future appears to have moderated over the decades, if his novel *Ada* is any guide, though he certainly remained interested in the possibility that future events are not quite as indeterminate as they seem. But nature and art will, let's hope, continue far into the future, and following the links of the chain are a surety of that—even if, as he wrote to his friend Edmund Wilson, his own scientific contribution might be eclipsed in just twenty-five years.[15]

Just as Nabokov's artistic writings inspire with their precision and originality of expression, their sensuous beauty, and their perceptual intricacy, so his scientific studies surprise and energize with their discovery of minute detail, their attention to microscopic difference, and their profoundly sustained engagement with tiny corners of the natural world. Through this perseverance, the studies also uncover and explore new forms of natural beauty, and these Nabokov analyzed obsessively through his six-plus years at the dissecting bench.

Not everyone will find a direct connection to Nabokov's scientific passion by reading his research articles. Nearly everyone, however, is drawn to the beauty of butterflies, and Nabokov's own engagement with that beauty is especially visible in his comparative drawings of butterfly wings. We can follow his interest inward, zooming from the naked eye to the microscopic, as he examines wing patterns at finer and finer grain, ending with meticulous analysis of exact locations of tiny individual wing scales, each about one twentieth of a millimeter in width: pixels of color on the wing's broad palette (about 250,000 dpi resolution). This level of magnification then shifts to the insect interior and Nabokov's chosen evolutionary diagnostic tool: the male genitalia. These he figured in the hundreds, maybe even thousands. Their general morphology was often critical to establish genera, and the ratio of their component parts' measurements could distinguish entities at the species and subspecies level. Through this examination, Nabokov developed a keen ability to imagine the evolution of many of the Blues, North American and South American—to picture the flow of morphology in time and visualize evolution as if it were a film.

On seeing the drawings represented in this volume, one anonymous reviewer wondered: Should we be troubled that Nabokov devoted so much energy to the depiction of butterfly sexual organs? As late as his story "Father's Butterflies" (1939 or later), Nabokov created a fictitious scientist who mocked the "genitalic" approach to butterfly taxonomy. Yet at most two years later, he himself became a disciple of the practice, beginning in 1941 under the tutelage of William P. Comstock at the American Museum of Natural History in New York. In his renowned 1945 article "Notes on Neotropical *Plebejinae*," he expressed his view that, for species with highly changeable and even unpredictable wing patterns—not to mention those practicing mimicry—the genitalia were likely to evolve more slowly and offered the most reliable clues to long-standing taxonomic relationships.[16] Is it coincidental that Nabokov penned his first treatment of pedophilia at around the same time ("The Enchanter" [*Volshebnik*], 1939 [published posthumously in 1986])? Human sexuality is central to at least three of his American works of fiction, *Lolita*, *Pale Fire*, and *Ada* (and in the unfinished *The Original of Laura*), and a few of his Russian ones (*King, Queen, Knave*, *Laughter in the Dark*, and "The Enchanter"; one could add, controversially, *Despair*, especially in its revised translation). There is scant evidence to support the idea that Nabokov was obsessively interested in sexuality, deviant or otherwise; more consistent with his life, career, and works is the idea that he was intrigued by the important role of sex in human lives—as in any sexually reproducing species—and the dominant place of sex in modern (Western) commercial culture. Even without considering popular Freudianism and its broad influence, how could anyone ignore the world of Western sexual culture in the twentieth century and consider themselves serious observers of life and humanity? It may be, in fact, that Nabokov's profound study of butterfly evolution through genitalic mutations heightened his already acute artistic interest in the complex, troubling, often disruptive place of sex in human affairs. If sex leads evolution (at least among sexually reproducing entities), what does human sexuality say about the future of the species? These would be pressing questions for the artist-scientist with a passion to know and understand the chains linking the past to the present and the future.

The drawings themselves vividly demonstrate the connection between aesthetic "seeing" and taxonomic "knowing"—some with almost startling clarity. Although a few of the images appear essentially utilitarian and some are not so carefully drawn, an aesthetic principle dominates in most. We see his attention to detail and also to form and perspective, with its attendant variations on the perceived asymmetries. The often entrancing shapes have formal beauty in themselves, but many of the drawings accentuate that beauty in a way that seems unlikely to have been quite so clear through the microscope (for example, Figures 26 and 66 and Color Plate 19). They represent not just seeing things that have never been seen before but also seeing them *in a way* most things have never been seen. Nabokov appears to suggest (as in his comment about "inseparability," above) that viewing with an eye toward both structural exactitude and beauty can lead to a more complete form of knowledge.

In the laboratory, Nabokov studied natural phenomena and their historical perception by others, recording this information onto his note cards. He magnified things; he drew them; he broke them into component parts and measured them. As an artist, he studied the details of worlds that came into being through a mix of observation, mental effort, and inspiration, added to these details the study of relevant documentary facts and philosophical opinions. Beginning after his time as a professional scientist, he recorded all of this artistic material on numbered cards. This coincidence of the material of recording indicates an important link, a commonality, between the two kinds of information being recorded. Such a leap will provoke objections: index cards alone do not produce the unification of science and art. But Nabokov's belief in the specific reality of his artistic worlds emerges in his explanation of the use of obscure vocabulary: "The main favor I ask of the serious critic is sufficient perceptiveness to understand that whatever term or trope I use, my purpose is not to be facetiously flashy or grotesquely obscure but to express what I feel and think with the utmost truthfulness and perception."[17] The very fact that he sees creativity as a matter of truth and of perception immediately suggests parallels with scientific work. At the very least, the formulation indicates a scientific attitude toward what one may "feel and think." The idea of "utmost perception" reminds us that perception exists on a sliding scale: it can be performed casually, even passively, and it can be performed to the utmost, actively, to the extreme limits of human mental and physical capacity. As Nabokov acknowledged in his "lily" quotation, even "utmost perception" in science is a paradoxical, tentative perception, because "reality is an infinite succession of steps, levels of perception, false bottoms, and hence unquenchable, unattainable."[18] Not surprisingly, the things of the mind—thoughts and feelings—are no different.

Considering the close bond Nabokov saw between art and science, it seems inevitable that his index card method would transfer laterally from his scientific research into his artistic process. This major shift in work habits (at age forty-nine!) offered the ability to transcend linearity, to change and manipulate chronology, to rearrange in order to highlight connections between small, widely spaced sections. It also gives the manuscript the feeling of a large rectangular block, yet one that evokes the systematized knowledge of the scientific card file.

Ultimately, these drawings demonstrate attention to detail as a vital part of existence in the world—and attention is the less noted but equally vital part of the formula. Just as in nature, Nabokov insisted in art on perceiving "such combinations of details as yield the sensual spark without which a book is dead."[19] These combinations abound in his work: think of the swaying stalk of grass on page 344 of *The Gift* that connects to another from page 133 and last memories of Fyodor's father, and to Tolstoy; or the thirty-one tree species named in *Pnin* and the surprising patterns they seem to form; or Brian Boyd's discovery that references to typing corrections in *Ada* encode a message from the beyond; or Dieter Zimmer's revelation that *Pale Fire*'s Queen Disa refers not only to a genus of orchids *and* to one of butterflies but *also* to a tragic Queen Disa of Scandinavian mythology; or Alexander Dolinin's discovery that the jars of jelly in "Signs and Symbols" form a particular pattern that correlates with the letters on a rotary telephone's numbers.[20] Or, of course, we can also simply concentrate on the beauty and aesthetic force of these minute pieces of natural engineering, just as we can choose to pause and revel in the sensuous flow of Nabokov's language, the precision of his descriptions, and the intricacy of the compositional structures to be found in his art.

NABOKOV WAS OBSESSED WITH TIME. He entered into adulthood simultaneously with the full fame of Einstein's theory of general relativity, an era saturated by modernist interest in time through the works of Marcel Proust and Henri Bergson, and his own fascination can only have burgeoned in response. Thrilled by the sensuous feeling of elapsed centuries in the objects around him, he wandered the streets of Cambridge like a time traveler.[21] The living natural world and the artistic world are both particularly ripe for temporal exploration: the first through the lens of evolution, the second through a similar lens that follows the development (Nabokov even called it evolution) of artistic expression over human history. Looking at his surroundings, he wanted to see into the temporal dimension: into the past of such objects as trees, buildings, streets, and pencils. Working with thousands of butterfly specimens in the 1940s, he turned the same attitude to their diversity and its evolution. To figure out what a creature is, one often needs to have a good idea of what its ancestors were, and where, in the distant past.

Although as a child and even as an adult amateur Nabokov dreamed of discovering and naming a new species, once he took up professional work he quickly accomplished that goal and recognized richer, more significant tasks that a zoologist could undertake. When Nabokov looked at a group of species and studied their geographic diversity, he saw closely related yet significantly varying details in their anatomy—patterns of similarity and difference, presence and absence. The more he worked, the less he focused on species as fixed entities: he came to see them as a "relative category" within generic variation, with "peaks" and "valleys" of morphological structure allowing more or less confident classification. He was also interested in the related species or genera as conveying the appearance (illusion) of consecutive development, imagining spatial variety as if it were temporal development (see Figures 80–83). To this end, he compared anatomical features presumed to be of more recent origin with those considered archaic or primitive (for example, in "The Nearctic Members of the Genus *Lycaeides* Hübner").[22] Modern species sporting what appear to be primitive structures served proxy for creatures from millions of years ago, as Nabokov imagined morphological transitions and elaborations through time. That is, he tackled the diachronic problem from two perspectives: through variations among modern, nonprimitive species and along an imagined continuum extending from the appearance of forms in the deep evolutionary past right up to the present. The first approach creates imaginary analogies of evolution, while the second emphasizes spontaneous appearance of characters (mutations) and their gradual elaboration and diversification.

Speaking to his students about the practice of good reading, Nabokov emphasized the kind of deep familiarity that comes from repetition and review. Repetition allows the imaginative departure from strict chronological time. "Rereading" in his lectures refers to an activity that builds up layers of reliable and useful memory—perhaps it is knowledge—of an increasing body of facts and details.[23] The benefit of this approach is that it allows one to consider the linear work along a new, atemporal axis—or even freed from any axis whatsoever. If first-time reading ties the reader to the text's unidirectional flow and whatever sequence of events and description it provides, the mental artifact produced by multiple rereadings creates what Nabokov might have called a hypertext if the word had existed, but which instead he compared to the perception of a painting: one can view a painting in many ways, attend to its details in a variety of sequences, and move from part to whole and back again in an instant. A text can be converted into a basis for that kind of mental experience by mapping it into the brain through memory, where it can be manipulated at will, its parts explored in large or small magnification, and its patterns and layers perceived more readily. Nabokov's analogy imagines the initial temporal experience of narrative as a one-dimensional line and explicitly transforms it into something two-dimensional, although the addition of still further dimensions is also implicit in the analogy. Some of these extra dimensions appear through hidden allusions to literary, musical, or visual artworks or through the discovery of yet more layers of encoded pattern that extend beyond the traditional locations of artistic meaning—such as anagrammatic typos, date-keeping mistakes, "polygenetic" allusions (coined by Pekka Tammi) and "doubled italics" (James Ramey).[24] The use of index cards for writing novels demonstrates the creative side of the same concept: the cards and their nonsequential creation represent the author's own liberation from textual (and implicitly temporal) linearity.

Nabokov's drawings invite the same kind of multiple experience: the more we look at them and revisit them, the more their details coalesce into meaningful shapes and patterns, sometimes with comical results. A nonscientist who invests the time can imagine the researcher studying these drawings in multiple series, determining which features are most important, which seem primitive and which more advanced, and how one structure might have diversified into several variations or, possibly, evolved through several iterations. Nabokov was looking at groups of specimens to determine how many genera and species they represented—the classificatory question. But he felt that to do this well, he also needed to imagine them as evolved and evolving forms, as mobile features in time. As a spur to this imaginative attention, he composed fanciful phylogenetic trees from groups of modern species, noting that these are "but the shadow of a [phylogenetic] tree on a plane surface."[25] This is a characteristic move: Nabokov pushes a common heuristic tool or empirical concept out of its standard cognitive mode. This "tree," traditionally a two-dimensional diagram, printed on any flat surface, is escorted back across the threshold from two dimensions to three (the world of real trees), in which it can cast a shadow that has a definite, but not necessarily obvious or predictable, relationship to the original. The transition also crosses a temporal boundary, reimagining a synchronic series as if it were a diachronic one. In the absence of a complete fossil record, his ersatz tree serves as a useful surrogate for looking at real diachronic series. Taking that shadow-tree seriously may or may not lead to accurate conclusions about evolution, but it does, like Nabokov's description of re-reading, create circumstances for discovering previously neglected perspectives on the organisms under study. This troubled, unsure relationship between shadows and objects, between the things in the world (including artworks) and their shadow in consciousness, is analogous to the difficulty of surmising the evolutionary history of taxonomic groups—the problem that fascinated Nabokov in his preparation of "Notes on Neotropical *Plebejinae*" and other studies.

These creative heuristic tools also allowed Nabokov to imagine following a pathway of evolution as one might watch a movie and to run the imaginary film forward or backward in order to explore appearances and disappearances in both positive and negative directions:

> A modern taxonomist straddling a Wellsian time machine with the purpose of exploring the Cenozoic era in a "downward" direction would reach a point—presumably in the early Miocene—where he still might find Asiatic butterflies classifiable on modern structural grounds as Lycaenids, but would not be able to discover among them anything definitely referable to the structural group he now diagnoses as *Plebejinae*. On his return journey, however, he would notice at some point a confuse adumbration, then a tentative "fade-in" of familiar shapes (among other, gradually vanishing ones) and at last would find *Chilades*-like and *Aricia*-like structures in the Palearctic region.[26]

This echo of time-lapse photography (moving here from the two- to the three-dimensional but implicitly also to four-dimensional space-time) offers the possibility of an entirely unfamiliar sense of *narrative* and *character:* the protagonists are shapes (forms) and diversity, the themes are ecological change, migration, population growth or decay, and mutation (leading to biological success or failure). No wonder Nabokov told Wilson that one of his research papers read like a *roman d'aventures*.[27]

There is also something to be considered in the care and detail invested in Nabokov's scientific drawings. The drawing process itself—whether by camera lucida or freehand—demonstrably translates the structural forms into an aesthetic realm. The drawings show scientific precision, to be sure, but they also show a level of detail, refinement, and compositional care—especially the ones that are colored—that appears to go beyond the scientific utility of shape-recording and comparison, although Nabokov must have felt that anything he could notice and record might become scientifically useful one day. Especially in those drawings that clearly include more visual information than was the subject of Nabokov's research, one wonders: Did the extra contemplation, the surplus attention to shape, color, and shadow, and the time spent reproducing them, deepen his ability to think about their interrelations, their change and purpose across time and genera? Inevitably, such intensive looking must have enabled him to see new synthetic features—like "pugilistic Klansmen" and platonic triangles.[28] It is also worth exploring whether, and how, such contemplation might have added dimensions to the way he conceptualized and undertook artistic creation as a writer. At any rate, as he perhaps jokingly considered artists to be "God's little plagiarists," as he wrote to his mother, he must have thought about the ways that strange structures in nature might find analogs in strange variations in artistic form.[29] If insect genitalia can look like pugilists "of the old school"—a very funny analogy, when one thinks of the "struggle for existence"—surely a novel can mimic any biological apparatus one might choose—even, as James Ramey shows, the parasitic maggot of a bot-fly.[30]

WHEN NABOKOV LEFT the American Museum of Natural History to take a position as research fellow and de facto curator of Lepidoptera at the Museum of Comparative Zoology at Harvard, major questions were being debated regarding taxonomic methods and evolutionary theory. Nabokov's notes indicate that he was reading Bernhard Rensch's theories on the coloration of birds (1925), Theodosius Dobzhansky's *Genetics and the Origin of Species* (1937), Julian Huxley's 1938 essays on evolutionary biology and his book *Evolution: The Modern Synthesis* (1942), and Ernst Mayr's *Systematics and the Origin of Species* (1942). Nabokov's musings regarding the biology, taxonomy, and evolution of Blues were particularly linked to the writings of Rensch and Mayr, both ornithologists. In his notes, Nabokov often compares the views of ornithologists and entomologists, especially regarding matters of close species resemblance and mimicry. Blue butterflies, like many birds, showed confusing and contradictory patterns of external resemblance, geographic distributions, and differential interbreeding, the study of which Rensch had pioneered in his work on the "Rassenkreis," or circle of races phenomenon. This lack of clarity in the relationships of Blues was precisely what had left many of them widely unstudied across most of the world's continents.

For the insect taxonomist in the laboratory tasked with identifying, naming, and creating classification schemes for his subjects—in Nabokov's case, butterflies—the obvious questions were about species and speciation. What is a species, and how do species become distinct through time? But more than this, how does a working scientist gather and decipher information on his subject creatures—both in the laboratory and in nature—so that there is a reasonable chance of recognizing the path of lineages that had actually occurred through historical time? What made this a difficult matter for the incipient systematists and taxonomists of Nabokov's day was that quite different kinds of phenomena and data were informing the process, and all at once. From before the time of Darwin's work, and sealed within Darwin's concept of natural selection, everyone had agreed that some kind of interaction between environment and essence informed what was happening in nature: organisms had basic internal fingerprints—what had loosely been called "genes" for decades; then nature, by the processes of competition and survival of the fittest, selected which organisms would survive and pass on the information of their germ plasm to succeeding generations.

Varying perspectives on such questions brought many variables into the discussion. On the one hand, data and understanding surrounding genetics were constantly expanding. The melding of the emergent Mendelian view of genetics with Darwin's view of evolution by means of natural selection was creating the context for what would be a new or neo-Darwinian synthesis. It was defined variously then, but driving the synthesis was the marriage of genetic study (microevolution) with the study of wider patterns in nature (macroevolution). The scientists Nabokov was reading were exploring many pivotal questions. Further, Nabokov himself, solidly confident of his own capabilities at such a transformative moment, was contributing his views, including occasional new terms and definitions.

The questions vexing evolutionists all turn up in Nabokov's notes. It was generally agreed that a species was an entity that could not interbreed with a different species—but what might be the reasons for the incompatibility? Was it just in the germ plasm (the genes), or might there be other factors in the natural environment, too, such as factors creating isolation—geography, subtle ecological conditions, or behavior of the organisms? Entomologists certainly knew there was a rub here. Even in butterflies, entities never known to interbreed in nature sometimes interbred voluntarily in the laboratory. Further, if you hand-paired butterflies, you could get interbreeding which itself would never happen naturally even in the lab. Still more unfortunate for entomologists of the time, the situation and conditions in nature were sometimes at least as enigmatic as in the laboratory. Lepidopterists had known for decades that across a wide distribution of what was generally agreed to be a species, if you looked at the scattered populations of some butterflies—including the Blues—as if they were strings of beads, some of the populations making up those beads interbred with others and some didn't, and even some populations that interbred in one place didn't interbreed elsewhere.

Known in Europe for decades before Nabokov's departure for the United States, this enigma was known as the *Rassenkreis,* or circle of races dilemma. It created an even bigger question for insect taxonomists. If the matter of interbreeding was so plastic, reflecting so many contradictions, what was a good definition of species in the first place? Was the interbreeding criterion (the biological definition of species) enough? Or, if that data became a muddle, was the taxonomist doomed to create morphological—anatomical and wing-pattern-based—definitions for species? The taxonomist in a museum or university might not actually know what was interbreeding with what in nature, and so the question arose: Weren't morphological definitions of species perhaps better simply because they could be more universally consistent?

Unfortunately, there were just as many fallibilities in attempts to find a consistent structural definition for species. What characteristics would one use—external? internal? . . . both? Everyone who knew groups like the Blues—often neglected simply because they were so perplexing—realized that a structural definition of a species by external characteristics could result in a hodge-podge of entities sharing a general resemblance, but often simultaneously including entities that most lepidopterists readily knew didn't interbreed in nature. Where to draw the line? Relying on internal characteristics might create more fine-grained groups, but what was known about these in nature? Did any of these interbreed? Or were some of them just radically variant populations at the extreme edges of a widespread range whose middle ground was simply unknown to lepidopterists?

MOST TAXONOMISTS OF NABOKOV'S DAY came to agree that a biological definition of species based on the criteria of differential interbreeding was desirable but also that it had to be informed by knowledge of structural characteristics. Such a combined definition could satisfy two needs. It could encourage discernment regarding the problem of inconsistent interbreeding in nature and give scientists in a laboratory or museum collection criteria available to the naked eye by which they could make at least some needed decisions.

Genetics itself, at the level of both the genetics of individual organisms and that of populations, was in ferment at the time, further confusing the landscape. Was genetic change itself enough to drive evolution down a path that would result in new species? Or did it always need an intervention from selection? If so, what kind(s) of interventions were involved?

Models of speciation had been appearing in the scientific literature for decades. First of all, if genetic change was in some sense driving evolution, at what point, and how, did a population become separate from another such that eventually it might become reproductively distinct? Based on scientific definitions that organisms in the same distributional range were sympatric and organisms with separate ranges were allopatric, this most obvious possible cause of differentiation over time—spatial separation leading to reproductive incompatibility—became known as allopatric speciation. But what did it take to cause it?

The allopatric speciation process was particularly enigmatic in Nabokov's day because what we recognize now as plate tectonics was both little known and generally dismissed.

As a result, the understanding so obvious today—that major cleavages of biological distributions break apart interbreeding populations on a massive scale—was simply not part of the discussion. Scientists saw, or envisioned, allopatric speciation in smaller theaters—mountain range changes, river course changes, and so on—but these were all equivocal enough to complicate the confidence scientists had in their causal role in species evolution. Since nearly all scientists of Nabokov's day believed that landmasses were static, they envisioned that somehow both plants and animals had moved around the planet somewhat. Thus, surely the obvious divisions of ranges apparent planetwide—as with the African and Indian Elephants—meant that allopatric speciation had occurred, although its causation was not well understood.

Another question remained. Could speciation occur with far less dramatic splitting of ranges? This was a big question, particularly for anyone who worked with more vagile (far-traveling) organisms, such as butterflies, birds, and fish. How much range division did it really take, and how long, for species to arise in far less dramatic fashion? Some biologists saw evidence for quite fine breakages in ranges that resulted in species—or no breakage at all—what was called sympatric speciation, speciation among organisms that were still sharing a general range. Others saw patterns of step-by-step, level-by-level speciation and from that deduced parapatric and peripatric speciation, much as astronomers envision planets slowly congealing within the rings of debris around the Sun.

In sum, taxonomic scientists in Nabokov's day were engaged in a pivotal debate about processes within organisms at the genetic level, genetics among groups of populations of organisms (population genetics), and genetics in the sphere of environment-organism interactions.

YET ANOTHER SET OF PRACTICAL PROBLEMS faced Nabokov and his fellow biologists. In studying evolution and making taxonomies and classifications, how should one retrieve, record, and arrange information? Further, having retrieved information, how should one systematically present it in a landscape of ideas and concepts? Even if there were agreement about definitions in matters like species, natural processes, and philosophical underpinnings, how should a scientist work to retrieve and use data consistent with these views? What if one disagreed with the consensus definitions? The problem of information retrieval (what constitutes data and how it is obtained, formulated, stated, and then compared) has always presented challenges.

Nabokov's notes show that he did not shy from these puzzles. For example, what data account for the biological definition of species? How does one get such data? How does one know what entities interbreed in nature? What does it mean when they interbreed in a lab or, more confusingly, interbreed freely in a lab but not in nature? Likewise, what structures define a species based on morphology? If many structures, do they all have equal weight? What about structures that accidentally look alike (analogy—analogous structures) versus structures that resemble each other because of an actual evolutionary relation (homology—homologous structures)? All of these dilemmas plagued the study of the Blues.

As a pioneer butterfly anatomist, Nabokov found himself uncovering new structures not previously used for differentiating species. In fact, he discovered arrays of structures in the genital apparatus of some Blues that were lacking in other Blues long assumed to be their closest relatives. In his 1945 publication naming new Latin American genera, he recognized previously undiscerned structures at the terminus of the male clasper (valve). He also described a structure (near the aedeagus, or penis) that he called the sagum, which occurred in some, but not all, Latin American Blues that were otherwise confusingly alike externally. He was confident that such structural differences meant that these Latin American Blues could not actually be in the same genus.

Unfortunately, Nabokov's contemporaries and subsequent lepidopterists ignored these data for nearly fifty years. If contemporaries and later critics actually read his explanations of the sagum and the structures at the terminus of the male valve, they appear to have made an arbitrary decision not to consider these as data. Historically, in Blues, wing patterns and (if anatomy was even consulted) the general overall shape of the male clasper were considered the sources of data. In fact, lepidopterists never discussed these newly discerned features of Blues' anatomy—especially the differential occurrence of the sagum—until the 1990s. Until then, the judgment of whether Nabokov's classifications were right or wrong revolved around evaluations of the known characteristics of traditional taxonomy.

Such anomalies exemplify the historicity of Nabokov's work as judged by his critics, of course, but there were larger general issues as well, relating to the determination of actual evolutionary relationships. For example, once homology is distinguished from analogy, what characteristics help a scientist recognize which organism is most closely related to another? And, once that is determined, how are the relationships best portrayed? Historically, science has portrayed relationships of organisms in branching diagrams, often generally called dendrograms, or evolutionary trees. It began to occur to scientists, at about the time of the books Nabokov was reading in the early 1940s, that science had not taken sufficient care to differentiate whether a particular branching diagram was a actually a statement about characteristics or a statement about a lineage of organisms. And these are two very different things. A statement about lineage is clearly a statement about evolutionary history. A statement about characteristics may, or may not, reflect the historical relationship of organisms (like butterflies, birds, and bats all having wings). So, at about the time of Nabokov's active work, scientists began to differentiate between diagrams about characteristics (sometimes called phenograms) and diagrams about evolutionary relationships (variously called phylograms or cladograms), as well as to debate about when the two were synonymous and when not.

All of these questions were pivotal to the developing science of systematics (the theoretical construct of which taxonomy is a part). Two major transitions in systematics, during and after the time of Nabokov's active research career, would further affect and characterize both his work and his legacy.

The first was the international process of establishing a generally accepted rule book for how to do taxonomy. What became the modern International Code of Zoological Nomenclature (published in French and English in 1953, updated 1958, and then in many subsequent editions) evolved from a number of international rules books from as early as 1905. Like other taxonomists of his era, Nabokov followed general protocols drawn from these earlier works, but there was no standard work clearly acknowledged as internationally normative. Like the resolution of other upheavals in systematics and taxonomy, the universally accepted international code postdated Nabokov's active years. As it would turn out (see below), this would affect the historical status of a number of the major names for Blue butterflies he proposed.

The second factor was the progression of systematics through at least three major, and quite different, methodologies as knowledge of evolutionary science increased decade by decade. Without some understanding of these, there is no way a lay Nabokov enthusiast (or even literary scholar) can understand the landscape of biology during Nabokov's active years or what became of his work afterward.

SYSTEMATICS BEGAN WITH the pioneering work of Carl Linnaeus (1707–78), who created a binomial nomenclature. He recognized a kind (a species) connected to a more general category (a genus), which provided some understanding of their larger relationships. Above the genus are a number of larger categories (each known generically as a taxon and together as taxa)—family, order, class, phylum, kingdom—which are called obligatory categories, to be used by all scientists. Several other taxa are not obligatory but are found to have utility and thus are often used by many systematists. Examples common in butterflies are tribe, subfamily, superfamily, supergenus, and subgenus, Nabokov's now famous bunch being the tribe Polyommatini and many of the related groups (often called outgroups) in the larger subfamily Polyommatinae (see especially figures and color plates under the heading "Old World").

Before much was known about evolution, classical systematics allowed scientists to group organisms into the categories of taxonomy by whatever criteria they felt were useful—obviously a source of great argument. With the advent of Darwinism and the various neo-Darwinian views, systematics moved into a mode usually called either traditional or evolutionary systematics. It was part science and part art—because there was some information that could be gotten by scientists (like structural, biological, or ecological data) and some that couldn't (like the grand patterns of change through time). It is unfortunate that through historical happenstance the traditional school took the name "evolutionary," because it would actually turn out to be less allied to concerns about real evolution than subsequent schools. How to retrieve and reconstruct the actual path of evolution was the challenge of the competing schools of evolutionary theory and practice. We find Nabokov musing on this challenge, in his scenario for the origin of the New World Blues, when he imagined using "a Wellsian time machine" to see what really happened.[31]

Accordingly, the science within evolutionary systematics was the hard data assembled by the scientist, but the art was in the intuition by which that scientist then made a statement about a lineage or a tree—some kind of a branching diagram. However, in the imprecision of the day, it was often not apparent whether the proposed tree or lineage was really a statement about the similarity of characteristics or a statement about the actual relationships of organisms. This was no small matter, because if one made a statement about a lineage, that was a statement about evolution. But if one made a statement about characteristics, the statement was perhaps no more than a statement about organisms sharing this or that apparently similar trait. That was information from which an inference about evolutionary history might be drawn but was not a direct statement about evolutionary history itself.

Further, the question of whether a branching diagram should be a statement about organisms or a statement about characteristics produced a critical choice affecting the entire meaning and enterprise of systematics and taxonomy. Should a taxonomy reflect evolution, or is it enough to have a taxonomy reflect character relationships and not even claim to be about evolution? Should taxonomy simply help scientists sort stuff, or should it actually reflect the course of evolution? These questions would end up determining the successive changes in systematics that ensued during Nabokov's life. But in evaluating Nabokov's work the situation is even more complicated, because Nabokov was a visionary, and in the 1940s he was already speculating privately and in print about the very questions that would end up determining which direction systematics would take.

Early on Nabokov placed himself solidly in the camp declaring that taxonomy and classifications should reflect actual evolution. He also asserted his belief that the necessary methods were available. Here he placed himself on the right side of history, a position that makes some sense of his declaration in 1950 that "natural science is responsible to philosophy—not to statistics."[32]

The advent of computers, and the desire of systematists to have a more empirical way of doing taxonomy than the combined science and art of the evolutionary taxonomic method, led to the development of what is today called phenetics and its often computer-based offspring, numerical taxonomy. Based solely on various measures of resemblance, the development of this paradigm and its ambitious competition with the prevailing evolutionary taxonomy spanned the later years of Nabokov's active scientific career. Numerical taxonomy paralleled the timeline of the computer age, arising with World War II and the second and third waves of computer development into the 1960s. The computer became a major methodological tool in taxonomy (including lepidoptery) in this latter decade. Whether involving computers or earlier hand calculations from various statistical packages, the larger issues—concerning whether phenetics was the better methodology—were the same as before: What is a relationship, and should it be based on traits or on evolution?

Working with Blues, so extremely alike in superficial resemblance, Nabokov was obviously not primed to favor a phenetic approach. Simple resemblance could be misleading. In his notes, Nabokov differentiates the deceptively similar Latin American Blues into his new genera by detailed attention to anatomy. Confusing superficial resemblances brought Nabokov to anatomy in the first place. Keenly aware of the difficulties, he scoured the scientific literature to see how others workers had approached the problem. On one card, Nabokov evaluates all the terms he has read in the literature possibly pertaining to the ephemeral superficial resemblance he sees in the Blues.[33] Consequently, he distinguished a unique kind of enigmatic resemblance, common to the Blues, which he described with a nuanced term, "homopsis," later recounted in his studies of *Lycaeides*.[34]

For a taxonomist attuned to the problem of superficial resemblance, phenetics had particular weaknesses. Those who wanted taxonomy to reflect actual evolution knew that a fundamental flaw in phenetics's reliance on simple resemblance was that species sharing primitive characteristics (those basal in a diagram), and not further distinguished from successively shared derived characteristics upward in the diagram, could make unrelated groups wrongly appear to be immediate relatives, creating a false monophyly—a false conclusion about lineage. For the lay reader, a crude example of this problem would be thinking that the shared characteristic of having wings makes butterflies, bats, and birds immediate relatives. It was such imprecisions that had made Nabokov critical of phenetic taxonomists in his notes regarding the Europeans' (and especially the Germans') penchant for confusing relationships based on simple resemblance and American lepidopterists whom Nabokov also singled out for such criticism (like Frank Chermock).[35]

Numerical taxonomy, phenetics's computer-based offspring, would have been even more offensive to Nabokov because it could mathematically generate taxonomies based on simple resemblance for increasingly large groups and samples. Nabokov believed in an evolutionary classification that reflected the actual evolution of real species. Contrarily, instead of "species," the jargon of phenetics often used OTUs (operational taxonomic units). Further, phenetics often stated unequivocally that the relationships of OTUs, and thus even of entire classifications, did not need to have anything to do with evolution. Notes from Nabokov's years at the Museum of Comparative Zoology demonstrate that such utilitarian approaches to natural species appalled him.[36] These approaches also appear to have contributed to his unfortunate disdain for statistics, another rudimentary part of the early approaches to phenetic taxonomy.

Perceiving this weakness in phenetics, Nabokov employed what science now calls a transformation series. A transformation series recognizes the progressive derivation of traits along a path of evolution. In his *Lycaeides* publications, Nabokov illustrated these by showing anatomical characteristics and lineage diagrams (phylogenetic trees) side by side. With such a method, Nabokov could be reasonably confident of his statements about evolution.

We suspect, however, that Nabokov, later, would have appreciated some of the things that numerical taxonomy could do. For instance, by processing enormous data sets with computers, phenetic methods shed a far more comprehensive light on the clusters of populations that made up the enigma of the Rassenkreis phenomenon. This would have made sense to Nabokov because, in his Rassenkreis notes, he tried to group constellations of characteristics that would make sense of what was interbreeding with what. This was a complicated matter, since interbreeding would be expected to result in convergent character states whereas lack of interbreeding would result in divergent traits (see Figures 79 and 81).[37] In fact, Blues expert John C. Downey would employ these very methods to sort out the enigmas in Nabokov's taxon *Icaricia* (see Figures 30 and 31).

AFTER YEARS OF CONTROVERSY, the kind of systematics that Nabokov sought began to arise: a phylogenetic taxonomy that could retrieve and reflect an actual evolutionary lineage. The breakthrough began in 1950 when, in German, Willi Hennig published *Grundzüge einer Theorie der phylogenetischen Systematik,* translated into English in 1966 as *Phylogenetic Systematics.*[38] It contained what Nabokov had been looking for: the skillful recognition that distinguishing nested sets of what Hennig called "shared derived characters" (synapomorphies) allowed reliable identification of an actual evolutionary lineage. This possibility was something Nabokov himself had raised. Indeed, he had employed the basics of the method, in rudimentary fashion, in his revision of the genus *Lycaeides* in 1944.

After a stormy period in the 1970s and 1980s, and aided by a revolution in the philosophy of science (the so-called Popperian Revolution), the modern phylogenetic paradigm of systematics became paramount. It is known also as cladistics and with its companion, vicariance biogeography, is based on a clear understanding of the implications of allopatric speciation. It understands that diagrams about the relationships of taxa (species, genera, and so on) comprise data sets parallel with diagrams about spatial relationships (the geographic distributions of species, genera, and so on) because the lion's share of species have arisen by outright spatial splitting. The phylogenetic paradigm became further buttressed by modern DNA-based methodologies that reflect the same hierarchical principles that cladistics established.

There is an irony in the fact that a modern phylogenetic paradigm, precisely what Nabokov had been looking for, arose just as Nabokov was leaving active work on the Lepidoptera—and so unfolded without his participation. By 1953 Nabokov wrote that he had "hardly any time" for butterfly work.[39] Had he remained active in the field, he might have helped further frame the new approach. Perhaps, in light of his critiques of German taxonomy, and like *The Gift*'s Fyodor on a Berlin streetcar, he would have been amused to learn that a German devised the paramount phylogenetic method.[40]

THE MUSINGS INSCRIBED on Nabokov's cards demonstrate a scientist quite ahead of his time in seeking a phylogenetic approach to systematics—one that could discover the actual lineages of evolving organisms. Nabokov's subject butterflies—the Blues—undoubtedly stimulated his quest. As a boy in Russia, Nabokov developed his knowledge of Old World butterflies. Poorly known groups like Skippers and Blues, resisting ready identification, particularly fascinated him. Before beginning professional work in the United States, Nabokov had collected specimens of a confusing Blue in southern France, publishing on this peculiar Blue and a puzzling Old World Skipper in 1941 (naming the Blue *Lysandra cormion*, new species).[41]

Differentiating both the Blue and the Skipper from possible relatives required dissection, a technique Nabokov is said to have learned from his mentor Comstock at the American Museum of Natural History.[42] A penchant for anatomical study was soon to become addictive for Nabokov. Anatomical characteristics, he found, pointed toward a far larger, yet hidden, diversity in cryptic butterflies. Further, dissection was a window into one of the long-standing enigmas of Old World butterfly lore—the peculiar reproductive patterns across far-flung and often erratically disjunct distributions of cryptic butterflies like the Blues. Traditionally, the Blues and Skippers were often nearly impervious to conventional methods of identification and classification. That Nabokov's first named species was a Blue (even though it would later turn out to be a hybrid of two other species) should not surprise. Blues and their many puzzles, ill-studied on nearly every continent, offered Nabokov a fertile landscape in which to create a legacy for himself in the field of lepidoptery.

Studying Blues, Nabokov was inevitably led to ponder the relationship of expanding distributions, distribution disjunctions, time, and speciation. In a number of drawings he tries one scheme after another to make sense of how entities known today—species, subspecies, and races—might have arisen across complex mazes of relationships (see Figures 80, 82, and 83). The complex circumstances surrounding the Blues led Nabokov to realize that study of his subject butterflies involved far finer distinctions than most. In his notes made on Ernst Mayr's *Systematics*, he remarks how butterflies are quite different from birds and require far more nuanced questions about evolution's ways. These musings led him to question the coherence of the dominant systematics of his day—and the reliability of the biological definition of species.[43] He wondered about its efficacy. If a biological species definition was the standard, how would it be determined—practically—in both the field and the lab? Further, was there a coherence between determining biological species and what was indicated by their physical traits—wing patterns and internal anatomy? Was there a coherent methodology for determining species that might rely solely on physical traits? These were, in fact, the major questions in systematics that would initiate the eventual appearance of Willi Hennig's purely phylogenetic (cladistic) method in 1950.

Like the later phylogeneticists, Nabokov sought a taxonomic method that could lead to the recognition of evolutionary lineages. He had little sympathy with phenetics or numerical taxonomy, which, surrendering to the difficulties of determining phylogeny, had abandoned efforts to reflect evolution. In contrast, Nabokov's view was consistent with the modern phylogenetic school of systematics, a consistency mirrored in at least four areas. Each is fascinating, and we review them briefly below.

NABOKOV REFLECTED ON the fleeting meaning of the individual or population in finite time, wondering how one would approach the flow of evolutionary change over far more expansive time in an ascending lineage; in his 1945 paper he used the metaphor of a train passing through successive stops, each having its own identity.[44] Later, Willi Hennig pondered the same phenomenon, even coining terms for the external appearance of a species today compared to its ancestral forms. This desire for a vertical view required taxonomy to move toward recognizing hierarchies of nested sets of successively derived characteristics—the key to a phylogenetic taxonomy. Nabokov's related musings led to his ten-million-year-long scenario for the wave-by-wave origin of the Neotropical Blues, verified by Roger Vila and colleagues in 2011.

In his *Lycaeides* study (1944), before Hennig's work appeared in any language, Nabokov began using what was later called a transformation series, in which the successive derivation of certain characteristics provides the information used to project the path of evolution. Nabokov began by hypothesizing the ancestral ground plan for the larger group (Figure 88 and Color Plate 38). He then recognized successive structural developments (through measurements according to his magic triangle method). He subsequently furthered this protocol into a projected phylogeny, illustrated with drawings of the character transformations (Figures 84, 85). Elsewhere, Nabokov's comments show that he clearly understood the challenge of reliably discerning ancestral characters, later worked out and codified by Hennig.[45]

Nabokov's note cards demonstrate his awareness that spatial disjunctions meant the disjunction of populations (and thus of gene pools). With these would come changes reflected in differing, analyzable characteristics, and these in turn created parallel data sets that, together, could disclose the paths of evolution. This was what the later phylogenetics would perfect in the cladistic methods of Willi Hennig and his followers.

At the time Nabokov was active in research, the scientists he was reading (including Ernst Mayr and Julian Huxley) were also struggling with the metaimplications of allopatric speciation. Their vision was clouded by a major obstacle: their belief in a stable earth in which the landmasses themselves had not moved. Without the modern awareness that even the most massive of landmasses is always in motion, the implications of spatial splitting could simply not be seen. The truth is that the major systematists and biogeographers of the time rejected the idea of continental drift. They had generated the prevailing dispersalist biogeography, and their writings were full of arrows showing how plants and animals had moved among the static continents. When plants and animals showed disjunct distributions, the dispersalists posited how dispersion might have happened, including "trans-oceanic land-bridges in other parts of the world," as Nabokov ironically noted in "Notes on Neotropical *Plebejinae*."[46] In that paper, Nabokov questioned the efficacy of such explanations, saying he doubted such possibilities would adequately explain the origin of the New World Blues. History would record the essential irony that though Ernst Mayr, George Gaylord Simpson, and P. J. Darlington established the allopatric speciation model, they could not comprehend its ultimate implications.

A card Nabokov subtitled "A Tricky Question," in which he ponders the crux of the riddle presented by allopatric speciation over time, is a boon to historical study of these matters. Like the later phylogeneticists, Nabokov boils the matter down to a realization that the fundamental challenge facing systematists was establishing how, when presented with three entities, one could determine which two are most closely related.[47] This deceptively simple conundrum needed to be resolved if there was to be a methodologically sound phylogenetic approach underlying scientific systematics. Known as "the challenge of the three taxon statement," the conundrum is complex, and the answer is simple, but it was unavailable until Willi Hennig's method enabled a solution.[48]

Consider that if one imagines a branching diagram (or tree) of relationships of any size, the hierarchy of the diagram includes any number of end branches that are then joined at sequential nodes downward through the diagram. Each of these levels (nodes) explicitly corresponds to a different level of entities (or taxa): species, genera, tribes, and so on. Thus any diagram can be separated into any number of three taxon relationships, wherein lies the vexing evolutionary question: Which two of the three are most closely related?

In taxonomic lingo, answering this question resolves the three taxon statement into a diagram showing the most likely actual evolutionary (phylogenetic) relationships—and thus the history—of the three organisms or groups. When a scientist is unable to answer the question, the evolutionary relationships are unresolved. Thus, when the systematics community finally came to debate whether a phylogenetic taxonomy was possible, and if so, what it might look like, it ended up comparing the ability of the different taxonomic schools to resolve a three taxon statement. In doing so, the systematics community was echoing Nabokov's Wellsian time machine anecdote—by proposing how to model what had really happened.

Obviously, neither traditional evolutionary taxonomy nor phenetics (numerical) taxonomy (which relied on simple resemblance) could solve this problem—except perhaps by luck. However, Willi Hennig's phylogenetic systematics, using the same (transformation series) approach published by Nabokov in 1944, could do it. The new phylogenetics could reliably make the decision based on distinguishing shared ancestral (primitive) characteristics from shared derived characteristics and building the nested sets of relationships on that distinction. Yes, Nabokov did it, too, and even earlier—using this method to create his *Lycaeides* phylogeny—but he did not himself state any overarching taxonomic principle associated with the method, perhaps because he abandoned active research soon after finishing his revisionary work on *Lycaeides*. Such recognition is what was unique to Hennig and the early phylogeneticists who followed him.

It is interesting that Nabokov calls the enigma "a tricky question." He then tries to frame the question about how one might resolve, in a coherent fashion, which two of three taxa are most closely related. His handwritten text says:

A tricky question: if it can [be] proved that species A is the ancestral form of B, the latter dwelling in a remote S.E. locality, and now specifically distinct from A; and if C, in a remote S.W. locality (thus, with B and C, in no contact, past or present, whatever and similar in all such characters as are subject to variation in the given group), both B and C sharing a definitive, constant, biologically important character (structure of the male genitalia) which is a perfectly obvious specific indication in the members of the group, can B and C be considered subspecies of one another?[49]

With hindsight, in light of how exacting, albeit simple, that question needed to be to solve this riddle, Nabokov's statement falls into the same trap as many of his early peers. He does not state the situation of each of the three branches in precisely the same way. Further, he intermixes, and thus confuses, comments about the spatial data and the taxon data. Thus, with the resulting verbiage about all three of the branches in something of a muddle, he never states or resolves the matter with perfect economy. Typical of all precladistic views of systematics, without simply and clearly differentiating shared primitive characters from shared derived characters, there could be no consistent criteria by which to resolve the three taxon dilemma.

IT HAS NOW BEEN WELL OVER a dozen years since attention during the 1999 Nabokov Centenary made it well known that Nabokov's previously controversial classifications for Latin American Blue butterflies had, after over a half-century of malign neglect, been universally recognized as correct.[50] *Nabokov's Blues: The Scientific Odyssey of a Literary Genius* contains a detailed account of Nabokov's studies of Latin American butterflies and the work of Kurt Johnson, Zsolt Bálint, and Dubi Benyamini in rehabilitating the reputation of Nabokov's New World classifications and securing their recognition. Since then, two significant additions have been made to that record: the 2010 promotion of the Karner Blue to species status, and the surprising 2011 validation of Nabokov's major evolutionary and biogeographic hypothesis concerning the origins of South American Blues, corroborated by sophisticated DNA and computer analyses.[51]

Systematics acknowledges today that, in 1945, Nabokov successfully recognized ten of the eleven natural groups (or genera) of the polyommatine Blues (tribe Polyommatini) in the New World, including the Blues belonging to the old name *Hemiargus* Huebner 1818. Of these eleven groups, five still bear his original 1945 generic names: *Cyclargus, Pseudochrysops, Pseudolucia, Echinargus,* and *Paralycaeides.* Two others (*Parachilades* and *Pseudothecla*) were replaced after 1945 because of technical rules of the modern International Code of Zoological Nomenclature. There was also one genus that Nabokov was unaware of simply because he never saw specimens (*Eldoradina* Balletto 1993), and there were two for which he used older names incorrectly (at least by the modern nomenclatorial rules). These two are *Itylos* Draudt 1921 and *Madeleinea* Bálint 1993 [*Itylos* sensu Nabokov 1945]. There is also a non-polyommatine Latin American Blue genus (*Leptotes* Scudder 1876) in which modern workers have named butterflies after Nabokov. Accordingly, in common parlance, these Blues comprise the Latin American component of Nabokov's Blues.[52]

Examining firsthand the hundreds of four-by-six note cards making up the bulk of Nabokov's laboratory records, we can now comprehend the nuances involved in Nabokov's taxonomic classification of these Blues, as well as his hypothesis articulating their origins. We see not only the details of his meticulous work on Latin American Blues but now appreciate the meaning of scores of cards devoted to specimens of species from the Old World (especially those areas abutting the Beringian region) and the letter-sized sheets on which Nabokov recorded his research strategy (see illustrations of Old World Blues, Figures 7, 9–16, 18, and 19; Color Plates 2–16; and Table 1, below).

Overall, some 70 percent of Nabokov's laboratory illustrations of the anatomy of Old World Blues center on his suspicion that northeast Asian and Beringian region Blues held the key to understanding the origins of the New World faunas. He outlined this direction on his research strategy sheets. It was also germane to virtually all of his Old World laboratory cards aiming at deciphering the relationship of the then-current Old World taxonomies for Blues, especially those taxonomic categories that were in use for both New and Old World Blues.

For example, multiple cards aim at deciphering the relationships of perennially used Old World generic names like *Agrodiaetus* Huebner 1822, *Plebejus* Kluk 1780, *Polyommatus* Latrielle 1804, *Albulina* Tutt 1909, and *Cyaniris* Dalman 1816. Nabokov's cards often record his doubt regarding what name to apply (Figure 13) or list options from the old nomenclature (Figures 7–10). This work became even more complicated after Nabokov had to factor in generic names that had been used in both the Old and New Worlds, such as *Glaucopsyche* Scudder 1872 (Figures 17, 28, 29), *Scolitantides* Huebner 1819 (Figure 27), *Polyommatus* (Figure 19), and *Plebejus* (Figure 30). Indeed, even today the Old World national and regional lists differ widely in their use of these generic names. Only in the past decade have DNA studies begun to sort out which names should be used (Figure 12 and Color Plates 2–4, 6, 7, 10, and 16).

Nabokov also selected individual specimens for dissection with an eye toward Beringia, even if only from cursory locality data sometimes available on nineteenth- and early twentieth-century specimens. He chose specimens from Korea (Color Plate 12), China (Color Plate 14), trans-Siberia (Figures 7 and 29 and Color Plates 8 and 18), Manchuria (Color Plate 19), and the Lake Baikal region along the Siberian and Mongolian border (Figure 17).

WE CAN SENSE THE EXCITEMENT in Nabokov's scientific process as we watch him move from his thorough examinations of the Old World Blues and into his discoveries concerning Blues of the New World. His comments in laboratory notes reflect those in a 1945 letter to his sister Elena: "To immerse yourself in the wondrous crystalline world of the microscope, where silence reigns, circumscribed by its own horizon, a blindingly white arena—all this is so enticing that I cannot describe it."[53] Once Nabokov entered the domain of the New World Blues, his realization that the entities he was studying were new to science was immediate. Seeing peculiar anatomical structures evident in new genera and species of Blues, Nabokov shifted from the style of notation typical to his work on Old World Blues (often abbreviated drawings and sparse notes) to the construction of significantly more detailed drawings, often intricately shaded or in color, and with profuse notes about the strange structures, accompanied by expansive lists of measurements and means.

Nabokov's laboratory drawings, presented in the book's next two sections, describe a story in itself for each of the genera that Nabokov studied or named across Latin America:

Cyclargus: Figures 44, 46, Color Plate 32
Echinargus: Figures 49–55, Color Plate 33
Parachilades (now *Itylos*): Figures 67–70, Color Plate 36
Pseudolucia: Figures 65, 66, Color Plate 35
Paralycaeides: Figures 63, 64
Pseudochrysops: Figure 78
Itylos (now *Madeleinia*): Figures 75, 76
Pseudothecla (now *Nabokovia*): Figure 74
Hemiargus Huebner 1818: Figures 45–48, 56–62, Color Plates 13, 32, 34, 51

Immediately evident is Nabokov's immersion in the discovery that New World Blues were far more diverse than scientists had suspected. On the cards listed above for *Cyclargus* and *Hemiargus*, one can readily trace his documentation that Huebner's old catch-all name *Hemiargus* 1818 (applied by lepidopterists to the lion's share of New World tropical Blues) was not a single group (Figures 57–61 and Color Plate 34), a fact buttressed by his subsequent drawings of what would become his new genus *Echinargus*. In both *Cyclargus* and *Echinargus*, Nabokov documented the salient anatomical feature he called the sagum, making the genitalia of these two groups of Latin American Blues unmistakable (Figures 44, 47, 50–53, and 62 and Color Plates 32 and 33).

When Nabokov saw new entities previously unrecognized by science, he immediately noted new species or genera ("n. sp.!" or "n. gen.!"), jotted new names that he might propose, inserted exclamations about unusual-looking characteristics (see Figures 44, 61, 63, 65, and 67–69), or constructed tables of measurement data for his triangulation of genital shapes (his magic triangles: Figures 30 and 88 and Color Plate 37).

One can palpably feel Nabokov's surprise and excitement as he ran across new entities and novel structures. Further, one sees Nabokov's comprehension of key evolutionary questions and puzzles—like the coincident relationship of evolutionary lineages and geographic distributions. It is these that set the stage for his biogeographic hypothesis that would become the subject of twenty-first-century study by scientists using DNA analysis and complex computer simulations.

The cards also tell us about the rapid pace of his work and his confidence in it. A new example, constructed by comparing dates on research cards, manuscripts, and page proofs, shows us now that Nabokov created his new, at the time controversial, New World genus *Echinargus* Nabokov 1945, not during the article's original research and writing, and not even during editorial revisions, but when he was correcting galley proofs (having then to correct printed errors in that section by hand when mailing reprints later that year)![54] We see, overall, a scientist who could quickly change his mind and radically contravene current convention when confronted by significant new data. We also see that his meticulousness is probably what allowed him to create correct classifications from extremely small samples (in his 1945 study, one hundred twenty specimens, compared to two thousand in his study of *Lycaeides*).[55] Such boldness may have been viewed as capriciousness or lack of rigor, with negative consequences for his work's reception by his contemporaries and even many later lepidopterists. In retrospect it is both a surprise and a marvel that Nabokov, knowing fewer than twenty Latin American species in his day, constructed the classification for some one hundred known today. Kurt Johnson and Steve Coates have listed twenty-nine Blue butterfly species named for Nabokov or characters from his fiction.[56] Since the publication of *Nabokov's Blues,* at least a dozen more Latin American Blues have been named, one of which carries a "Nabokovian" etymology.[57] Another surprise from the modern studies of Nabokov's Blues was the discovery of a mimicry ring in Nabokov's own South American genus *Pseudolucia* Nabokov 1945, recounted and diagrammed in *Nabokov's Blues* and included herein: see Figure 65 and Color Plate 35; see also Color Plate E1, which shows several of these species in recent color photographs by Dubi Benyamini.[58]

ALTHOUGH THEY APPRECIATED the scientific value of Nabokov's work, even the authors of *Nabokov's Blues* and the editors of *Nabokov's Butterflies* failed to recognize the importance of three paragraphs in *Psyche* in which Nabokov presented his view of precisely how the Latin American Blues had evolved.[59] Whatever the historical context, the public's ongoing interest in whether Nabokov was a serious and important scientist likely explains why, when the 2011 DNA results were announced in the media, news of "Nabokov vindicated" went viral on the Internet.[60] Such a vindication provided a capstone for Nabokov's lepidopterological legacy. This ultimate confirmation of his far-reaching contribution to science was fortuitous because, even after Johnson, Coates, Bálint, Benyamini, and other scientists had proved the extent of his taxonomic skills and achievements, many reviewers of *Nabokov's Blues* rejected the notion that Nabokov, still an "amateur" to most, had done any truly "important" science.[61] Some asserted that claims of his work's significance, beyond being simply respectable taxonomy, were most likely hyperbole driven by his celebrity.[62]

One possible cause for this disregard is the myth of Nabokov's supposed anti-Darwinism, especially as expressed in his nonprofessional statements on mimicry.[63] This myth received further credence thanks to those who interpreted "Father's Butterflies" (an addendum to *The Gift* published posthumously in *Nabokov's Butterflies*) as a scientific essay instead of as the fiction it is. No reader of Nabokov's scientific articles can doubt his belief in natural selection as a regulator of beneficial adaptations in nature. However, he clearly believed that other mechanisms or mysteries were also at work, and he placed his early hope for discovering them in the phenomenon of mimicry and object resemblance (crypsis). Ever the iconoclast, Nabokov appears to have believed that it was scientifically more valuable to doubt and challenge natural selection in valid, testable ways than to become an accepting disciple of Darwinist orthodoxy. He left science too early to follow through on his instincts, but based on the strength of his work, we believe that Nabokov would have accepted (and perhaps did eventually accept) that mimicry is always or almost always a survival-oriented adaptation. We also believe that he would have mastered, and welcomed, all the new advances of the modern synthesis (including statistics). Any other assessment seems unrealistic in light of the comprehensiveness and capacity of his intellect. Nevertheless, we do not doubt that he would have found other ways to challenge orthodoxy and tug at Darwin's beard. Had he been more of a conformist, public dismissals of his scientific work might have faded away sooner.

CURIOUSLY, NOTWITHSTANDING NABOKOV'S sophisticated musings on the mutual interplay of spatial and phylogenetic relationships, it was not these that led to his major theoretical statements. What enabled Nabokov's boldest claims was his apparent ability to create a virtual zoetrope in his head to see ten million years of temporal and spatial dispersion across Beringia. Indulging in his Wellsian time machine, he went on to propose this

> series of events in the evolution of these butterflies. (i) From Asian ancestors, a first colonization event of the New World across the Bering Strait, followed by dispersal southwards to South America. This first stock would produce the current Neotropical taxa, but would subsequently vanish almost completely from North America. (ii) A second crossing of the Bering Strait made by the ancestors of the *Icaricia—Plebulina* clade. And finally, more recently, the dispersal of (iii) *Lycaeides*, (iv) *Agriades* and (v) *Vacciniina* (explicitly in that order) from Asia to North America following the same route.[64]

The results of Roger Vila and colleagues' DNA studies confirm that "he got every one right," as coauthor Naomi Pierce of the Museum of Comparative Zoology puts it.[65]

In the same paper, Nabokov was the first lepidopterist ever to state that the anatomy of the Blues of Latin America showed that they were not immediate relatives of those in the north. If he was right, the current distribution of New World Blues was not a simple and recent result of species moving north to south but instead a far more ancient and multifaceted, even layered, evolutionary process. Here we see Nabokov's finest hour as a scientist: these theoretical contributions to evolutionary and biogeographical thought were as significant as his descriptive ones in systematics and taxonomy. As noted above, those of us appraising Nabokov's work on the occasion of his centenary completely missed the importance of that one-page scenario. It looked like a throwback dispersal scenario from the discredited old-time dispersalist biogeographers. Fortunately for Nabokov's legacy, Naomi Pierce and her colleagues at the Harvard DNA labs, on reading about Nabokov's speculations in *Nabokov's Blues*, recognized that the cryptically audacious hypothesis could finally be tested.

Nabokov's luck carries with it a historical lesson. In the long run, history favors the taxonomic splitter rather than the taxonomic lumper. There are several reasons for this bias. As with the Blues, and particularly with any butterflies (if not most biota) of the tropical regions, it is likely that there are far more species in existence than were imagined by earlier scientific models. For the Blues, there were fewer than twenty species at the time of Nabokov's writing, whereas today the figure approaches one hundred (with new species being described regularly). Moreover, fine-grained taxonomic methods (like DNA analysis) usually reveal more and more species separated from interbreeding by subtleties of behavior or ecology. The dangers of historical lumping are particularly consequential for today's DNA studies. If the conventional taxonomy before DNA study is lumped, it is more likely that real species will be passed over—when the sampling list is made—simply because they are already lumped with something else. De facto, they do not exist. But with a split taxonomy predating DNA studies, all the entities are there to be studied and the chips can fall where they may following DNA analysis.

Nabokov's notes demonstrate his fine-grained approach. We find letter-sized sheets on which Nabokov created a wide column to the left and a narrow column to the right (Table 1).

Table 1. After Nabokov's research plan, written on 8.5-by-11-inch paper, BA7-Folder (miscellaneous notes)

date and init. genus of
    cyna Edw. Trans. Amer. Ent. Soc?
    tulliola

See Forster: *Azamus*
See Lorkovi[ć]: uncus of *Everes*
Dates of *Azamus, Brephidium, Lucia*
Examine *isophtalma, pseudofea,
barbouri,* and another specimen of *exilis*

Dissect an *Azamus*

Find and examine the "specimen in cigar
box" with [pear] shaped *aedeagus* as in
exilis

-->cp. (measur. and figs.) *felicis, galba,
filenus.*
----> check
Dissect some more *Hemiargus*

1 filenus, 1 antibubastus, 1 gyas,
2 Mexican 1 Haiti 2 Central American
2 Brasil [sic] ----at least 10 in all

Note outline of proc. inf., angle of
rostellum, body of valve

Dissect another *koa* and ♀ if any
and another *emygdionis*

check Field's papers (2) on *isola.*

Examine Zool. Rec. and S.Amer
publications for papers on the Andean
etc. "Blues"
Examine cleotas pattern also *cigar box.*
    lajus or *laius?*
      trochilus or *trochylus?*

Is "Cupido (Lycaena) speciosa Stdgr" a
homonym of *Lycaena speciosa* Hy.
Edwards?
Find it what sense Stdgr employed it.

Ought not *Azamus* (? With *Brephidium,* ? with *Lucia*) to be separated into a different subfamily from *Catochrysopinae*? (the "bouclier" see card of *Azamus*)

I do not doubt that several more Plebejinae await actual or revisional discovery in South Central America

The "*cnejus*" of Chapman genit. photo (also, apparently the *cnejus* of Bingham) is not Euchrysy. cnejus but a *chilades* extremely like galba. His *pandava* is like *lajus.*

The prep. of the aedeagi of *pheretes* --------------------------------> and *isaurica* have been interplaced on the plates and appar. in the descript.
Hanno from Venizaula [sic] and elsewhere in S.America belongs to a different species than *antibubastus* (with *filenus*), but what about the central amer. forms?

Study the striking resemblance between *hanno* and *galba* (homopsis N:137 , or "retained" mimetism). Falx differentiated in type from chilades, but derived from the latter (shrinkage) The tendency of the membrane of the valve of hanno, galba etc., bornoi to expand and be excised distally –or in Cyclargus, dorso-distally (under and along the neck of the rostellum). This has led B.B. into an error when distinguishing the *Chilades* forms.

The stretch of land from *Asia minor* (galba) to Bering Strait is exactly equal to the stretch from the latter to Peru. Several "horseshoes" hanging from the nail of Nome. But on the other hand Chilades species are also found in Senegal (*Eleusis,* test. Stempffer)

Miles from the Andes, the Marquises Schw[?] (?*clestus*) both of which are equidistant from the Andes each stretch being about ¼ of the minimum stretch via the north (from w. China – Cajus forms) It is difficult to regard in the same light . . .

On the left he recorded the views and conclusions of the currently reigning conventional taxonomy and his questions about them. On the right, he listed the steps he would need to carry out to investigate this orthodoxy or ask new questions. The sheets trace his exploration of the Asian Blues and the Blues of the New World, listing the various species he must dissect and study in detail to pursue this problem. This sheet is like a tourist's guidebook to the parade of cards collected here, showing the European and Asian Polyommatini tribe and larger Polyommatinae subfamily, and then the New World groups, one by one. Looking over the related cards, one sees the new Latin American groups all appearing, one by one, as Nabokov continues his work—startling even him!

Nabokov's fine-grained approach also served him well with regard to his studies of his beloved Melissa Blues—*Lycaeides melissa* and *L. m. samuelis*. Nabokov felt for some time that his *samuelis* was most likely a species on its own, an inkling he reiterated in his 1975 letter to Robert Dirig.[66] Already burned by *Lysandra cormion,* he originally named it as a subspecies, continuing to speculate about its actual status for many years. As Matt Forister and colleagues demonstrated in 2010, Nabokov's bolder private view actually proved the correct one. For further discussion, see the essay in this volume by Lauren K. Lucas and coauthors, who were part of the research team that established *Lycaeides samuelis* Nabokov as a distinct species.

Overall, relative to his peers, Nabokov adopted audacious approaches. Working before the Popperian Revolution's more liberal stance toward (testable) radical hypotheses, taxonomists lacked alternatives to the generally accepted accumulation of evidence methodology. Seldom would they jump to unorthodox, new (but testable) theories that the deductive mode enabled. Nevertheless Nabokov, even as a lesser-known figure, did propose very different, surprising conclusions and predictions that might be tested. The boldness apparent in Nabokov's articles is reflected in his quip that "natural science answers to philosophy, not statistics." This subtle iconoclasm also stands out in his phylogenetic view of evolution and systematics and in his consistent arguments against taxonomic methods based on resemblance alone. It is likely that Nabokov's extensive knowledge of philosophy helped make him a more innovative and pioneering scientist, as compared to many of his contemporaries.

Another unusual aspect of Nabokov's taxonomic work was his prowess as an artist. If Nabokov was not the only publishing scientist of his day also doing his own illustrations, he was one of very few. The sheer number of drawings in Nabokov's archives (well over a thousand) suggest that his intimate, tactile knowledge of structural variation over space and time propelled his evolutionary and biogeographic views. Eventually, the era of a causal relationship between the work of gifted illustrators and research results probably ended with the development of computerized methods for assessing evolutionary lineages (numerical cladistics) in the 1990s. Today, although there are several prominent lepidopterists who do superior illustrations (such as James Miller at the American Museum of Natural History), their biological discoveries come to a great extent from computer analysis, not from the kind of brain-powered activity that characterized workers in Nabokov's era. With his transformation series, his cinematographic time travel, and his mastery of an enormous diversity of data within a brief research career, Nabokov demonstrated how advanced the human brain's power of discernment can be. And yet, even in the midst of such intensive labor, he continued to see and portray his butterflies' anatomy as things of great aesthetic beauty, and their discovery as one of life's "most intense" pleasures.

The essays included here address Nabokov's scientific work, his artistic creativity, and the fascinating connections between them. Foregoing the one-by-one summary typically appended to such introductions, we simply note that they range from the purely scientific to the personal and to the decidedly literary-critical, with ample hybridization in between. These authors contextualize Nabokov's achievement and legacy, but they also present many new discoveries that he and his work have inspired. We hope that these drawings and their companion studies will inspire historians, scientists, and readers, as well as budding naturalists and artists, to carry Nabokov's adventurous spirit far into the future.

Notes

1. *NB*, 529.

2. *SO*, 16–17, 179, 132.

3. September 26, 1966, in *SelL*, 391.

4. *SO*, 11.

5. See Flower, "Scientific Art," for a valuable early critical treatment invoking the importance of Nabokov's science in his art.

6. D. Zimmer, *Guide*, 73–81. Some are no longer valid, and one has been promoted; see below.

7. Barabtarlo, "Nabokov's Trinity," 134.

8. *NB*, 387.

9. *SO*, 17.

10. *NB*, 312.

11. See letters to Edmund Wilson, May 8, 1944, and to Elena Sikorski, November 26, 1945, in *NB*, 312, 386–87.

12. "Deeply enjoying the profusion of fascinating figures provided by [Eimer, Kuznetsov, Schwanwitsch, and others]; and of course Kusnezov's [sic] masterpiece (1915, Insectes lépidoptères (Nasekomye cheshuekrylye) I (I), *in* Faune de la Russie) is unsurpassed by any other general survey of the morphology of Lepidoptera." Footnote to "Notes on the Morphology of the Genus *Lycaeides* (Lycaenidae, Lepidoptera)." In *NB*, 324.

13. *NB*, 399.

14. *NB*, 399.

15. October 11, 1944, in *NB*, 346.

16. See Babikov, "'*Dar*'"; Nabokov, "Notes on Neotropical," 5.

17. *SO*, 179.

18. *SO*, 11.

19. *SO*, 157.

20. Boyd, *Nabokov's "Ada"*; "*Erebia disa*," in D. Zimmer *Guide*, 147; Dolinin, "Signs," 257–59.

21. "[I thought] of Milton, and Marvell, and Marlowe, with more than a tourist's thrill as I passed beside the reverent walls. Nothing was closed off in terms of time, everything was a natural opening into it, so that one's mind grew accustomed to work in a particularly pure and ample environment, and because, in terms of space, the narrow lane, the cloistered lawn, the dark archway hampered one physically, that yielding diaphanous texture of time was, by contrast, especially welcome to the mind." *SM*, 269.

22. "An ancestral type of *Lycaeides* male armature was deduced by me from a preliminary study of the variation in the genitalia of the palearctic and nearctic forms . . . and was later discovered to have survived in a butterfly still inhabiting the mountains of Peru." *NB*, 411.

23. *LL*, 3.

24. Tammi, *Russian Subtexts*; Ramey, "Parasitism."

25. *NB*, 280.

26. Nabokov, "Notes on Neotropical," 44.

27. March 26, 1944, in *NB*, 311.

28. *NB*, 322.

29. Letter to his mother, E. I. Nabokova, in *VNRY*, 245.

30. James Ramey suggests that in *Pale Fire*, Nabokov imagines the novel as analogous to a parasitic botfly, whose eggs are delivered to mammalian hosts by mosquitoes. Ramey, "Parasitism."

31. His scenario was later corroborated by the now well-known DNA studies from the Harvard DNA lab published in 2011, and discussed by its lead authors, Roger Vila and Naomi Pierce, in this volume. See also Nabokov, "Notes on Neotropical," 44.

32. Nabokov, "Remarks," 75–76.

33. BA2-22b.

34. Nabokov, "Notes on the Morphology," 137–38.

35. *NB*, 309, 341, 347.

36. For example, "Reflections on Species Concept," in *NB*, 335–37.

37. See also the card at BA6-17c.

38. See these works; many of the authors were

part of the early debates in *Systematic Zoology* and at Nabokov's first research home, the American Museum of Natural History, and subsequently elaborated the Popperian Revolution in the philosophy of science and the arising of a phylogenetic methodology for systematics: Cracraft, "Phylogenetic Models and Classification" (1974); Nelson and Platnick, *Systematics and Biogeography: Cladistics and Vicariance* (1981); Hull, *Darwin and His Critics: The Reception of Darwin's Theory of Evolution by the Scientific Community* (1983); Eldredge and Cracraft, *Phylogenetic Patterns and the Evolutionary Process: Method and Theory in Comparative Biology* (1985); Hull, *Science as a Process: An Evolutionary Account of the Social and Conceptual Development of Science* (1990); Hull and Ruse, *The Cambridge Companion to the Philosophy of Biology* (2007); and Schuh and Brower, *Biological Systematics: Principles and Applications* (2009).

39. *NB*, 497.

40. *NB*, 309.

41. Nabokov, "*Lysandra cormion.*"

42. *VNAY*, 24.

43. For example, *NB*, 335–43.

44. *NB*, 335–43.

45. BA2-33b, BA2-33d.

46. *NB*, 378.

47. BA2-22b.

48. Also called "the search for the sister group" or "the search for the nearest neighbor."

49. BA2-22b.

50. Today they are recognized in all professional systematic literature and websites (e.g., ButterfliesofAmerica.com) and the current authoritative West Indies book (Smith, Miller, and Miller, *Butterflies of the West Indies and South Florida*). At this writing, only Wikipedia's pages (see *Hemiargus, Cyclargus*) and one enthusiasts' association (NABA.org) still favor the old classifications.

51. Vila et al., "Phylogeny": this study became a major biogeographic model regarding New World biotic origins via Beringia. Karner Blue: Forister, "After Sixty Years."

52. Historical counts of "genera" for Nabokov's Blues may vary from our numbers here if one includes both old generic usages far earlier than Nabokov's work (especially for South America) such as *Lycaena* Fabricius 1807 and how many times one counts a name like *Itylos* Draudt 1921, which early taxonomists applied to quite different groups of Blues.

53. *NB*, 387.

54. Conclusion from (1) March 16, 1945, Nabokov reply to W. P. Comstock ms. (later published in *NB* as "Note on the Male Genitalia of *Hemiargus hanno* Stoll and *huntingtoni* Comstock, with Two Figures," *NB*, 349–51), wherein Nabokov enumerates status and members of his then current view of Latin American Blues classification; compared to (2) Nabokov, "Neotropical *Plebejinae*," published October 26, 1945; compared to (3) subsequent reprints of (2), error-free except for inked-in Nabokov corrections and additions on pages 27, 28, and 39, clearly indicated as corrections of printer's errors after the addition of *Echinargus* in galley proofs. The corrected reprints are at the American Museum of Natural History lepidopterology [Dept.] reprint library (maintained by the last Lepidoptera curator there, F. H. Rindge).

55. *NBl*, 24.

56. *NBl*, 341ff.

57. Pelham, *Catalogue*, All-America pages, Lycaenidae.

58. *Nbl*, center section plate, "Mimicry in Nabokov's Blues."

59. Nabokov, "Neotropical *Plebejinae*," 43–44; *NB*, 377–78.

60. Reference to Carl Zimmer's *New York Times* article "Nonfiction: Nabokov Theory on Butterfly Evolution Is Vindicated" on Google will show nearly forty thousand entries under "Nabokov vindicated" and over one hundred thousand entries under "Carl Zimmer Nabokov," attesting to the number of international media that carried this story.

61. This neglect continued a tradition: Nabokov was undoubtedly a victim of the unfair assumption that he was an overzealous splitter. A further problem for Nabokov was his scientific writing style. William P. Comstock also wrote to Nabokov about his dense scientific prose (*NB*, 348). Even a friend of Nabokov's, Charles L. Remington, remarked, during the years when no one seemed to know if Nabokov had been a "good" or "bad" taxonomist, that "a lot of people have been uneasy about how well his work would stand up under the scrutiny of good professionals" (Coates, "Nabokov's Work").

Much of this confusion resulted because lepidopterists' skepticism was buttressed in 1975 by the publication of *Field Guide to the Butterflies of the West Indies* by the British Museum expert

Norman D. Riley, which summarily ignored Nabokov's Latin American work. We know today that Riley evidently either didn't read Nabokov's 1945 treatise on Latin American Blues or, if he did, overlooked the details of text and figures. See *NBl* 26, 92ff., 100ff. As Johnson and Coates record in *Nabokov's Blues,* and as readers can judge in reviewing the illustrations herein, the distinctions in Nabokov's new genera (especially that of the structure he called the sagum and also in the terminus of the male genital valve) make Nabokov's new genera unmistakable—as DNA studies later showed. For decades, however, Riley's negative assessment of Nabokov's work was given the benefit of the doubt, and thereafter, lepidopterists simply believed what they had come to hear. Nabokov was a known lepidopterist, but he was considered quite nonauthoritative, and most of his classifications were ignored until the 1990s.

62. *NBl* details comments (chapters 1, 4, 14), most salient from *Harvard Magazine* (Zaleski, "Nabokov's Blue Period": Nabokov was an "amateur") and *Boston Globe* (Taylor, "Nabokov Exhibition": Nabokov held "a bottom position").

The tradition continued in reviews of *Nabokov's Blues:* "There is no reason to inflate Nabokov's scientific contribution; . . . He achieved distinction in another arena" (Berenbaum, "Blue Book Value"); "The reader needs to feel confident about Johnson's science. But he seems to be hitching his star to Nabokov's a bit too assiduously" (Conniff, "Vlad the Impaler"); "This scientific work was trivial. . . . They make some rather grandiose claims . . . Nabokov's seems a respectable but minor contribution to science. . . . Nabokov's celebrity lend[s] an aura of importance" (Grice, "Review").

More recently, things have turned around: of the Transitional Nabokov conference in 2007, the organizers wrote, "This is one of the most highly contested issues. . . . There was intellectual debate over the validity of Nabokov's scientific views" (Norman and White, "Transitional Nabokov"); "Gould's parable doesn't work quite so well if the literary critics Gould dismisses—the ones who sought to explain away Nabokov's scientific efforts as unrecognized genius—were in fact *correct.* . . . Was Gould wrong to dismiss the 'Nabokov was a scientific genius too' argument so readily?" (sources quoted in C. Zimmer, "Nabokov 2.0").

63. Among many: "Nabokov's rejection of Darwinism" (D. Zimmer, *Guide,* 47, also at www.dezimmer.net/eGuide/Intro16%20%28Species Concept%29.htm); Ahuja, "Nabokov's Case against Natural Selection"; "He dismisses categorically the standard scientific, or Darwinian, explanation" (Alexandrov, "Note," 239); Klinghoffer, "Vladimir Nabokov, 'Furious' Darwin Doubter." Contrarily, see Johnson, "Lepidoptera" (presents a timeline), and Blackwell, *Quill,* 29–30.

64. Nabokov's hypothesis as paraphrased in Vila et al., "Phylogeny."

65. C. Zimmer, "Nonfiction."

66. *NB,* 713–14.

Black and White Figures

*Except as noted, captions are by Kurt Johnson
(following the current international taxonomic code)
and edited by Stephen H. Blackwell.*

Old World

Special Taxa

Figure 1 [*cormion, coridon*]

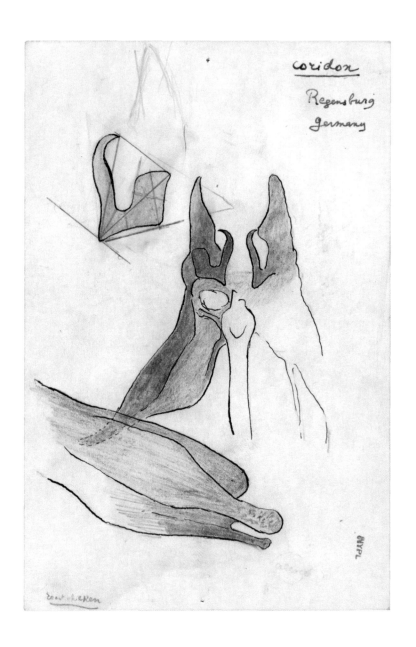

Figure 2 [*cormion, coridon*] Figures 1 and 2: Nabokov's first adventure into naming new species or subspecies was in the Old World *Lysandra coridon* (Poda 1761) complex, wherein he named a "new species" *cormion* Nabokov in 1941. Figure 1 shows the aspects of the male genitalia of *Lysandra coridon* from which Nabokov distinguishes his new name, *cormion*, marked with the basic triangulation lines for a "magic triangle" computation. Notice "roast chicken" at lower left of Figure 1. See the Introduction to this volume and Figures 84, 85, and 88. See also Color Plate 1.

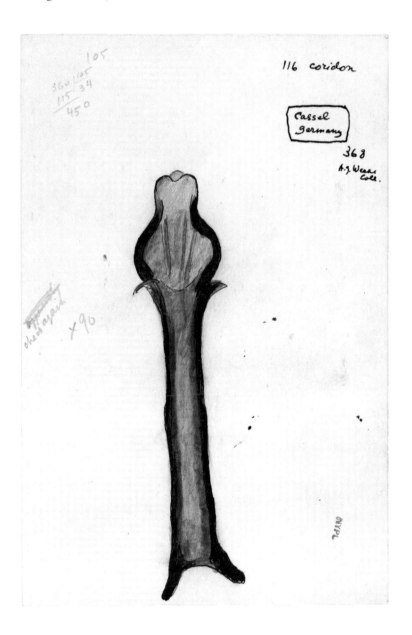

Figure 3 [*cormion, coridon*] Figures 1–3: In July 1938, Nabokov netted two unusual Blues "on the flowery slopes above Moulinet" in the Maritime Alps of southern France. The specimens resembled no known species, and he wondered whether they might be "the freakish outcome of such evolutionary gropings which fashioned a few specimens in the season of 1938, never to bring out that particular make again." Nonetheless, in 1941 he described them as *Lysandra cormion*, a name compounded from two species whose traits they shared, *L. coridon* (the Chalkhill Blue) and *Meleageria daphnis* (Meleager's Blue). Experiments later showed this "new species" to be a hybrid of those species after all. The first drawing depicts the male genitalia of *L. coridon*; the second, just its aedeagus, or penis; and the third, the male genitalia of "*L. cormion*."—Robert M. Pyle

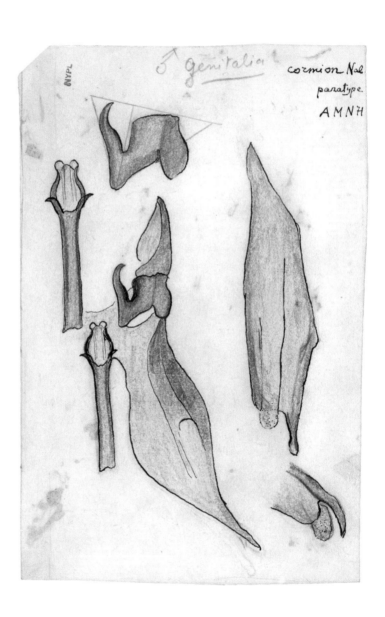

Figure 4 [*menalcas*];
see also Color Plate 5

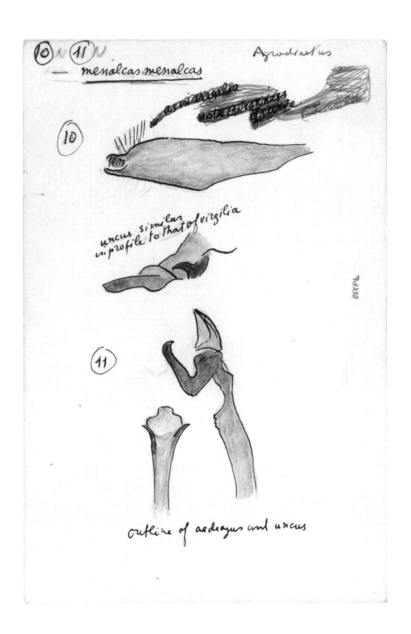

Figure 5 [*iphigenides*];
see also **Color Plate 5**

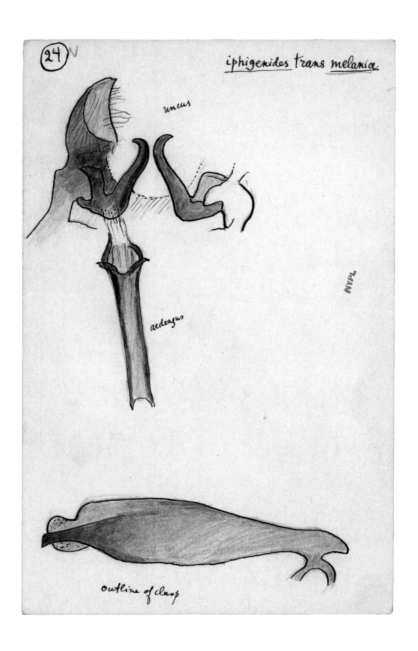

Figure 6 [*admetus*];
see also **Color Plate 11**

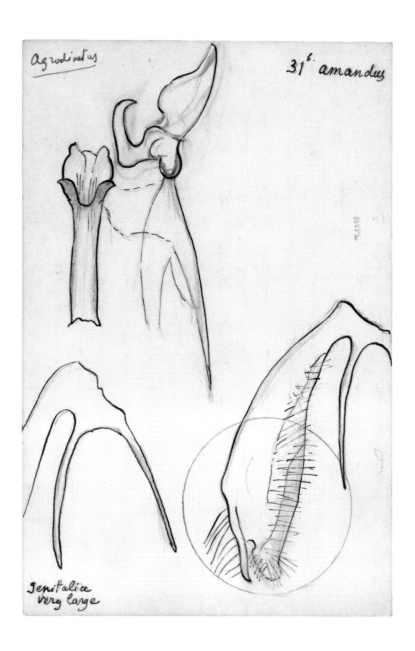

Figure 7 [*agestis*]

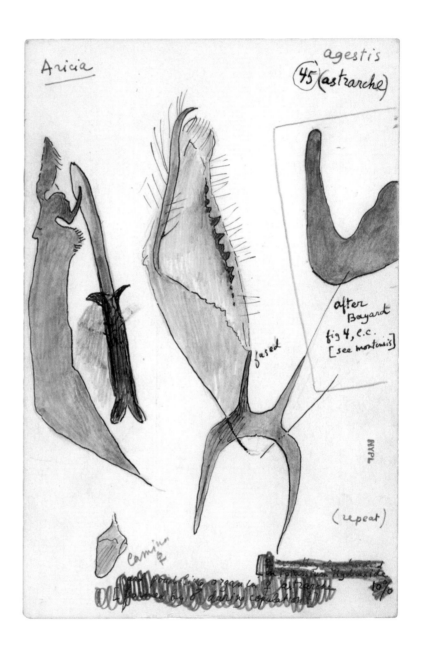

Figure 8 [*agestis*] Figures 7 and 8: Genital characteristics of specimens identified as *Aricia agestis* (Denis & Schiffermueller 1775), the Common Brown Argus, which is distributed across Eurasia to the Bering Strait. Mindful of characteristics of the female (often hard to find in collections), Nabokov appears to draw the terminus of the female genitalia (labeled "lamina" by him—also often called the lamellae). A note says "repeat," evidently a reminder to check additional specimens. See Color Plate 14 and the essay by Pierce and colleagues in this volume.

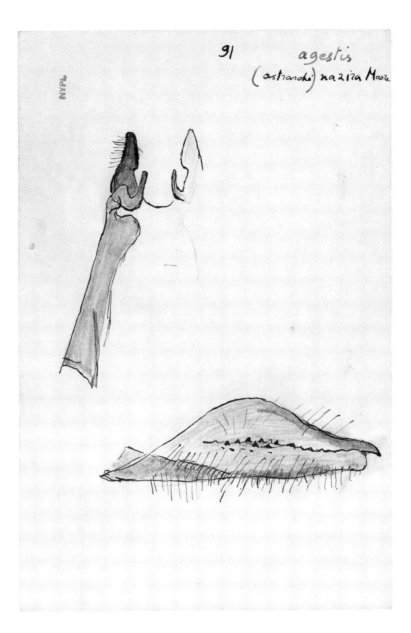

Figure 9 [fifty-four specimens]

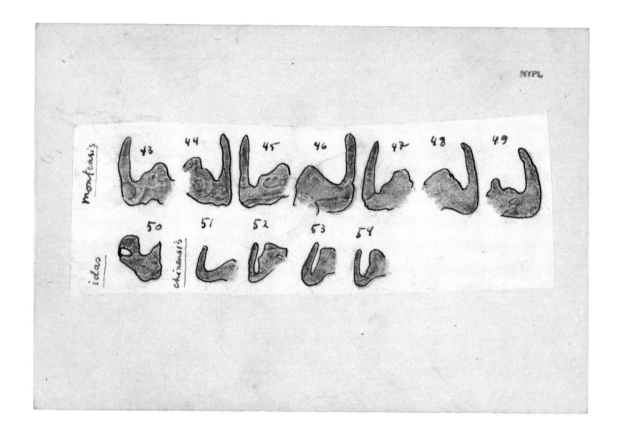

Figure 10 [fifty-four specimens, continued]

Figures 9 and 10: Meticulous about comparative detail, Nabokov sought to determine which characteristics of blues genital anatomy were most informative about relationships. On these cards Nabokov draws lateral views of the falx/humerulus from fifty-four specimens historically assigned to the genera *Plebejus* Kluk 1780 or *Aricia* Reichenbach 1817 (see Color Plate 14). The falx/humerulus in lycaenids has since been determined to be quite variable and only somewhat informative about relationships. However, Nabokov drew additional, and extremely useful, information by deriving triangulated measurement data from the falx/humerulus combined with the other structures of the male genital's dorsal terminus (his "magic triangle" method): see Figures 84, 85, and 88 and Color Plate 37.

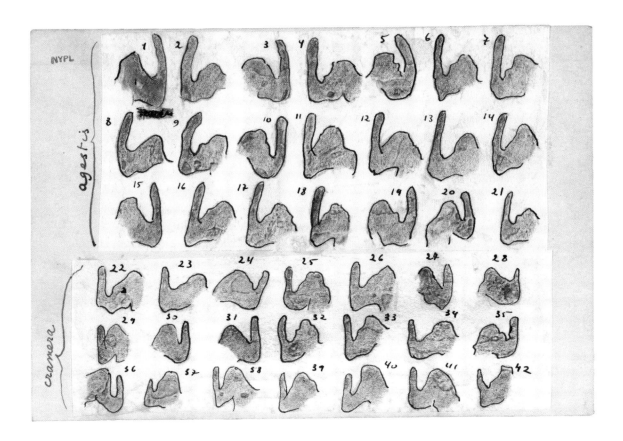

Figure 11 [*cramera*]   Study of the unclear relationship of widely distributed Blues of *Aricia* Reichenbach 1817 or *Plebejus* Kluk 1780. After studying the Eurasian Common Brown Argus (see Figures 9 and 10), whose distribution reached the biogeographically pivotal Bering Strait, he turned his attention to the Southern Brown Argus—species *cramera* Eschscholtz 1821. Usually placed in *Aricia*, this Blue extends into North Africa and westward onto nearby Atlantic islands.

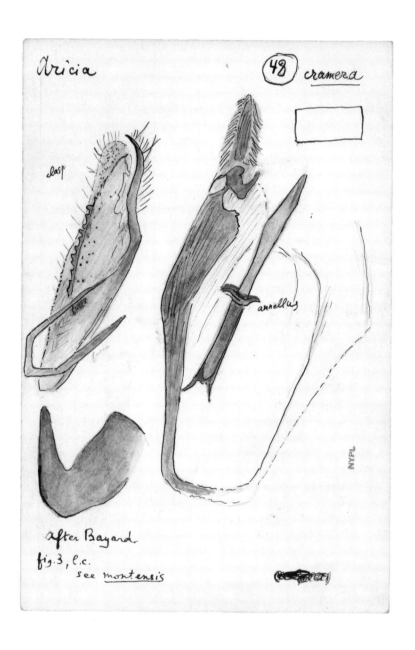

Figure 12 [*semiargus*] Characteristics of the genus *Polyommatus* Latrielle 1804. It can be considered the flagship of the tribe of Blues Nabokov studied—the Polyommatini. Here we see the species *Polyommatus semiargus* (Rottemburg 1775), the Mazarine Blue, well known across Eurasia and even reported in some transplanted populations in Oceania. The entire genital apparatus is shown at the bottom left in three dimensions, something uncommon in Nabokov's lab drawings. We connect falx and humerulus here because they represent, respectively, the bent forearm and base of a contiguous structure. *Polyommatus*, extremely divergent and including many

species names, had historically been used only for Eurasian Blues. However, members of the wider tribe associated with *Polyommatus* make up precisely the sample Nabokov was faced with in trying to decipher the identity and relationships of New World Blues. Many other genera Nabokov studied, including *Agrodiaetus* Huebner 1822, *Lysandra* Hemming 1933, and *Cyaniris* Dalman 1816, have, over time, been considered by some lepidopterists to be the same as *Polyommatus,* with DNA studies as recent as 2010 only beginning to show them as distinct.

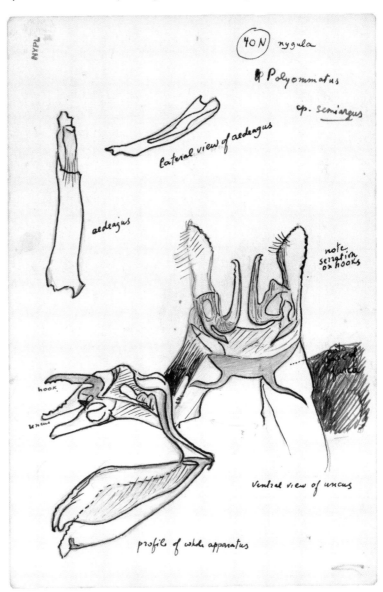

Figure 13 [*vogelii*] *Aricia* Reichenbach 1817 / *Plebejus* Kluk 1780 investigations (see Color Plate 14). Characteristics of a specimen identified as *vogelii*—*Plebejus vogelii* (Obertheur 1920). On the card, Nabokov wonders about the truly proper assignment for this entity—"? Plebejus" (and, below that "... Aricia ... ?")—and its relationships to a number of other species-level taxa listed at right and below.

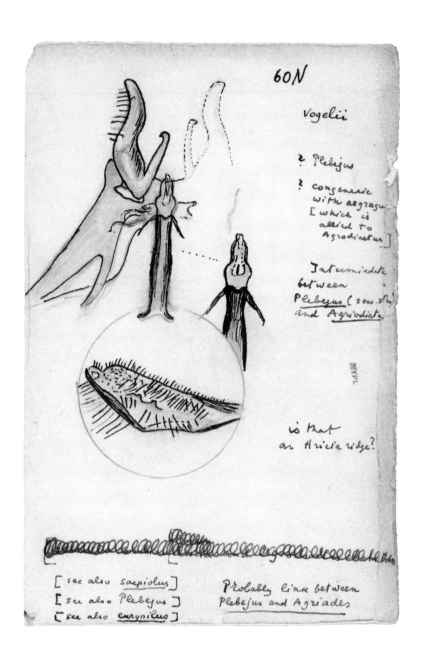

Figure 14 [*Iolana*]   Genital features of a distinctive but often poorly known genus of Old World Blues—*Iolana* Bethune-Baker 1914. These are distinctive Blues because they are larger than most others but the handful of species or subspecies making up the genus are also poorly known, extending in reported range from Europe across the Middle East and into the Siberian region. This probably explains Nabokov's single card on which many details about these odd Blues are recorded.

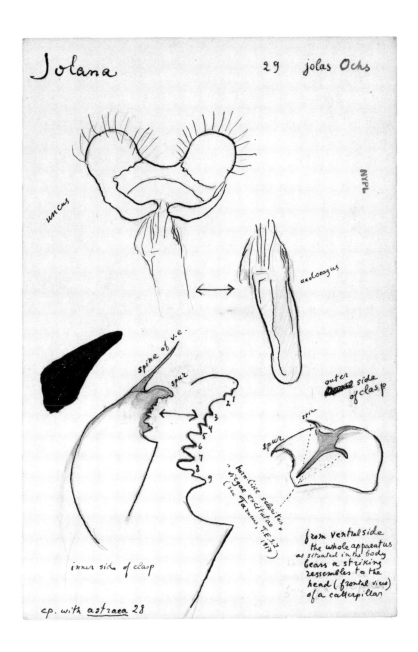

Figure 15 [*helena*] Whether Nabokov has suspicions about the species *helena* figured on this card cannot be determined because he makes no notes. However, *helena* is an Old World Blue of the genus *Cyaniris* Dalman 1816 and is today placed as *Cyaniris semiargus helena* (Staudinger 1862). *Cyaniris* is a genus confirmed only in 2010 to be different in DNA than the Blues of *Polyommatus* Latrielle 1804. Nabokov notes here that *helena* was placed in either *Polyommatus* or *Cyaniris* (see Figure 12), depending on the lepidopterist.

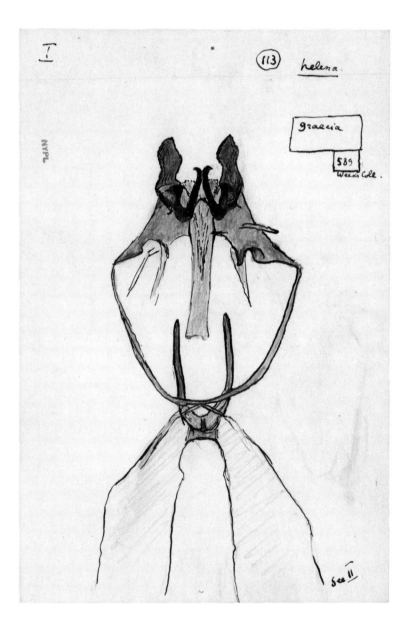

Figure 16 [*hyrcana*] Genitalia of a poorly known Blue from the region of Turkey east and northward into Asia. It exemplifies the problems of taxonomists with poorly known entities from puzzling geographic areas. The specimen is identified as "hyrcana," referring to *Lycaena hyrcana* (Lederer 1869). Since then, it had been placed in *Cupido* Schrank 1801, *Albulina* Tutt 1909, *Plebejus* Kluk 1780, *Polyommatus* Latrielle 1804, and several other genera! Although Nabokov does not note the locality, he likely would have tried to find a specimen from close to the locality from where *hyrcana* was named (the type locality), which is generally recognized as Astrabad, Persia (today known as Gorgan, Iran). On the edge of Asia, this was certainly an area that interested Nabokov when it came to Blues. The distinctness of the recorded traits suggests that it is a highly unusual Blue. Indeed, a Russian taxonomist has placed *hyrcana* in a new genus, *Farsia* Zhdanko 1992.

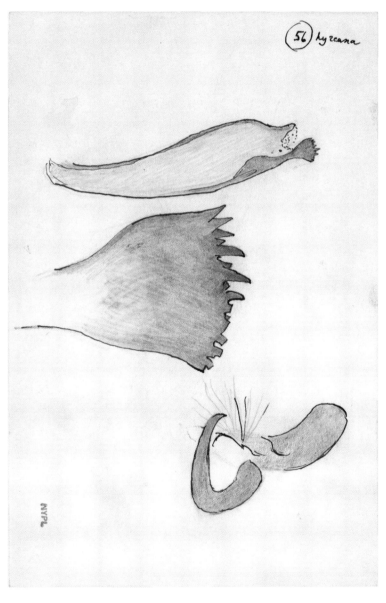

Figure 17 [*lycormas*]   *Glaucopsyche lycormas* (Butler 1866) is a little-studied Old World Blue generally known from only a few localities in the Lake Baikal region of central and southeast Siberia. Nabokov wanted to ascertain its relationship to *G. cyllarus* (Rottemburg 1775) (today considered the same Blue as an earlier name, *G. alexis* (Poda 1761)). Stating here that the two are, to his view, "allied," he reflects the modern view that *alexis* is considered a distinct species, with a distribution extending into both the heartland of Europe and the more eastern regions of Siberia. The relationship was thus germane to Nabokov's considerations of Beringia as a migration corridor in the evolution of Blues. *Glaucopsyche alexis* is endangered across most of its European range, recalling that *Glaucopsyche* hosts the only extinct New World species of these Blues, *G. xerces* (Boisduval 1852), which was lost in the 1940s due to the decimation of its dune habitats in California.

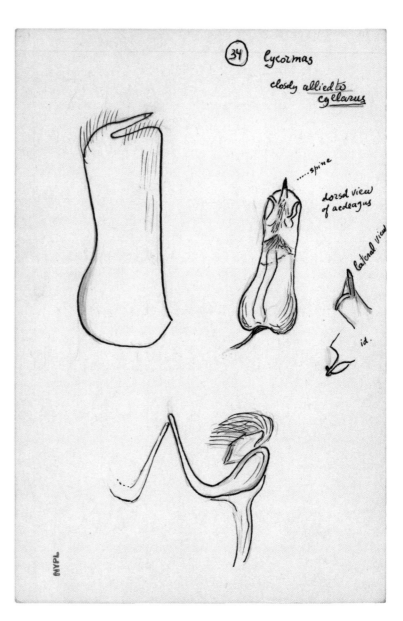

Figure 18 [*persephatta*] Genitalia of an extremely confusing Blue from the Trans-Altai Mountains on the border of Tajikistan and Kyrgyzstan. Nabokov suggests many generic assignments for this Blue *persephatta* Alferaky 1881, today generally placed in a separate genus, *Eumedonia* Forster, 1938. He remarks that the butterfly has the "body of a Lycaeides!" At bottom are details of the valve's hooklike terminal structures, which might have been key to his view of Beringian migrations to the New World.

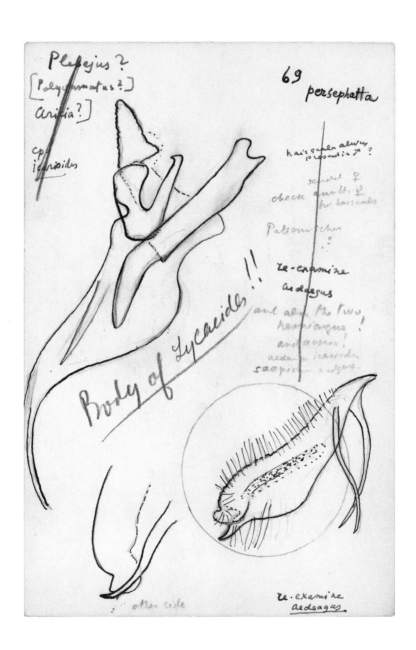

Figure 19 [*icarus*]   One of the world's most common Blues, *Polyommatus icarus* (Rottemburg 1775), the Common Blue. It occurs across Eurasia and North Africa and has been introduced in northeastern North America. Nabokov chooses a specimen from Manchuria, a northeast Asian extreme for this Blue's distribution, relevant to his view that New World Blues originated through waves of migration across the Bering Strait.

Figure 20 [dorsal genital terminus comparative, Old World Blues]

Line drawings of the dorsal terminus of the male genitalia of several Old World Blues.

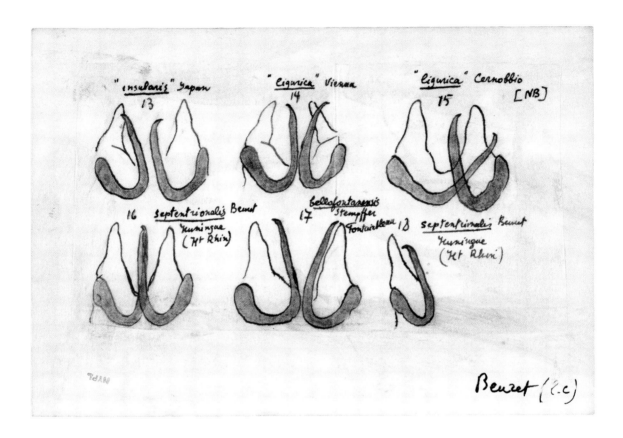

Figure 21 [*argyrognomon, idas*] Blues of the *argyrognomon* Bergstraesser 1779 and *idas* Linnaeus 1761 complexes, two very old names with possibly broad Old and New World distributions, were quite a puzzle in Nabokov's day. Here he resorts to detailed comparisons of the coxcomblike structures at the terminus of the male valve to show differences across a number of names associated with this complex to determine which are conspecific (the same species). Unfortunately, modern DNA studies have shown that in these complexes, identification by wing pattern and genitalia is often futile; the actually evolutionary relationships are far more complex and resolvable only by DNA analysis.

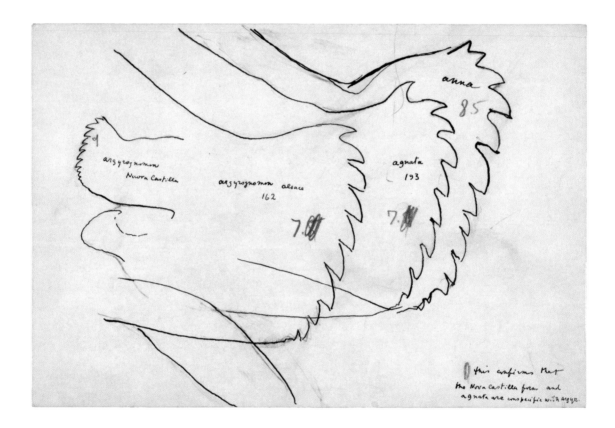

Figure 22 [*argyrognomon, idas*]   Another avenue Nabokov investigated to try to find reliable characteristics by which to distinguish Blues otherwise confusingly alike in wing pattern was study of wing scale anatomy. Here he depicts two wing scales from the *argyrognomon* Bergstraesser 1779 and *idas* Linnaeus 1761 complexes. However, it has turned out that in these transcontinental complexes of Blues, only DNA analysis is really reliable for identification.

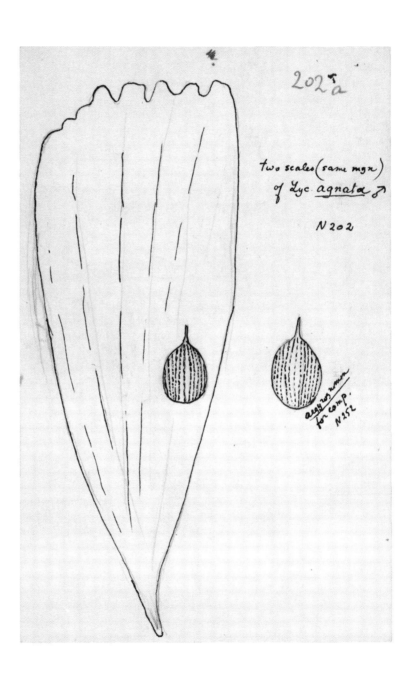

Figure 23 [*piasus*] When Nabokov confronted the New World Blues he had to evaluate the uses of the generic Old World names *Scolitantides* Huebner 1819 and *Glaucopsyche* Scudder 1872. He figures the species *G. piasus* (Boisduval 1852), the well-known and widely distributed Arrowhead Blue, associated by various taxonomists with both of these old generic names.

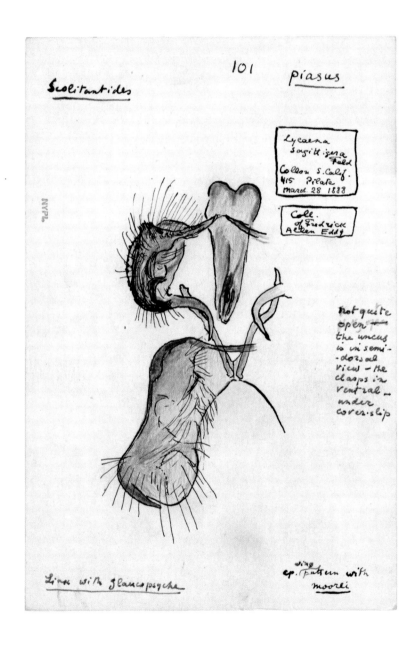

Figure 24 [*cleotas*]   In looking at Blues from the far reaches of the world Nabokov here draws a lateral view of the New Guinean Blue *cleotas* Guérin-Méneville 1831, which he places in "Chil," the Old World genus *Chilades* Moore [1881]. Inset: a figure showing spots on the undersurface of the hindwing. Modern New Guinea lists put this species in *Luthrodes* Druce 1895.

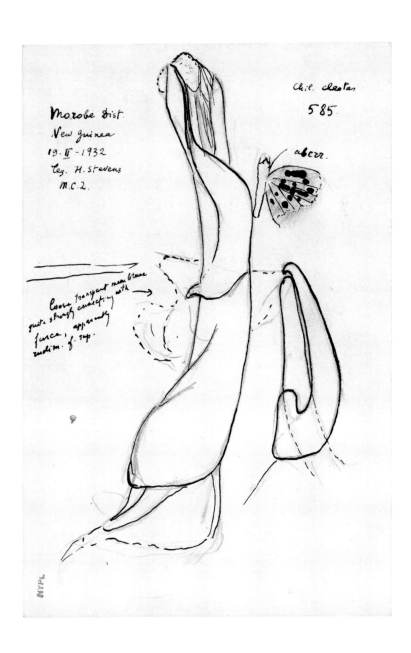

Figure 25 [furci of *Plebejus, Lycaeides*] A structure called the furca in male butterfly genitalia. The shape of the furca is compared among Blue butterfly species: (1) *Lycaeides idas*, (2) *Plebejus insularis*, (3) *Plebejus cleobis*, (4) *Lycaeides melissa*, and (5) *Plebejus argus*. *Lycaeides idas* and *L. melissa* are found in the New World, whereas *Plebejus insularis*, *P. cleobis*, and *P. argus* are found in the Old World. All these Blues figured in his studies of *Lycaeides* Huebner 1819, often seen now as a subgroup within the wider distributed worldwide *Plebejus* Kluk 1780 assemblage.—Lauren K. Lucas, Matthew L. Forister, James A. Fordyce, Chris C. Nice, and Kurt Johnson

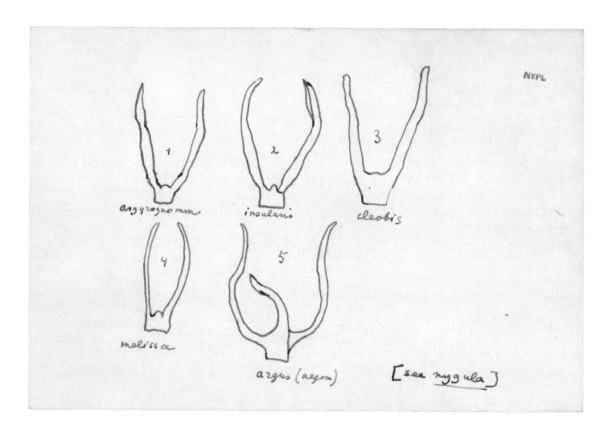

Glaucopsyche

Figure 26 [*Turanana*] Odd genital structures of the Old World Blue group *Turanana Bethune-Baker 1916*, exploring its status and relationship in relation to genera named in the New World—*Scolitantides* Huebner 1819 and *Glaucopsyche* Scudder 1872. Nabokov figures the species *T. panagea* (Herrich-Schaeffer [1851]), a species name often confused and combined with other Old World Blues names because of its geographic distribution along the cusp of Europe and the Middle East. The structures are obviously far more robust than in many Old and New World Blues, with the broad and club-ended male valve at center left, enclosed by a narrow genital ring attached upward to a lobate uncus and a very narrow and curvate falx/humerulus.

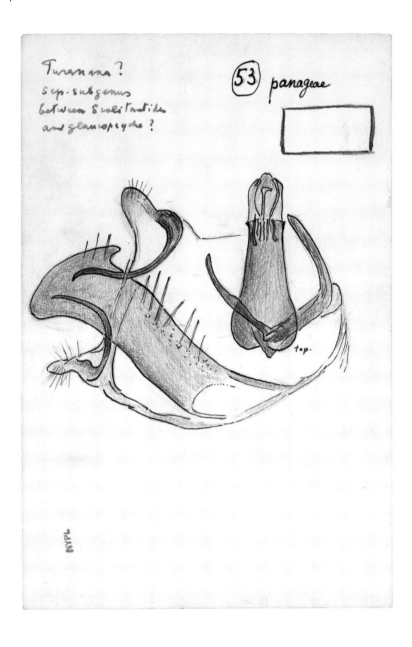

Figure 27 [*Glaucopsyche* wing pattern] Once Nabokov begin studying the Blues of the New World, they presented the same nomenclatural problems as those of the Old: there were many old names, but what did they apply to beyond the species originally named with them? *Glaucopsyche* Scudder 1872 and *Scolitantides* Huebner 1819 were two good examples, the best-known species being common and widely distributed. This complicated their relationship to names available from the Old World. The simple figure is adequate because species of this genus have a generally unmarked light ground color with prominent dark orbs running across the wing.

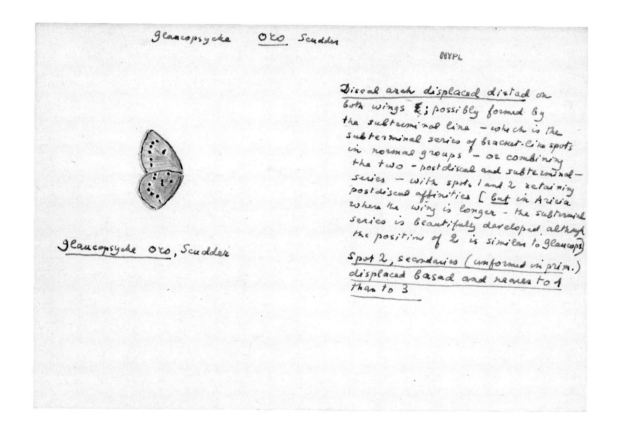

Figure 28 [*oro*]   Studying species of the New World genus *Glaucopsyche* Scudder 1872, Nabokov sought to sort out the relationship of available Old World and New World names to a transcontinental evolutionary view of the Blues. Following the wing pattern of *oro* Scudder 1876 (see Figure 27), we see here the genital anatomy of *oro* (today considered a subspecies of *lygdamus* Doubleday 1842). The drawing shows the dissection teased open with all the parts still contiguous.

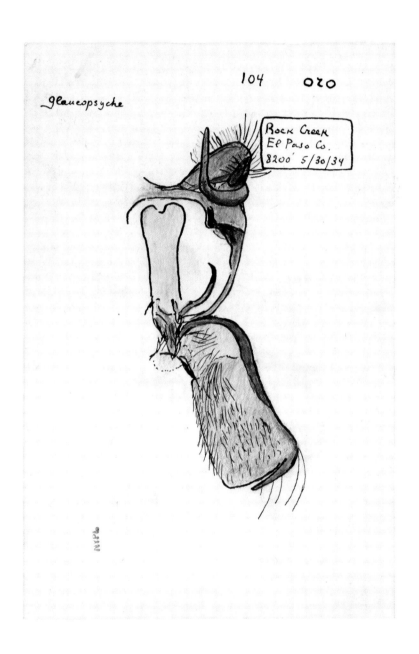

Figure 29 [*laetifica*] *Glaucopsyche laetifica* (Puengeler 1898) interested Nabokov because of its few specimens hailing from where the mountain regions of today's China abut the Lake Baikal region and central and eastern Siberia. As such, its evolutionary identity was part of the larger question of the proposed Beringia corridor to the New World.

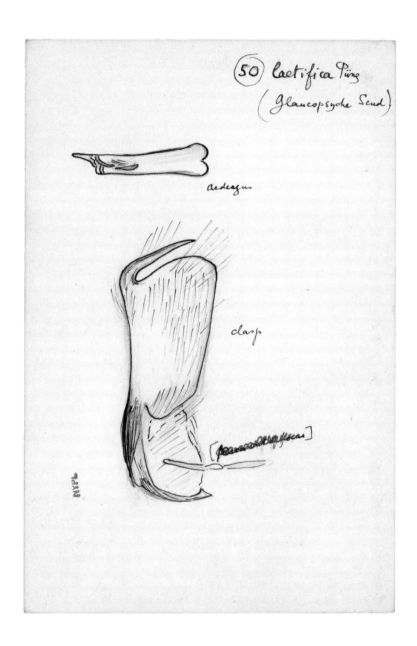

Figure 30 [*shasta*] The conventional nomenclature for North American Blues used some generic names from the Old World and others coined by earlier American lepidopterists. Mirroring the situation with Old World *Agrodiaetus* Huebner 1822, *Plebejus* Kluk 1780, *Polyommatus* Latrielle 1804, and others, entire complexes of species needed to be evaluated and classified, and Nabokov was soon to add new names as he came to note the striking difference among North American Blues both in the field and in the lab. Here he examines the well-known species *shasta* W. H. Edwards 1862 (the Shasta Blue). His notes indicate he suspects this specimen represents a new subspecies of *shasta:* "ssp. nov."

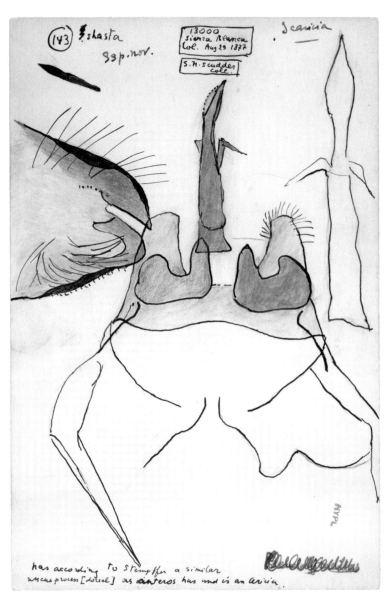

Figure 31 [*shasta*]   In a more detailed drawing of *shasta* W. H. Edwards 1862, Nabokov indicates that he will place it in his new North American genus *Icaricia* Nabokov 1949.

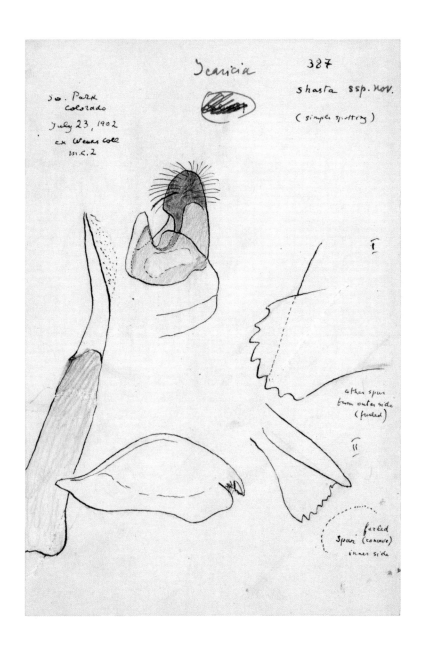

Figure 32 [*icarioides*]   Nabokov's genus *Icaricia* 1949 was named to include some of the best-known but taxonomically confusing North American Blues. One of these—*icarioides* Boisduval 1852—showed a widespread distribution full of the kinds of circles of races (*Rassenkreis*) that had become notoriously befuddling to lepidopterists of the Old World. It was the *icarioides* complex, and others, to which Nabokov's young friend, the eventual blues expert John C. Downey, would apply numerical computer methods in the 1960s and 1970s to try and sort out the complexes of subspecies, races, and food-plant related strains. Today there are nearly thirty subspecies associated with *icarioides*, one of which is *pheres* Boisduval 1852, shown here.

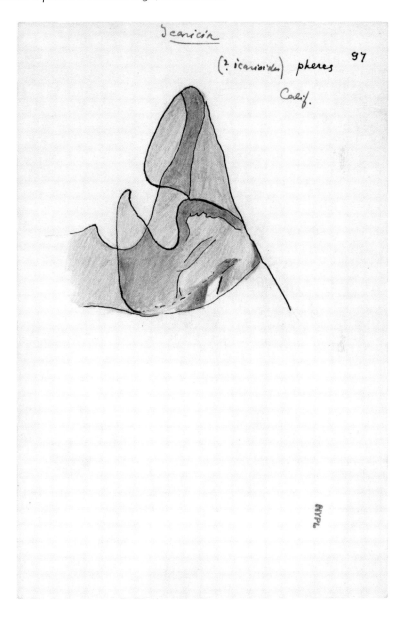

Figure 33 [*neurona*] Genital characteristics of the distinctive West Coast North American species *neurona* Skinner 1902 (the Veined Blue), which today is included in the genus *Icaricia* Nabokov 1949. Although there is still some disagreement today, most lepidopterists include about a half-dozen North American species in Nabokov's *Icaricia*. For a number of decades, *Icaricia* was either not used or considered a subgenus.

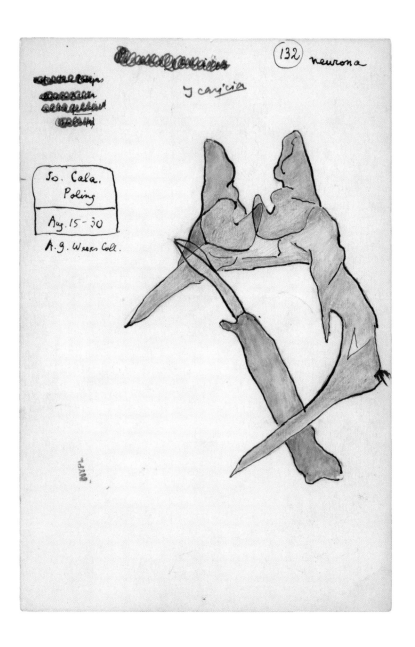

Philotes [North America]

Figure 34 [*rita*]   Nabokov labels this drawing *Philotes*, that is *Philotes* Scudder 1876 and cites the species *rita* Barnes & McDunnough 1916. Since Nabokov's time, however, specialists on western North American Blues have restricted *Philotes* to a unique dune habitat species, *sonorensis* C. & R. Felder [1865], of which at least one subspecies is, because of habitat destruction, considered extinct. In 1978, American lepidopterists recognized *Euphilotes* Mattoni [1978] as the genus containing the taxon that Nabokov renders here rather crudely. It is likely that Nabokov considered the very restricted ranges of these butterflies—in desert washes and outcropping of American's Southwest—as less relevant to his concerns about the Blues of the Old and New World's northern forests and his questions about Beringia.

Figure 35 [*idas/longinus* hybrid] Eedeagus and pair of unci from a male, hybrid *Lycaeides idas longinus* specimen from Jackson Lake, Wyoming. This is an intermediate genitalia in size and shape compared to *L. idas* from Canada in Figure 37 (and referred to as "*argyrognomon*" in the drawing) and *L. melissa* in Figure 39.
—Lauren K. Lucas, Matthew L. Forister, James A. Fordyce, and Chris C. Nice

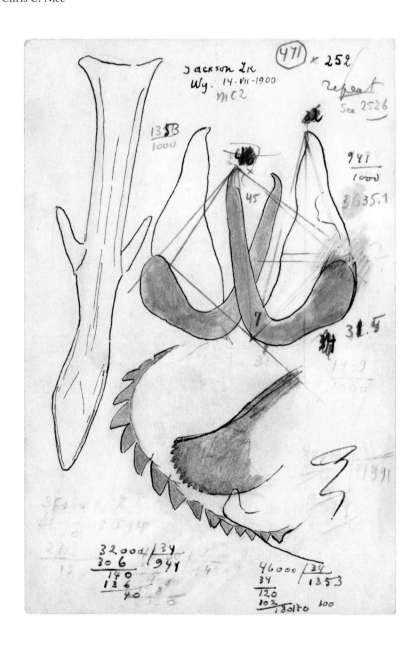

Figure 36 [*melissa*]

Dorso-terminal male genital elements from a specimen of *Lycaeides melissa* (W. H. Edwards 1873) from Wyoming.

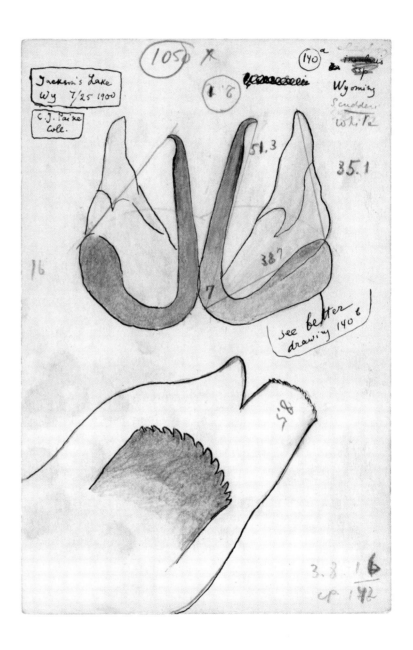

Figure 37 [*anna*] Drawing of a pair of unci from a male *Lycaeides anna* specimen from Calgary, Alberta. *Lycaeides idas* genitalia are characteristically shorter and wider in shape compared to *L. melissa* genitalia.—Lauren K. Lucas, Matthew L. Forister, James A. Fordyce, and Chris C. Nice

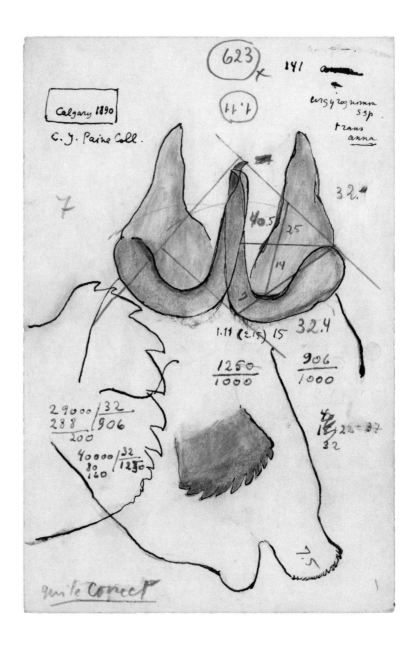

Figure 38 [unlabeled] In this unlabeled rendering, Nabokov presents an inked line drawing showing the lateral view of an entire male Blue butterfly genitalia, with the valve included only as dotted lines. Below it, he shows a detail of the terminal one-third of that valve.

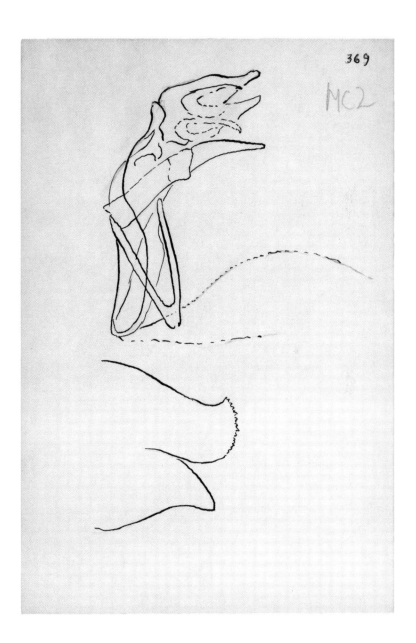

Figure 39 [*melissa*] A pair of unci from a male *Lycaeides melissa* (W. H. Edwards 1873) specimen from El Paso County, Colorado. The falx is colored more darkly than the rest of the figure and consists of the forearm (F); the long, narrow structures oriented up and down in the figure (the raised fists of the pugilists); and the elbow (E), or the bend at the bottom of the forearm that connects to the lateral humerulus (H). Nabokov also measured part of the uncus (paired structures with lighter shading in the drawing that he likened to Ku Klux Klan hoods) from the end of the humerulus to its tip. There is another male *L. melissa* genitalia illustrated in the lower left of the drawing. *Lycaeides melissa* unci are characteristically longer and skinnier in shape than *L. idas* (Linnaeus 1761) unci.

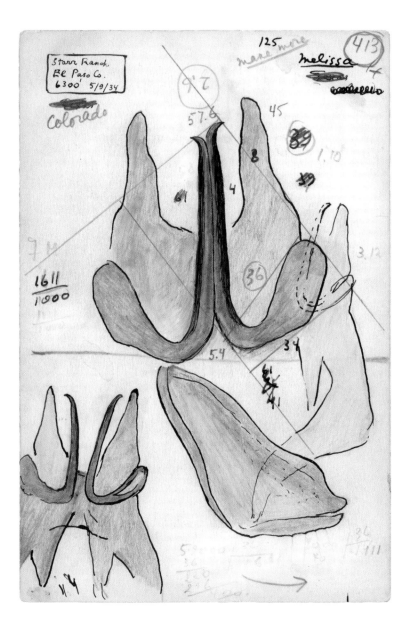

Figure 40 [*samuelis*] This drawing features an uncus (center) from a male *Lycaeides melissa samuelis* (Karner Blue) specimen from Albany, New York, similar to Nabokov's drawing of the holotype (Figure 88). The rest of the drawing illustrates other parts of the genitalia that Nabokov did not use for taxonomic designations of Lycaeides.—Lauren K. Lucas, Matthew L. Forister, James A. Fordyce, and Chris C. Nice

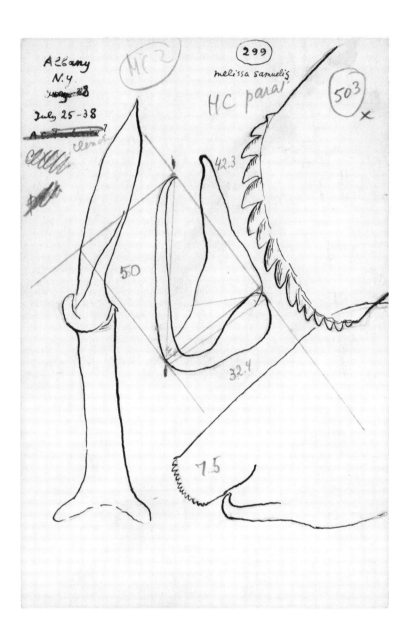

Figure 41 [*samuelis*] *Lycaeides melissa samuelis,* illustrating black spot position on a forewing and hindwing, elements that Nabokov considered defining characters of the Karner Blue. This drawing is similar to Color Plate 40, which emphasizes the alignment of the black spots on the forewing (left) and hindwing (right) of a different *L. melissa* specimen.—Lauren K. Lucas, Matthew L. Forister, James A. Fordyce, and Chris C. Nice

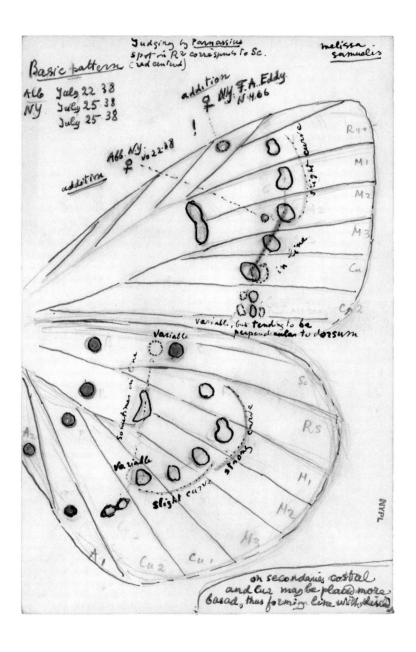

Figure 42 [*melissa*]

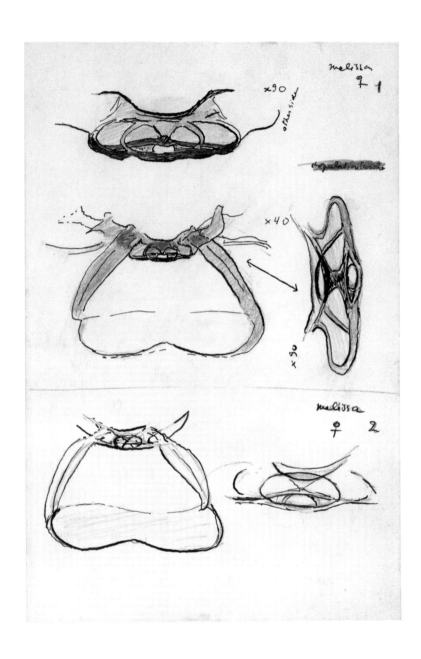

Figure 43 [*melissa*]   Figures 42 and 43: Although Nabokov dissected and drew structures found in female *Lycaeides melissa* genitalia, as shown in these two drawings, he did not use these female structures for taxonomic designations of *Lycaeides*.—Lauren K. Lucas, Matthew L. Forister, James A. Fordyce, and Chris C. Nice

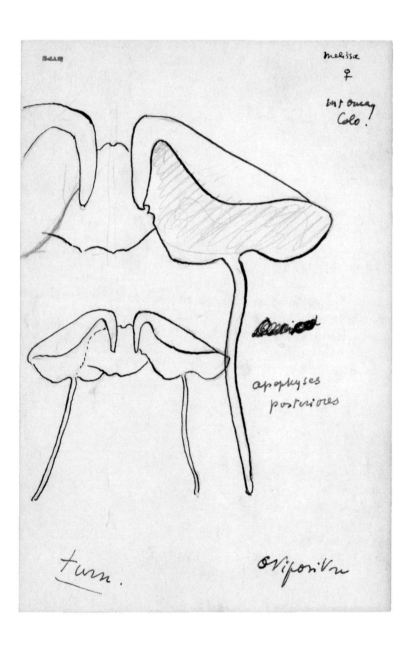

Latin America

[Cyclargus]

Figure 44 [*Cyclargus*] This card shows the burst of detail and activity in Nabokov's notes that reflect discovery of an unmistakably new taxon. Here he has already noted the new generic name "Cyclargus" and "n. sp.," and even suggests a species name, "caimanella," subsequently abandoned. This was because he later found, well beyond this Cayman Island specimen, other new Blues of this distinctive group across the entire Caribbean region. Ultimately, Nabokov published the new genus name *Cyclargus* Nabokov 1945 for this Caribbean group, distinguishing it from Blues long associated with Huebner's old name *Hemiargus* (1818), noting that they differed drastically in anatomy. Differing from *Hemiargus* Blues, *Cyclargus* Blues have a distinct structure around the aedeagus that Nabokov called the sagum (shown here as the winglike expansions on the aedeagus drawn in the lower right). Also unique are distinctive coxcomblike structures at the terminus of the male genital valve shown here in the lower left in both general rendering and detail. His 1945 paper on Neotropical Blues featured these data prominently, but the paper was not taken seriously by contemporaries and his names were officially ignored in 1975, when Norman Riley's guide book on the West Indies was published. The wealth of notes and measurements on this card show Nabokov's concentration on details when new entities were discovered.

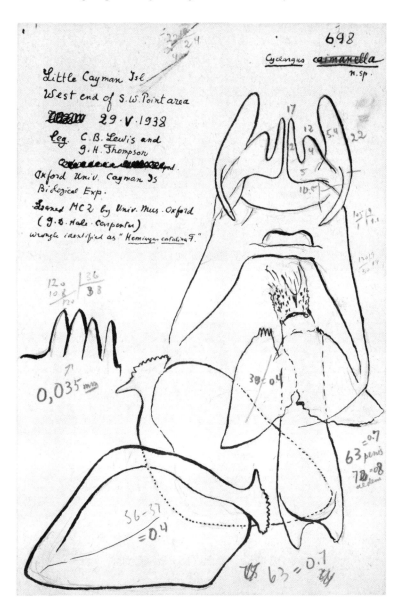

Figure 45 [*Cyclargus*]   In his 1945 publication on Latin American Blues, Nabokov prominently mentioned and illustrated the coxcomb-ended male valve, shown here at center in lateral view with the comb oriented left and then drawn in detail just below. It is inexplicable how other experts on Blues at the time overlooked such characteristics and considered *Cyclargus* Nabokov 1945 as indistinct from *Hemiargus* Huebner 1818. Given Nabokov's worldwide knowledge of the anatomy of Blues, it is probable that he remembered seeing similarly unique spines among Old World Blues in the outstanding genus *Iolana* Bethune-Baker 1914 (see Figure 14). These are not the same structures, of course (not homologous), but they most likely influenced Nabokov with regard to their significance as informative characteristics.

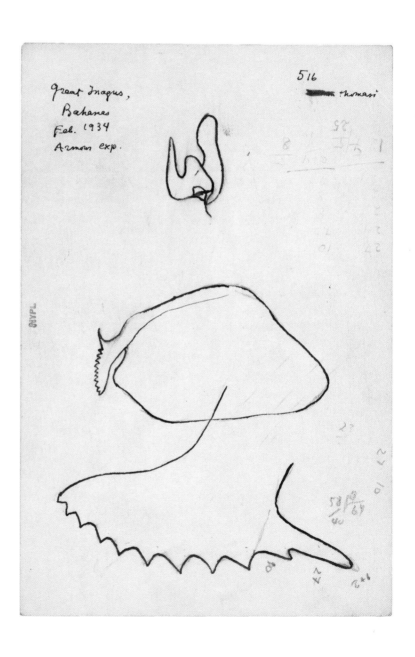

Figure 46 [*thomasi, erembis*]   Still studying what would become *Cyclargus* Nabokov 1945, Nabokov exhibits fascination with the prominent spinate and coxcomb structures at the terminus of the male valve that in *Hemiargus* (the old Huebner name) showed only a singular, nonspined, pencillate to curvate terminus. Nabokov likely remembered that such unique spines typify an outstanding genus in the Old World (*Iolana* Bethune-Baker 1914; see Figure 14). Here, Nabokov has recognized two species of *Cyclargus: erembis* Nabokov 1948 (published instead of "caimanella"—after his 1945 paper) and *thomasi* Clench 1941 (originally named by Clench in *Hemiargus*). Nabokov counts the teeth on the coxcomb clasper terminus and suggests triangulations on the major terminal spine.

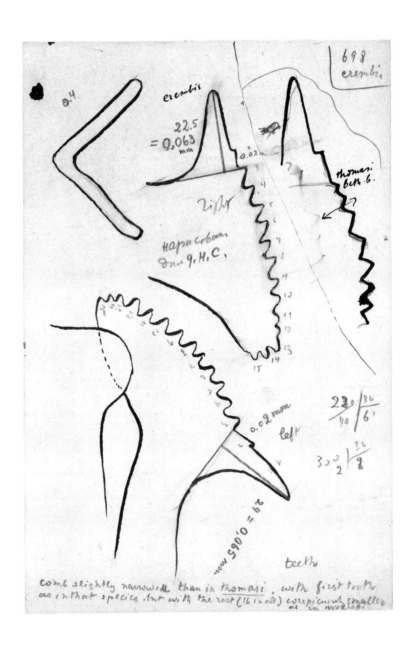

Figure 47 [*bethune-bakeri*] Confident of his distinction of *Cyclargus* Nabokov 1945 as a distinct genus, Nabokov includes in it a well-known Florida Blue that Comstock and Huntington had named in 1943 (as a member of *Hemiargus* Huebner 1818—*bethunebakeri* W. Comstock & Huntington 1943). It had become recognized as the southern United States subspecies of what Clench had named *Hemiargus thomasi* in 1941. Removing these common Blues out of *Hemiargus* to a new genus, especially when they looked so much alike, was a direct challenge to orthodoxy. Here Nabokov portrays all the genital traits that show that this Florida entity is a member of *Cyclargus:* the sagum, above left, and depicted again just below, connected to the aedeagus, as well as the furca and the distinctive male valve.

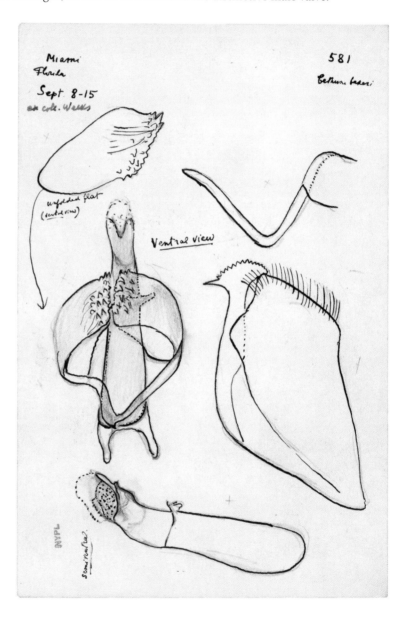

Figure 48 [*dominica*] Nabokov might have found it either ironic or humorous that even though lepidopterists doubted the reality of his genus *Cyclargus* Nabokov 1945 for decades, *Cyclargus* would eventually be understood to contain eight distinct Caribbean species and, within these, six regional subspecies distributed from the United States across the rest of the Caribbean region. What is more, its ranks would include another familiar Florida and Caribbean region Blue, the Nickerbean Blue (*ammon* Lucas 1857). Here Nabokov illustrates distinctive toothlike characteristics of the terminus of the aedeagus, known as cornuti in the distinct species *Cyclargus dominica* (Moeschler 1886) from Jamaica.

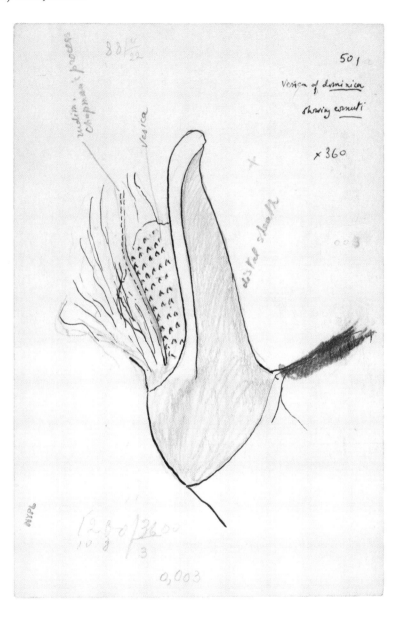

Echinargus

Figure 49 [*martha*] Nabokov assigned the South American species *martha* Dognin 1887 to his new genus *Echinargus* Nabokov 1945. After distinguishing the expansive spinate structure (sagum) surrounding the aedeagus (see Color Plate 33), Nabokov moved to drawing the elliptic male clasper with its hooked terminus (right), further details of the prominent dentate spines of the sagum (center), and the furca (below).

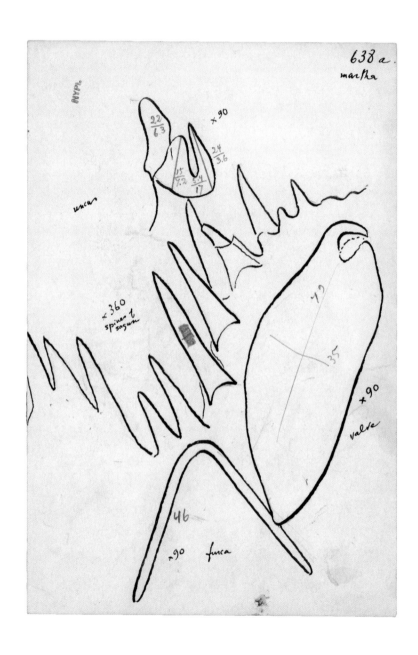

Figure 50 [*isola*] Nabokov's discovery that the common, widespread North American Blue *isola* Reakirt [1867] (known colloquially as Reakirt's Blue) showed the same radical internal structures as those in his new genus *Echinargus* Nabokov 1945 was unfortunate for his work's reception. For decades lepidopterists had been used to seeing this Blue placed in Huebner's genus *Hemiargus* 1818. The anatomic evidence, however, is clear, as Nabokov draws the expansive spinate sagum (bottom, center) wrapped about the aedeagus just above. The name *Echinargus* was not recognized widely until after the further work of taxonomists in the 1990s; old-time usages of *Hemiargus* persisted in some lepidopterist literature as late as 2012 (see the Introduction to this volume).

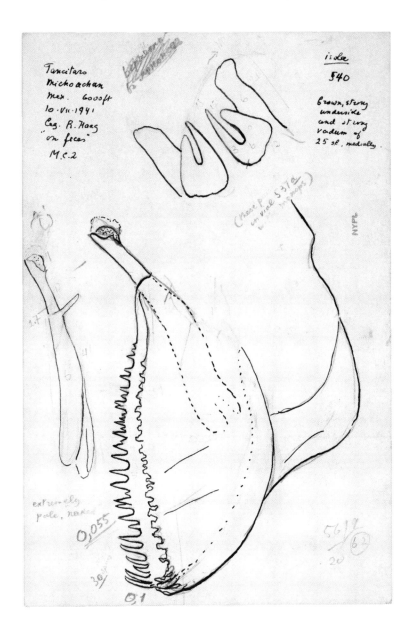

Figure 51 [*isola*]

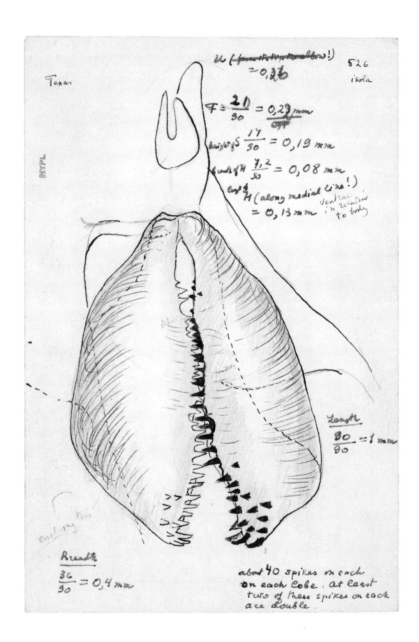

Figure 52 [*isola*]   Figures 51 and 52: Nabokov's dissection of a specimen of the familiar *isola* or Reakirt's Blue showed immediate evidence of a radically different genital apparatus than that of the other Blues historically often grouped with *Hemiargus* Huebner 1818. The gigantic spinate and bilobed structure surrounding the aedeagus, shown here (Figure 51) at center, dominated the genital ring and, perhaps, to some observers could be mistaken for the male valvae. In a line drawing in Figure 52, Nabokov again draws the dominating size of the sagum in the context of the surrounding genital ring (vinculum). Because the other species of this new genus were poorly known South American ones, most North American lepidopterists did not favor a new generic name for Reakirt's Blue, and *Echinargus* did not get traction for several decades.

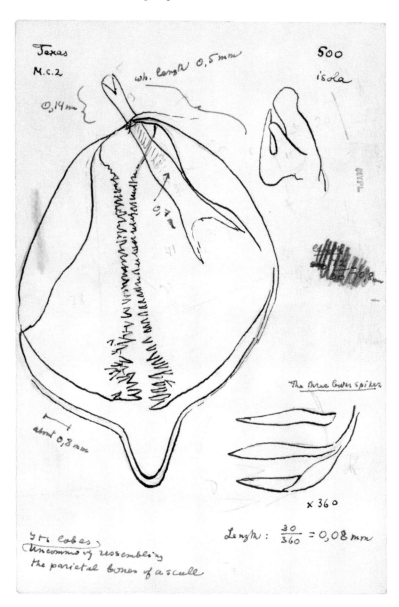

Figure 53 [*huntingtoni*] Studying Latin American Blues to determine which ones showed the distinctive anatomical characters of his new genus *Echinargus* Nabokov 1945, Nabokov draws the genitalia of the Trinidadian species that will later take the name *huntingtoni* Rindge and Comstock 1953. The prominent and rather bifurcate sagum (shown center), with a detail of the dentate terminus on one lobe, is less dominant than on other species he will assign to *Echinargus*—Andean species *martha* Dognin 1887 and widespread North American species *isola* Reakirt [1867]. For Nabokov, however, it marked *huntingtoni* as unquestionably a member of the *Echinargus* lineage. Nabokov did not name *huntingtoni*, out of respect for his friend William P. Comstock (its discoverer), who, he assumed, eventually would. However, because South American *huntingtoni* and *martha* so were little known to North American lepidopterists, Nabokov's name *Echinargus* seemed merely another for the common North American Reakirt's Blue. This made the name change for *isola* Reakirt [1867] unpopular and mostly unused for many decades.

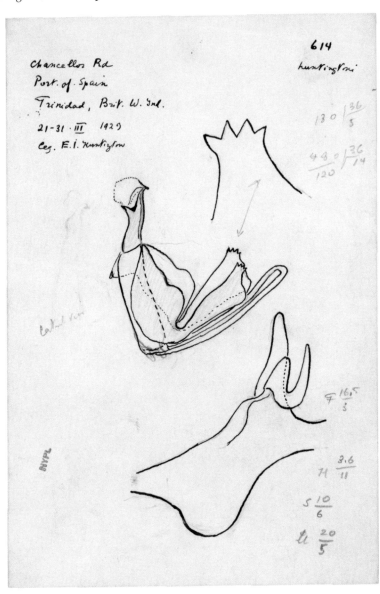

Figure 54 [*huntingtoni*]

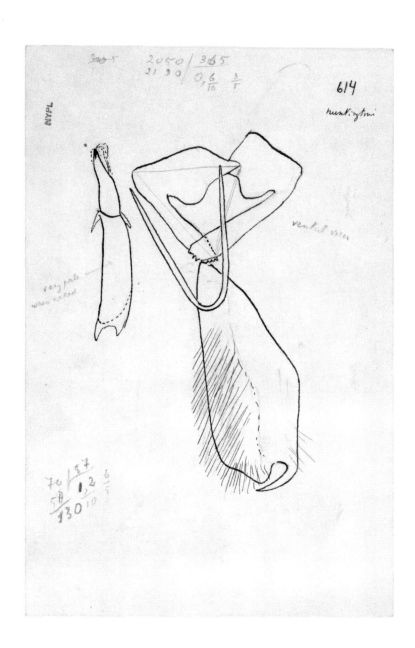

Figure 55 [*huntingtoni*] Figures 54 and 55: Although Nabokov did not name the species *huntingtoni* Rindge and Comstock 1953 (out of respect for Comstock, see Figure 53), Nabokov studied it thoroughly and referred to its existence (without the Latin name) in his 1945 publication on the Latin American Blues. Unfortunately, *huntingtoni* would not be named until after Comstock's death (by the subsequent curator at the American Museum of Natural History, F. H. Rindge, in 1953). This delay complicated lepidopterists' full appreciation of the distinctness of Nabokov's genus *Echinargus* Nabokov 1945. Always the meticulous worker, Nabokov here makes sure he has dissected and drawn the holotype, or defining specimen, of Comstock's new Blue. In Figure 54 he shows the base of the genital apparatus. In Figure 55 Nabokov labels the drawing as "holotype" and shows selected features, with measurements.

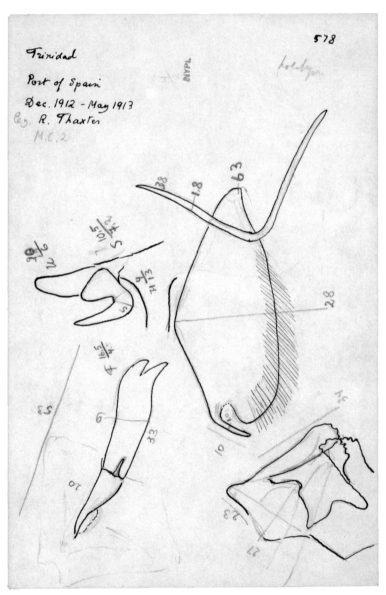

Hemiargus

Figure 56 [*Hemiargus*, redefinition] Nabokov's recognition that his new genus *Echinargus* Nabokov 1945, characterized in part by an expansive spinate sagum surrounding the male aedeagus, caused him to scour through the Latin American Blues then assigned to *Hemiargus* Huebner 1818 to determine how many other species actually belonged to *Echinargus*. One was the taxon *huntingtoni*, which had come to Nabokov's attention when proposed in a manuscript by his colleague William P. Comstock (see Figures 53–55). It is quite possible that Nabokov shared this card's contents with Comstock because the drawing distinguishes the slightly hooked male clasper of *Echinargus* (bottom left) from the elongately hooked terminus of the male clasper in *Hemiargus* (bottom center). Also drawn is the aedeagus typical of *Hemiargus* (above left) without a prominent sagum and, beneath it, the aedeagus and prominent sagum in *Echinargus martha* (Dognin 1887) and other comparative characteristics.

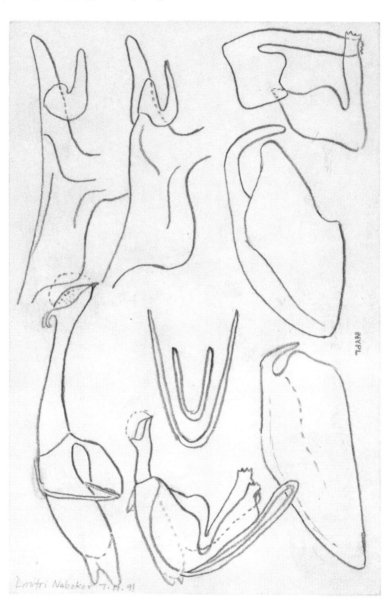

Figure 57 [*Hemiargus,* redefinition]

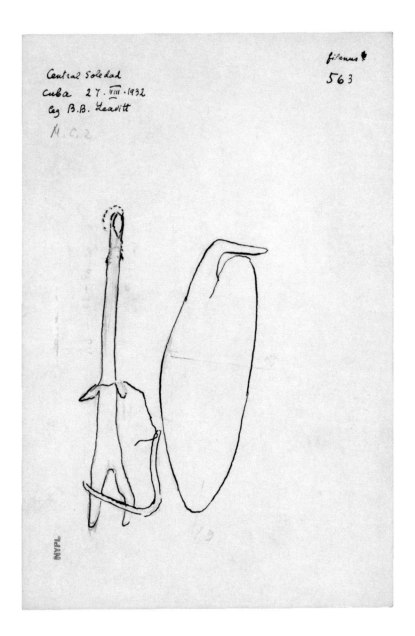

Figure 58 [*Hemiargus,* redefinition]

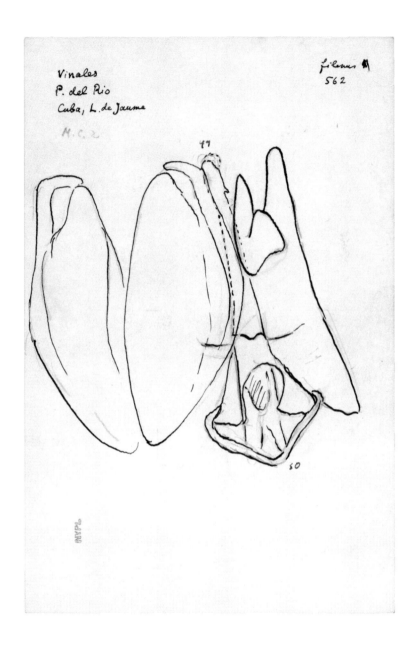

Figure 59 [*Hemiargus,* redefinition] Figures 57–59: Nabokov was the first to recognize that the elongately hooked terminus of the male valve and lack of a prominent sagumlike structure surrounding the aedeagus limited the number of species directly related to the defining species, or type species, of *Hemiargus* Huebner 1818 and thus the size of that genus. The expansive sagum (nearly dwarfing the aedeagus itself) in the lineage Nabokov would give the genus name *Echinargus* Nabokov 1945 further confirmed the reality that *Hemiargus* contained only a small number of species (two to five today, depending on the classification used). Accordingly, Nabokov's laboratory work led to him to distinguish three species of *Hemiargus* comprising some ten subspecific entities distributed across the New World: *ceraunus* Fabricius 1793, *gyas* W. H. Edwards 1871, *astenidas* Lucas 1857, *antibubastus* Huebner 1818, *filenus* Poey 1832, *hanno* Stoll 1790, *watsoni* W. Comstock & Huntington 1943, *bogotana* Draudt 1921, and *ramon* Dognin 1887, all listed here because they figure in Nabokov's drawings of the entities of this genus and, even today, are variously recognized as species or subspecies by modern lepidopterists, depending on the regional usage. Figures 57–59 depict specimens from Cuba (*filenus*) and Trinidad (*gyas*). The Trinidad (*gyas*) specimens were particularly important to distinguish this Trinidadian *Hemiargus* from Trinidadian members of Nabokov's new genus *Echinargus*.

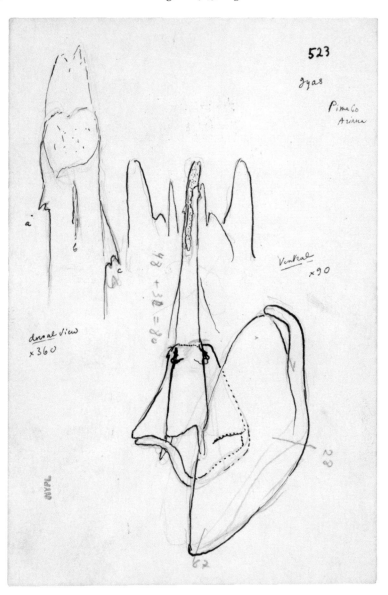

Figure 60 [*hanno,* neotype];
see also **Color Plate 34**

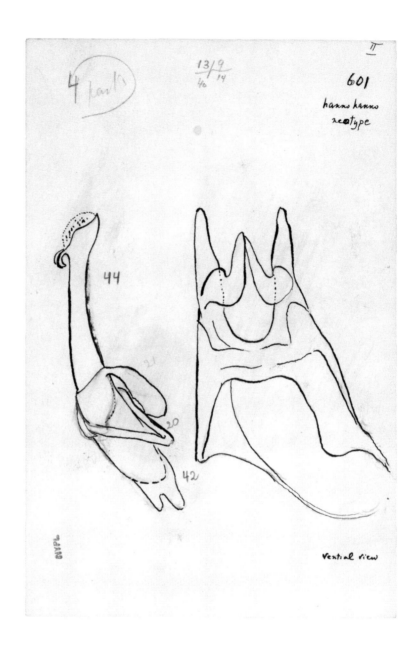

Figure 61 [*Hemiargus,* redefinition]   Figures 60 and 61: Continuing his survey of populations and taxa belonging to the evolutionary lineage of *Hemiargus* Huebner 1818, in Figures 60 and 61 Nabokov shows the genital anatomy of the defining specimen. In Figure 61 he elaborates an extreme for the *Hemiargus* complex, a specimen from Venezuela (closer to *bogotana* Draudt 1921) in which the elongately hooked terminus is very pronounced. See also Color Plate 34.

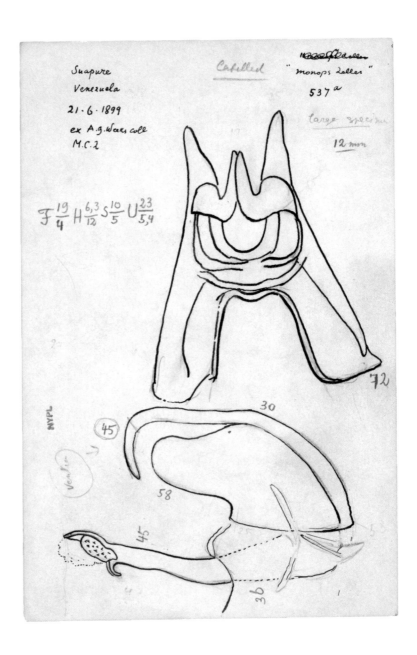

Figure 62 [*hanno*]   Here, in *Hemiargus hanno* (Stoll [1790]), Nabokov shows an extremely small, minutely dentate, structure attached to the far larger aedeagus, indicated at right and because of its larger size, shown only in part.

In other genera of the Latin American Polyommatini, which Nabokov distinguished in his 1945 paper, the minute structure shown here is often huge, sometimes more expansive than the aedeagus and also often densely dentate or spined. This structure was not distinguished, or named, before Nabokov's work. Nabokov called it the sagum and used its prominence in some lineages to distinguish such genera as *Echinargus* Nabokov 1945 and *Cyclargus* Nabokov 1945 (see section on *Cyclargus*, above). Here, the defining, or type, species of *Hemiargus: hanno* Stoll [1790] is shown. To distinguish further the importance of this structure in other Blues not directly part of the *Hemiargus* lineage, Nabokov also designated a neotype, a new definitive type specimen, for the species *hanno* (see Figure 61).

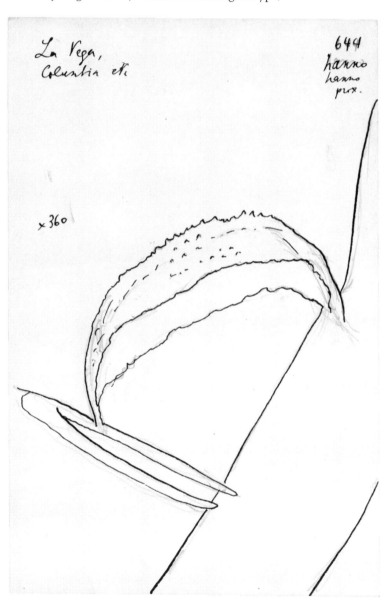

Paralycaeides

Figure 63 [*inconspicua*]   Here Nabokov expresses exclamation ("!") on discovering that the South American Blue *inconspicua* Draudt 1921 shows distinctive structures that warrant a new generic name. In pencil at the top he suggests a generic name "Palaeolycaeides," although ultimately he does not use this name but publishes, instead, the name *Paralycaeides* Nabokov 1945. Already thinking in an evolutionary (phylogenetic) framework, he exclaims in pencil, "! conforms to hypothetical ancestor."

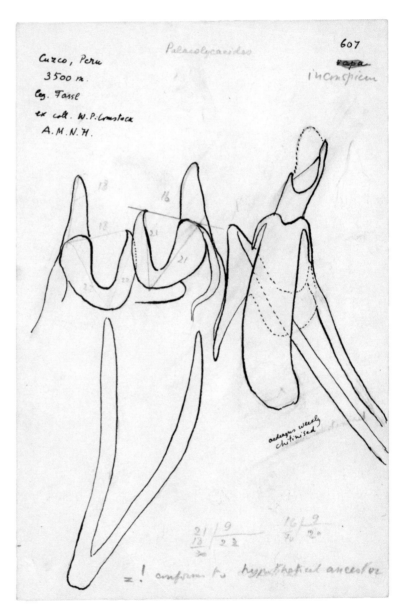

Figure 64 [*inconspicua*] Study of *inconspicua* Draudt 1921, continued, showing that it represents a distinct lineage and thus warrants a new generic name. Depicted here are additional line drawings of the entire genital ring vinculum with the furca.

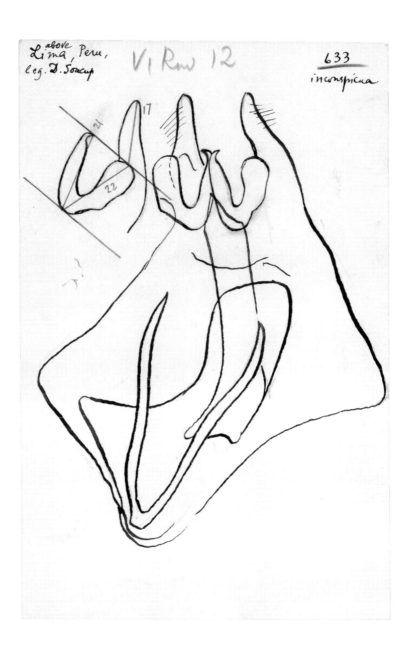

Pseudolucia

Figure 65 [*chilensis*]   Nabokov recognized a distinctively new mid- and southern South American genus of Blues, one member of which was the well-known and peculiarly brown-and-orange-marked Chilean and Argentine species *chilensis* Blanchard 1852. Here, with his usual flourish for adding detail and comment when a particular specimen piqued his interest, he notes a filamentous membrane apparently associated with the juncture of the furca that appears to wind around the base of the vinculum (shown larger and in color in Color Plate 35). There was a particular irony in this discovery. The *chilensis* Blue was fairly well represented in world butterfly collections and known to show patches of yellow to orange across areas of its wings. Because Nabokov could examine only a few species of South American Blues, he did not know that these yellow to orange patches occur in a number of then-unknown species allied to *chilensis* in southern South America. Further, he also could not know that similar patches of yellow to orange occurred in then-unknown species of several of his new genera. In the 1990s, when these new species were discovered—and the occurrence of the orange could be understood in a geographic context—it was shown that such Blues (along with some Hairstreaks and moths known as dioptines) apparently mimic the yellow-orange and orange patches of *chilensis*. This is because *chilensis* caterpillars feed on a toxic plant, *Cuscata*, which makes them toxic to predators. Across nature, hues of yellow-orange and orange are a common warning coloration, telling predators, "Don't eat me or you'll regret it!" Thus, ironically, unbeknown to Nabokov, his own Blues included a mimicry ring.

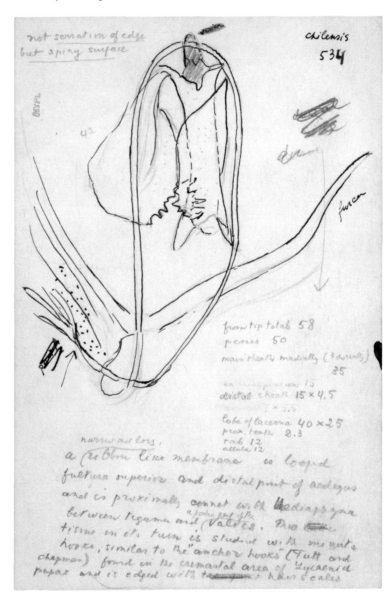

Figure 66 [*chilensis*] Continuing work on *chilensis* Blanchard 1852, from which Nabokov would name the new genus *Pseudolucia* Nabokov 1945, Nabokov recorded details of the ventral and dorsal aspect of the aedeagus. Fortunately for Nabokov, Blues of this primarily southern South American genus were distinctive enough that most lepidopterists readily accepted the new name *Pseudolucia*. No one suspected, however, how many more species would be subsequently discovered in this diverse group. Nabokov started with two, and today nearly forty species are known, many named for characters in Nabokov's fiction. See also Color Plate 35.

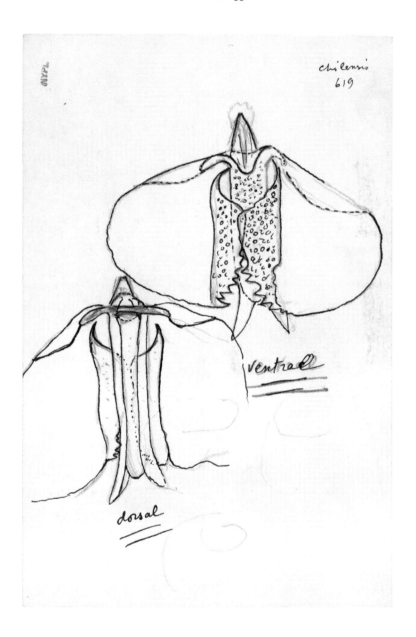

Parachilades (now *Itylos*)

Figure 67 [*titicaca*] Here Nabokov draws the genital features of the storied Titicaca Blue, *Itylos titicaca* (Weymer 1890). Well known for its small size and brilliant blue oblongate wings, this was the best-known Andean Blue to nineteenth-century lepidopterists. Nabokov recognized the singularity of the Blues related to the Titicaca Blue, which today are grouped under *Itylos* Draudt 1921. Their anatomy is outstandingly distinctive, as Nabokov immediately saw. He coined a new generic name for this group, *Parachilades* Nabokov 1945. However, *Parachilades* turned out to be a synonym; Max Draudt, a German lepidopterist, had already used the name *Itylos* for this group. Here Nabokov draws the peculiarly robust male valve with its distinctive double-pronged terminus oriented left. He remarks of the terminus, "!pointed mentum."

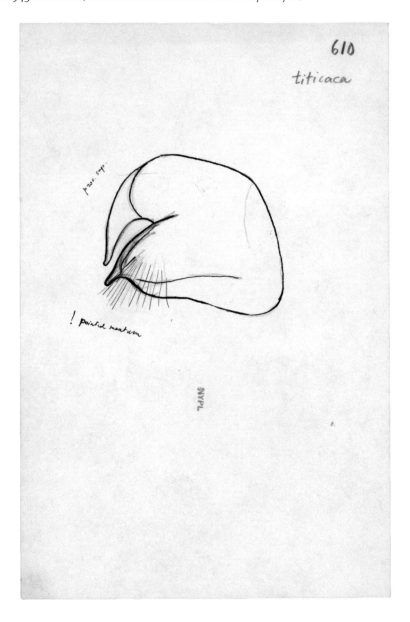

Figure 68 [*speciosa*] Nabokov found his (synonymous) genus *Parachilades* Nabokov 1945 to have at least two members—*titicaca* Weymer 1890 and *speciosa* Staudinger 1894. Modern workers, armed with a broader understanding of the diversity of *Itylos*, consider these two names to represent the same Bolivian species. Here, the entire genitalia of a *speciosa* specimen in the ventral view with the male valvae removed. He labels the "Chapman's Process," a term Nabokov coined for penile termini that show a sclerotized prong.

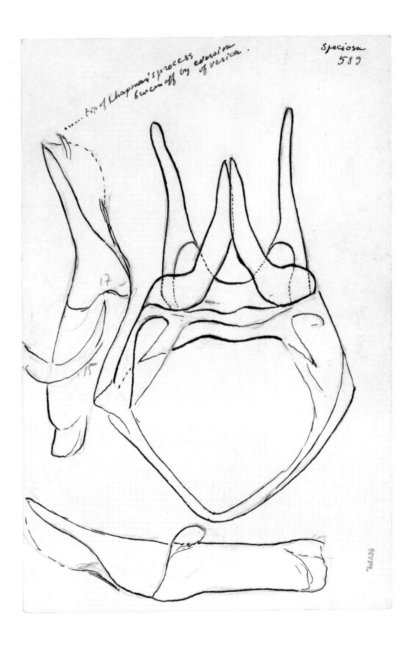

Figure 69 [*speciosa*]

Itylos speciosa (Staudinger 1894) (now considered the same species as the Titicaca Blue). Nabokov remarks that its male valve shown at bottom left (in lateral view) looks "wonderfully like an elephant."

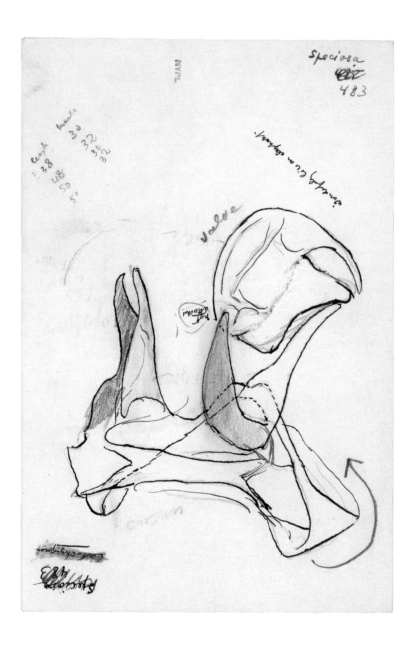

Figure 70 [*speciosa*]   Drawing the rest of the genital apparatus of *Itylos speciosa* (Staudinger 1894), Nabokov presents a rotated view of the complex parts of the dorsal terminus and vinculum (the aedeagus drawn within), all as if looking downward from the dorsal terminal elements. This may have been because he noticed that in lateral view the remarkably elongate elements of the uncus and falx/humerulus (shown shaded dark in Figure 69) can make the genitalia appear quite unlike other polyommatines and reminiscent of the dorsal termini in satyrids, or Wood Nymphs. See also Figures 71 and 72.

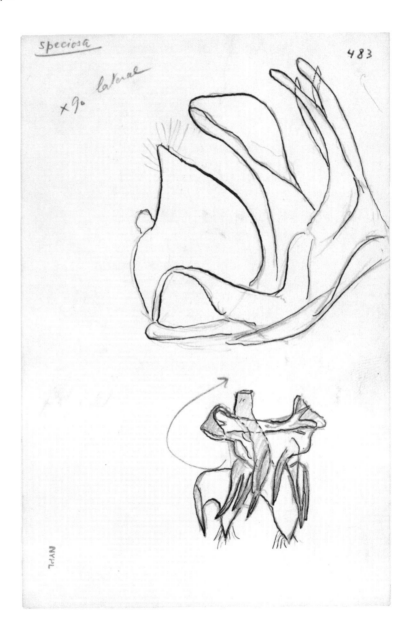

Figure 71 [*maniola*]

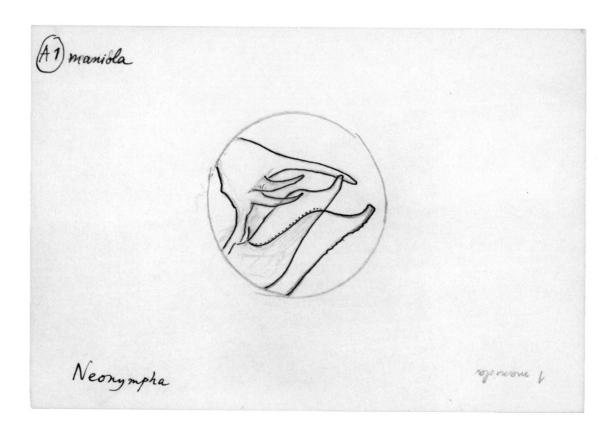

A 1 *maniola*

Neonympha

Figure 72 [*dorothea*] Wellesley graduate student Dorothy Leuthold drove Vladimir and Véra out west for collecting and lectures in June 1942. At the Grand Canyon, as the three hiked down Bright Angel Trail, Dorothy "kindly kicked up the first specimen" of the butterfly that would come to bear her name. In subsequent studies, Nabokov discerned two distinct types of gemmed satyrs in Arizona and Texas. He described these as *Neonympha dorothea* and *N. maniola,* the latter for its resemblance to the Meadow Brown (*Maniola jurtina*) of Europe. Both survive today as subspecies of more widespread types: *Cyllopsis pertepida dorothea* and *C. pyracmon maniola,* the Canyonland and Nabokov's Satyrs. The drawings show their male genitalia.—Robert M. Pyle

Figure 73 [*barbouri*]   Nabokov here figures the odd-looking male genital features of *Brephidium barbouri* (Clench [1943]), today considered the same subspecies of *exilis* Boisduval 1852 as subspecies *isophthalma* Herrich-Schaeffer 1862. The peculiar angle of the drawing, indicated by his curved arrow, makes it at first not readily recognizable with its elongate and terminally spined valvae, which are highly and irregularly sculptured at their bases. The radically Y-shaped furca is also outstanding, as is, to its left and shaded in gray, the robust form of the aedeagus. *Brephidium* Blues are not members of Nabokov's polyommatine Blues, but they include a number of Blues widely distributed and well known across the same distributions as Nabokov's Blues.

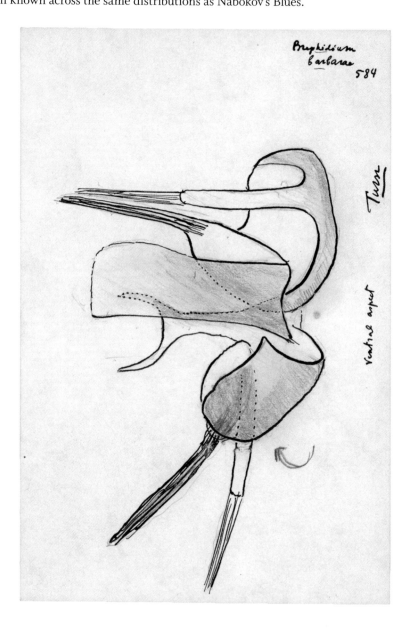

Nabokovia

Figure 74 [*faga*] Nabokov distinguishes here the uniqueness of South America's distinctive, brown-colored Faga Blue, for which he later created the genus *Pseudothecla* Nabokov 1945. This name ended up being technically invalid, leading British entomologist Francis Hemming to replace it with the name *Nabokovia* Hemming 1960. Nabokov probably chose the name *Pseudothecla* because the Faga Blue has short tails on its hindwings, thus looking somewhat like a member of the Hairstreaks, which were then called *Thecla* Fabricius 1807. Unfortunately, unknown to Nabokov, the European lepidopterist Embrik Strand had already used this name for another group of butterflies so, when modern nomenclatural rules evolved, Nabokov's name was obviously an unavailable homonym. As it turned out, another European entomologist, Francis Hemming, appreciated the irony and, knowing Nabokov's Lepidoptera work and his literary stardom, chose the replacement name *Nabokovia*.

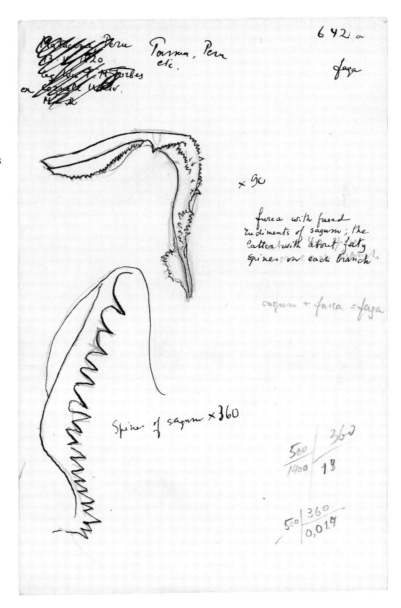

Madeleinea (what Nabokov mistakenly called *Itylos*)

Figure 75 [*koa*]   Nabokov here recognizes the distinctness of the South American Koa Blue and Moza Blue. However, he later mistakenly thought that these were the Blues that German lepidopterist Max Draudt had previously put in his genus *Itylos* Draudt 1921. As a consequence, although Nabokov was right about the uniqueness of this South American butterfly lineage, when *Itylos* was later shown to be the same as Nabokov's generic name *Parachilades* Nabokov 1945, this group ended up without a name. In 1993, forty-eight years after Nabokov published his New World Blues nomenclature, Hungarian Blues specialist Zsolt Bálint chose a name, *Madeleinea*, after Smithsonian Institution lepidopterist William D. Field, who had suggested this name but never published it.

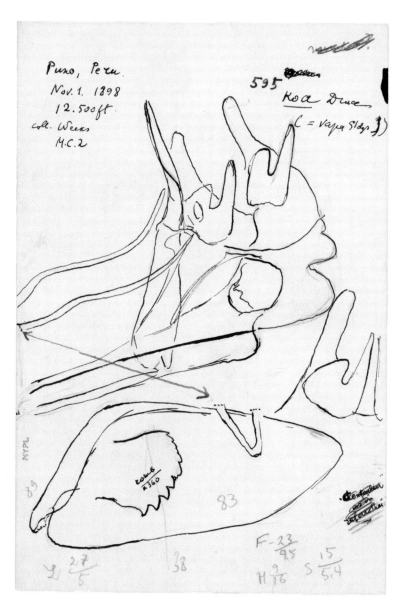

Figure 76 [*koa*] Here Nabokov adds detailed line drawings, with measurements, of the male genitalia of the Koa Blue, usually assigned to the genus *Madeleinea* Balint 1993. Some current lists of South American butterflies do not consider *Madeleinea* generically distinct and place it with *Itylos* Draudt 1921.

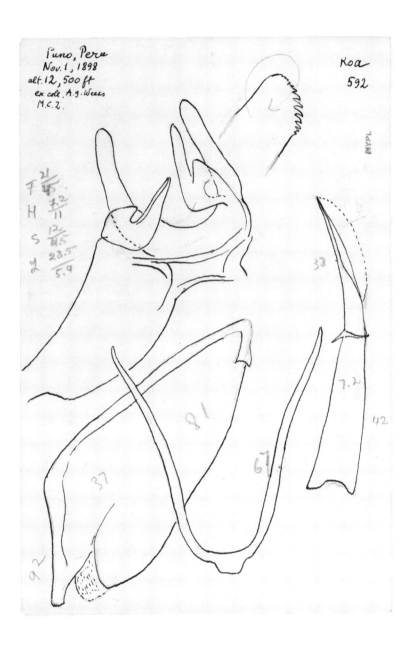

Figure 77 [*babhru*]   Here Nabokov draws the "papillae anales," or terminal elements of the female genitalia, in *babhru* A. G. Weeks 1901. Today this butterfly is considered the same Blue as *moza* Staudinger 1894, another well-known Andean Blue of the genus *Madeleinea* Balint 1993. Nabokov probably drew these female genitalia because female specimens of South American Blues were relatively hard to find in collections at the time of Nabokov's work and this specimen afforded him the chance to explore their anatomy.

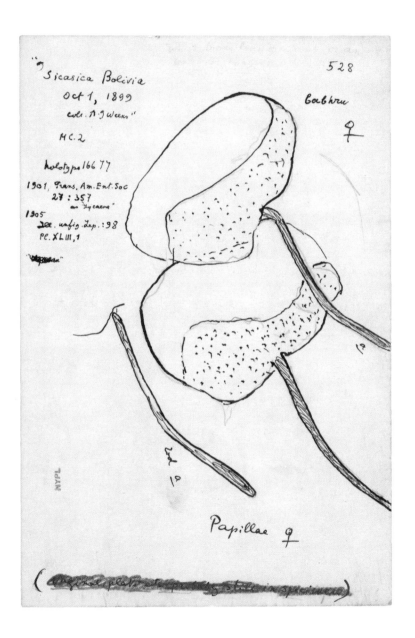

Pseudochrysops

Figure 78 [*bornoi*]   A uniquely outstanding Latin American Blue is the rare, colorful Bornoi Blue of the Greater Antilles. Nabokov found its anatomy to be remarkably distinctive and, making an exception to his usual practice, classified it in a separate genus, *Pseudochrysops* Nabokov 1945. No one has contested this group, and unlike the Blues of all the other Latin American genera, no other species of *Pseudochrysops* has ever been discovered.

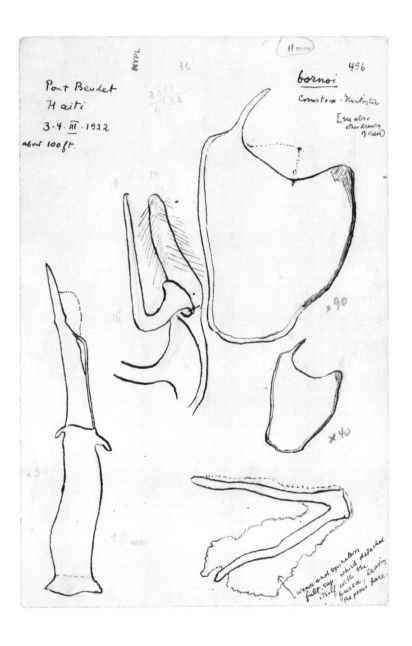

Evolution and Systematics

Figure 79 [locality map] On this card Nabokov seems to be plotting localities where he has either collected or plans to collect. On his journeys to the mountain areas of both Old and New Worlds, Nabokov was fascinated with the circle of races (*Rassenkreis*) phenomenon (see the Introduction to this volume).

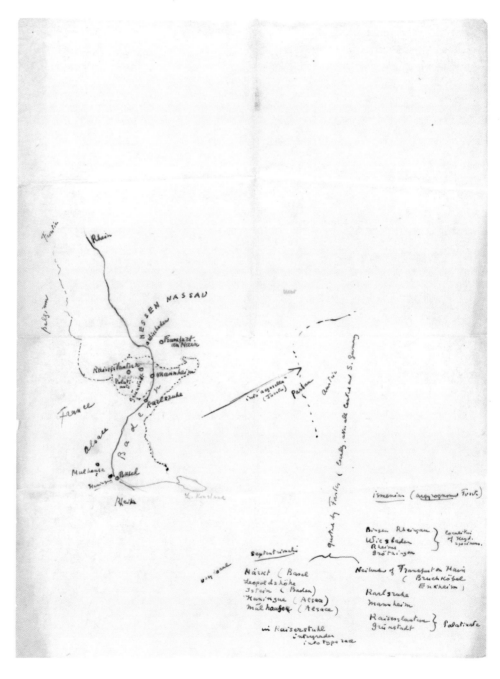

Figure 80 [phylogeny]   In these notes and a diagram, Nabokov points toward his eventual North American *Lycaeides* Huebner 1819 revision, envisioning an evolutionary tree, or phylogeny, for various Blues. He comments that speculation about New World Blues is difficult because of the unclear relationships of many of the Old World taxa—precisely the reason for his many cards dedicated to the Old Word species (particularly, Blues of *Aricia* Reichenbach 1817, *Plebejus* Kluk 1780, and *Polyommatus* Latrielle 1804). Already reflecting his growing anatomical knowledge, Nabokov notes which historically used generic names have been confused simply because no one looked at genitalia. Note that the species *saepiolus* Boisduval 1852 is at the base of the diagram, consistent with many cards Nabokov devoted to deciphering the members the genera *Plebejus* versus *Aricia* regarding the question of where *saepiolus,* a widespread species, belonged.

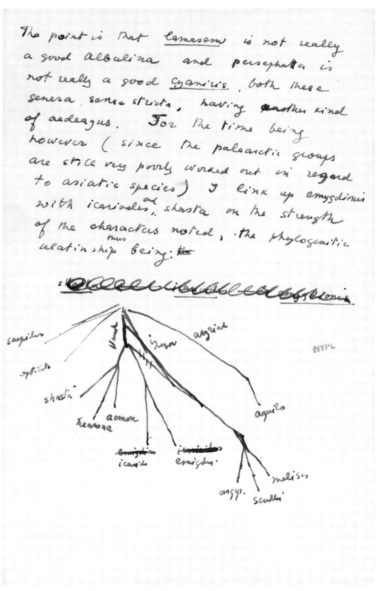

Figure 81 [*Rassenkreis* circle of races]   A complex of seven circles of races reflecting either Nabokov's own firsthand knowledge from fieldwork or, alternatively, perhaps specimens from a large number of disjunct localities he has examined in museums. Each circle stands for a race or subspecies, neither of which (by definition) are reproductively distinct except when separated artificially by time or space. Races are regarded as more ethereal in time than subspecies. Along the path of each circle Nabokov marks dots to refer to specific populations. The circles overlap at places Nabokov has placed a dot representing either (1) populations known or suspected to interbreed (or perhaps not interbreed depending on what he is noting!) or (2) populations where known complexes of traits that would usually be used to identify this or that race or subspecies are no longer clear (suggesting interbreeding).

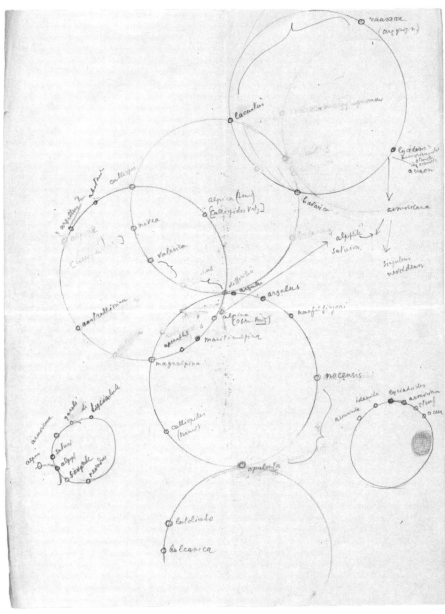

Figure 82 [geographic and phylogenetic relationships]

On this card Nabokov takes his reflections on the relationship of circles, races, subspecies, and species and their geographic distributions to a higher level. Ultimately, the relationship of an entity's geographic distributions and their corresponding phylogenies is the domain of biogeography. Because species are known to arise from spatial splitting followed by reproductive isolation, patterns of spatial information from butterfly populations in nature, as in Nabokov's circles here, can, when seen through time, suggest a companion branching diagram representing the group's evolutionary path. Here the circles could indicate geographic distributions or, perhaps, known patterns of divergence, inbreeding, or noninterbreeding. A branching line diagram, suggesting the evolutionary relations of the entities shown by his geographic distribution circles, is drawn crudely at the top. Its structure reflects that of the entities in the distribution circles, as evidenced by the number and position of terminal lines in the branching diagram, which match the number and position of the circles representing the distributional data. For example, there is a cluster of three lines (corresponding to three distribution circles) at the top of the upper branch of the diagram, versus two (usually) in the lower branch of the diagram. Obviously these were sketched quickly—most likely as Nabokov was thinking about spatial patterns and what they may mean regarding phylogenies. Ultimately, he is imagining something happening through time, which would become important later in his hypothesis that Blues arrived in the New World through successive waves of migration across Beringia. Of course, his diagrams are limited to two dimensions, and further, there is (as Nabokov was aware) a distinction between the kinship relationships of currently interbreeding entities in time and space (like complexes of races and subspecies) and that of different (reproductively isolated) species through time and space. No contemporaneous species is the ancestor of another contemporaneous species; rather, both are descendants from a no-longer-extant common ancestor. It is unclear whether Nabokov picked these cards at random from his files to doodle on or whether the note regarding the French entomologist Ferdinand Le Cerf (in the center of the card) is germane to the doodling. Le Cerf may have reviewed a complex of Blues similar to those Nabokov is considering here.

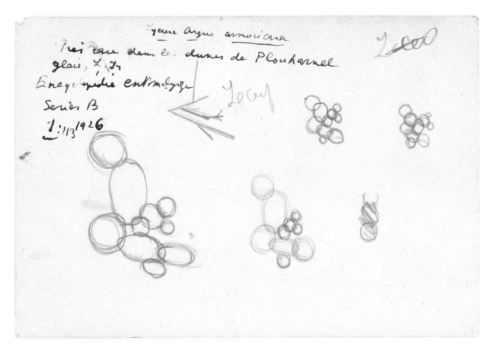

Figure 83 [geographic and phylogenetic relationships]

On this card Nabokov continues his thinking about the relationships of spatial areas and the kinship relationships of the butterflies inhabiting them. Fortunately, in contrast to the drawings of Figure 82, which have no labels, one of his diagrams here *does* contain taxon names. We see "anna" for *Plebejus anna* (W. H. Edwards 1861). Taking that as a hint we can make out at least "lotis" (*P. lotis* (Lintner 1878)), "aster" (*P. aster* (W. H. Edwards 1882)), and what appears to be "sublivens" (*P. sublivens* (Nabokov 1949)), "longinus" (*P. longinus* (Nabokov 1949)) and so on. What these represent today are the ten and six respective subspecies recognized by lepidopterists today for *P. anna* and *P. idas* (Linnaeus 1761) (a Blue occurring in both Old and New Worlds). Nabokov was thinking through the evolutionary relationships of Blues with regard to both spatial (geographic) and kinship (phylogenetic) data. The drawings here appear to be all about the same group, *and* he appears to have been thinking of the relationships of these names across two biological species. This is evidenced (especially at the left) by his distinction of the entities today associated with *anna* versus those today associated with *idas* (members of the *anna* complex are located basally in the diagram, members of the *idas* complex higher up).

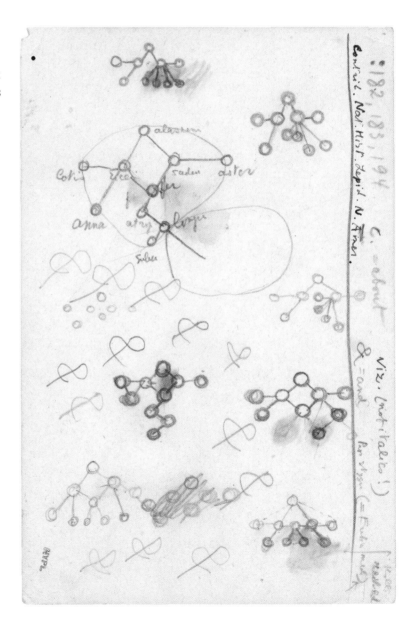

Figure 84 [transformation series, *Lycaeides*]

Figure 84: This image shows Nabokov preparing his data on transformation series of characteristics in *Lycaeides* Huebner 1819 and the branching diagram of evolutionary (phylogenetic) relationships he later presented in his 1944 study of the group. Figured are the dorsal termini of the genitalia (with triangulations, as in "2" above, center, for *anna:* the uncus ["28"] and falx ["38"]/humerulus ["33"]). He includes eight taxa in the group, including both Old and New World representatives (ARG meaning *argyrognomon* Bergstraesser 1779, AGN meaning *agnata* Staudinger 1889), a limited sample by modern standards and one that ultimately caused problems with the overall synthesis. The diagram readily shows the *lines* of evolution forming the branching phylogeny of these Blues and the transformations of anatomy characteristics reflecting that evolutionary pathway. In Figure 84 we see clean line drawings, and in the finalized figure in "Notes on the Morphology of the Genus *Lycaeides*," wherein he has moved his "Arg C." entity to another position, we see his hypothetical phylogenetic lines clearly, indicating the splitting events he attributes to the evolution of the group (see *NB*, 321). Although the evolutionary (phylogenetic) methodological concept is sound— companioning character transformation with lineage evolution—there are problems. First, Nabokov's vision is limited to anatomical data. With such data one can see apparently significant step-by-step transformations in characteristics. However, the sample limits what one may infer from these about evolution. Subsequent DNA studies have located the taxa of Nabokov's figure quite differently in branching evolutionary diagrams of these Blues. The species *argyrognomon*, for instance, is now generally considered European, and Blues from this group in North America are often placed in the genus *Plebejus* Kluk 1780, containing *anna* W. H. Edwards 1861, *idas* Linnaeus 1761 (including *sublivens* Nabokov 1949 and *scudderi* W. H. Edwards 1861), and with *melissa* W. H. Edwards 1873 and *samuelis* Nabokov 1944 considered distinct species. The ultimate definition for a species comes from the DNA of its "type specimen (holotype)," and this might be quite different from a specimen identified as that species by an early worker, even an expert.

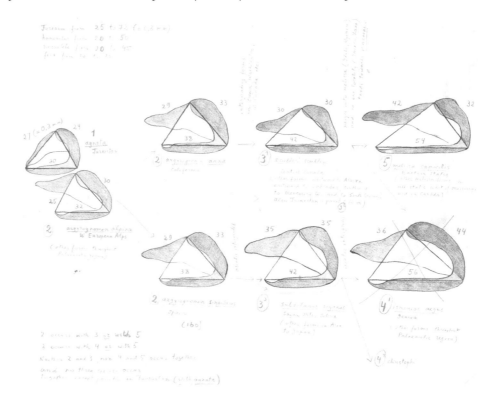

Figure 85 [taxonomic rank]   Sister species or varieties? This is a comparative drawing of several three-dimensional perspectives on the variable genital apparatus of *Lycaena collina* (Philippi 1859), which Nabokov included in his new genus *Pseudolucia* 1945. Disjunctly distributed across southwestern South America, *Pseudolucia zembla* Bálint & Johnson 1993 (named after the distant northern land in Nabokov's *Pale Fire*), and *Lycaena* lyrnessa (Hewitson 1874) are considered by some, but not all, lepidopterists to be varieties of *collina*.

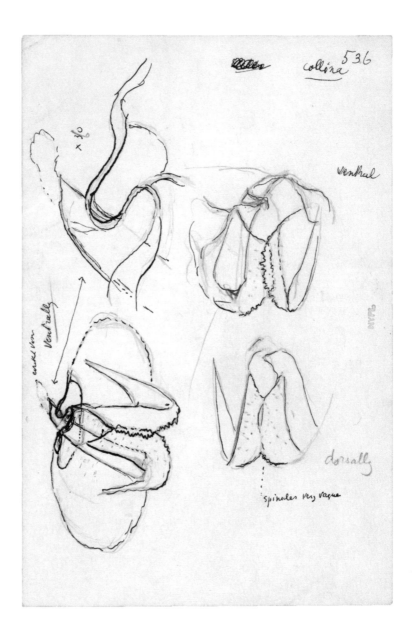

Figure 86 [outgroup comparison]

Contrasting Blues to a phylogenetic "outgroup," Nabokov depicts here, for comparative purposes, a lateral view of the aedeagus of a typical Hairstreak (subfamily Theclinae), one of the sister groups of the Blues (subfamily Polyommatinae). He used the species *Evenus candidus* (H. H. Druce 1907). See also Figure 88 and Color Plates 19 and 56.

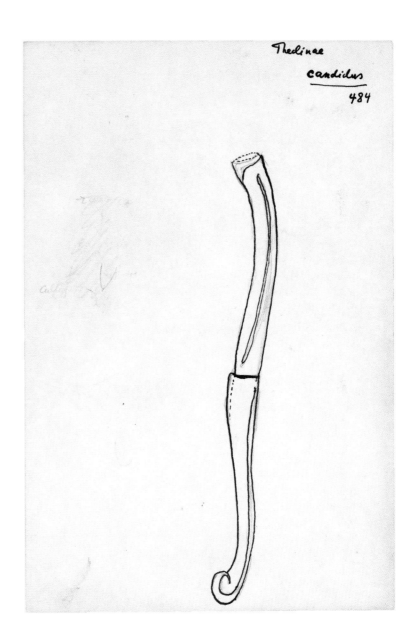

Figure 87 [magic triangles] Using the defining, or holotype, specimen of Nabokov's new Blue *samuelis* Nabokov 1944, this is a lateral view of the dorso-terminal elements of the male genitalia showing the elements of his magic triangles. Nabokov used these magic triangles for quantitative measure comparisons to characterize qualitative differences in general shapes of the uncus (U), falx (F), clarified as the "forearm of falx," and humerulus (H), clarified as the "humerulus of the falx." Above "U" (and near "40"), he draws another dotted line. This may be another measure; if so, it would be the length of the uncus's extension beyond where it triangulates with the humerulus. See Table 1.

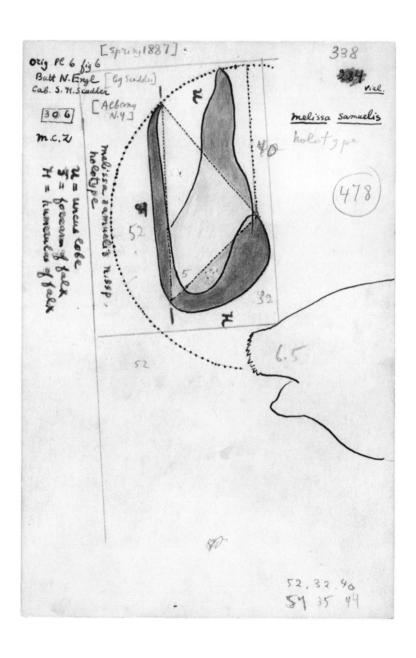

Wings

Figure 88 [comparative wing maculation]

Evidently for comparative purposes with his Blues, Nabokov here figures the pattern of the medial and postmedial spots that occur on the undersurface of the forewing in two other Lycaenid genera, *Turanana Bethune-Baker 1916* and *Lycaena* Fabricius 1807 (meaning in this case true Coppers, not Blues, for his model is stated as *phlaeus* Linnaeus 1761, a well-known Copper). Before more was known about Blues, eighteenth- and nineteenth-century studies often placed them in *Lycaena* simply because this was the genus early describers had used.

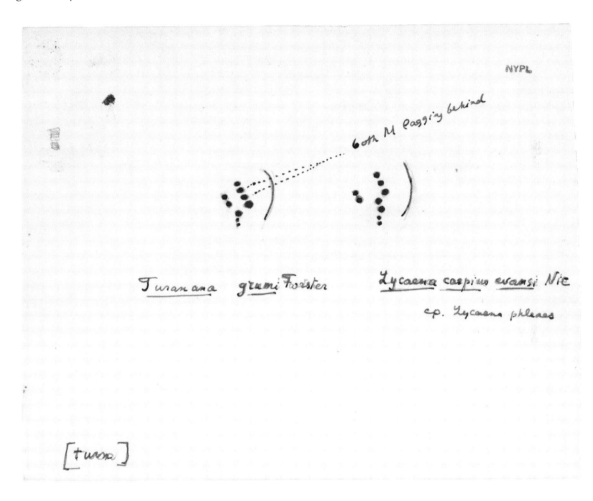

Figure 89 [comparative wing maculation]

In his efforts to track the occurrence and evolution of wing maculation, Nabokov reverted to using graph paper because the individual graph paper columns readily corresponded to successive cells on the butterfly wing. Thus, plotting the placement of macules on the wings in these vertical rows corresponding to the cells he can readily compare the relative positions across various species. He also then connects many of the figured macules with pencil lines so he can see generalities in the patterns, as reflected in various zig-zag patterns, and also patterns of alignment, which are readily read from the graph paper template. Here he portrays comparative macular variation in over two dozen specimens and taxa.

 prim. / sec. / prim. / sec. / sec. / prim. / sec

prim.	sec.	prim.	sec.	sec.	prim.	sec
1 aegagrus		7 persephatta			♀ icarus	
2 oberthuri	(spots)	8 coelestina			♀ icarus	
3 dardanus		9 semiargus	9a semiargus ⊗	12 antiochena helena intermedia ⊗	♂ icarus	
4 nygula omphissa galathea		10 semiargus			average icarus ⊗	
5 pheretes also camasem	M₃	11 amandus		11a amandus ⊗	Kenter icarus siticiae (one ♂)	
6 amphiroe				prim	devanica	
				Prim stolizkana (and amor)	sartha	
				eros (average)	abnorm. eros 2 ♂♂ and several ♀ have P. pattern	superba

Figure 90 [wing maculation schematic]

A detailed wing schematic showing Nabokov's skill as an illustrator. It shows the forewing's cells, venations, and general wing pattern regions, labeled in detail. The arc drawn with a line just right of center demarks the "postmedian" area where, in most Blues, the major macules, or wing spots, of the wing pattern occur. One of Nabokov's interests (see Figure 89) concerned the patterns of variation in the occurrence of medial and postmedial macules making up the basic ground pattern of Blue butterflies.

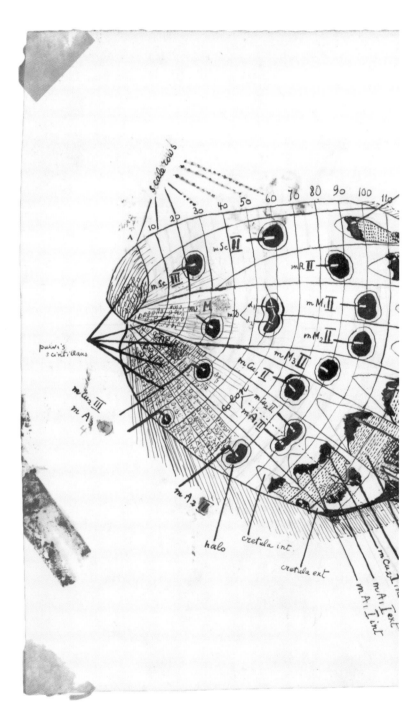

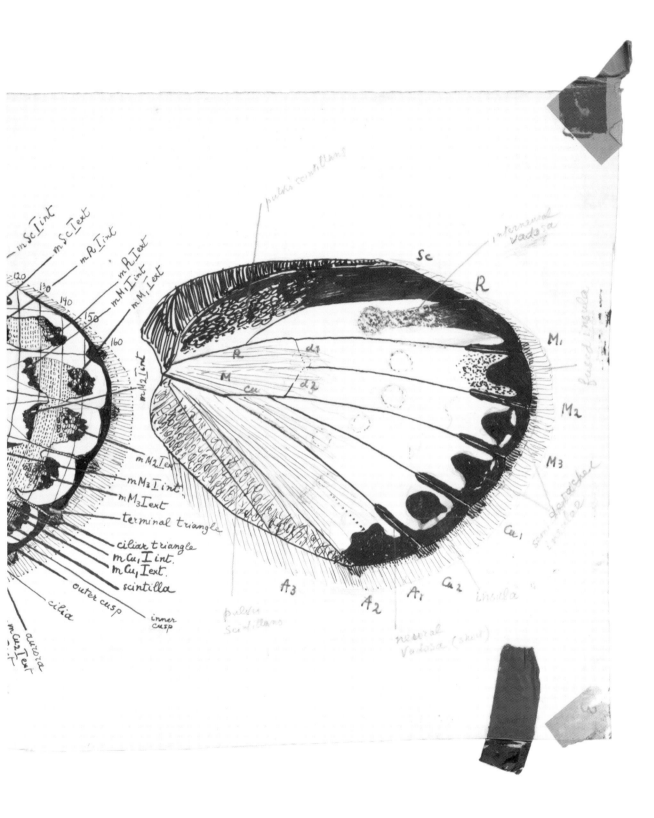

pulvis scintillans

interneural
vadosa

m Sc I int
m Sc I ext
m R I int
m R I ext
m M₁ I int
m M₁ I ext

120
130
140
150
160

m M₂ I int

Sc

R

M₁

fused insula

M₂

M₃

detached
insula

m M₂ I ext
m M₃ I int
m M₃ I ext
terminal triangle

ciliar triangle
m Cu₁ I int.
m Cu₁ I ext.
scintilla

outer cusp
inner cusp

cilia

aurora
m Cu₂ I ext
I int

R

M

Cu

d 1

d 2

Cu₁

Cu₂

insula

A₃

A₂

A₁

pulvis
scintillans

neural
vadosa (sheet)

Figure 91 [wing maculation schematic]

Nabokov shared with other lepidopterists of his day a passion to understand evolution through wing venation and pattern. Because wing patterns were the most used characteristics for identifying butterflies, a considerable scientific literature surrounded the search for understanding the evolution of butterfly wing patterns. Here Nabokov shows a schematic of venation and macular pattern (each wing spot is known as a macule) compared to a schematic (right) for butterflies in general (including variations on what veins and cells variously occur), with commentary on a generalized Lycaenid pattern of dark macules surrounded by "rims" of white.

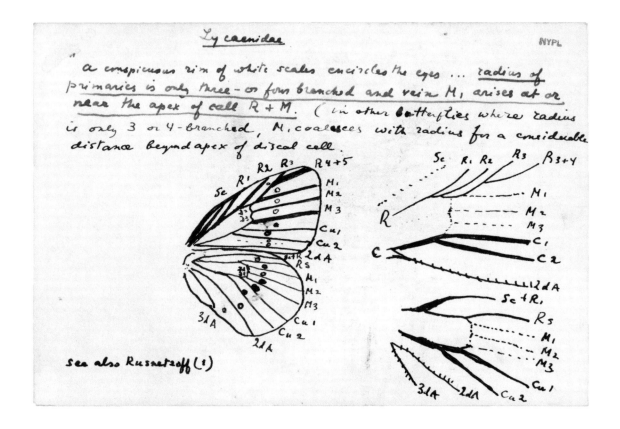

Measurements

Figure 92 [measurement chart]

Nabokov made morphological sketches and recorded his measurements for each collected series on index cards. Here he records measurements of male genitalia from six males of *Lycaeides idas scudderi* (W. H. Edwards 1861) (formerly *L. argyrognomon scudderi*) from several Canadian locales. Note that Nabokov's "means" are actually not the averages of the columns of numbers above them; instead, they are (unconventionally) the means of the two most extreme (large and small) values in each column. At the bottom of the drawing are notes about an unusual specimen from Ptarmigan Valley, Alberta. In this table Nabokov adds an additional measure to his usual "F" (falx"), "H" (humerulus), and "U" (uncus). It is "E" for "elbow" (see Color Plate E3), referring to the width of the "elbow" where the humerulus bends to become the falx fore-arm.—Lauren K. Lucas, Matthew L. Forister, James A. Fordyce, Chris C. Nice, and Kurt Johnson

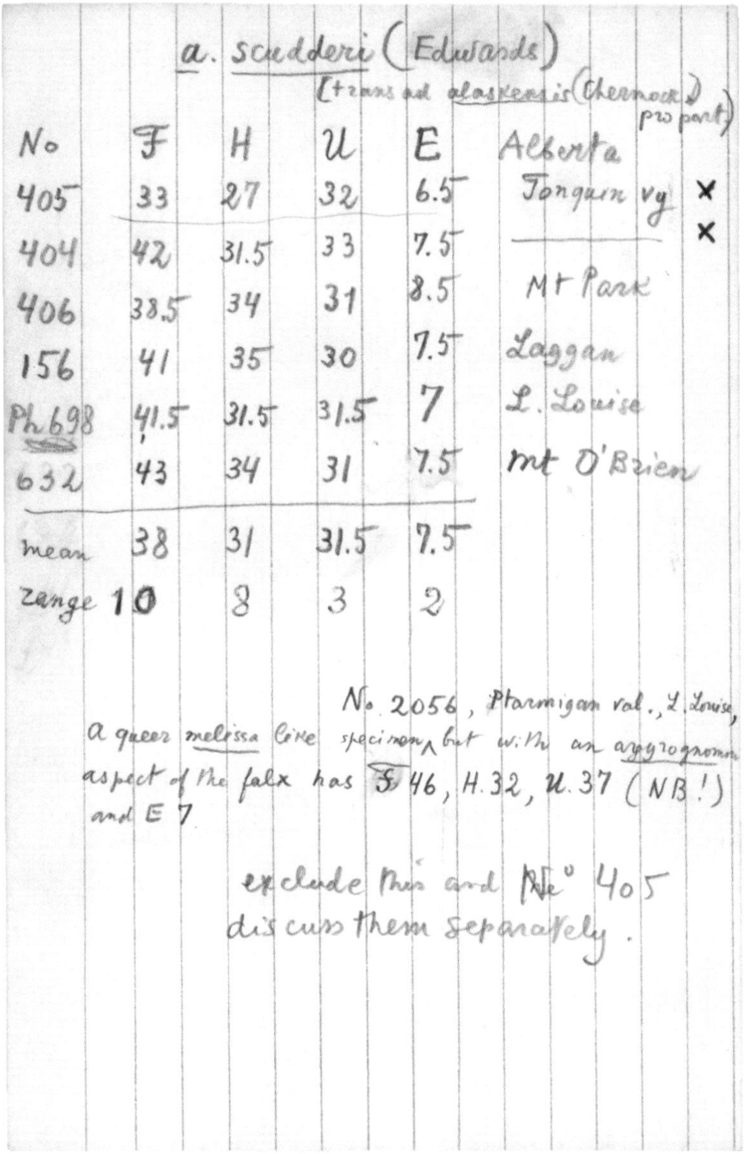

a. scudderi (Edwards)
[trans ad alaskensis (Chermock) pro part)]

No	F	H	U	E	Alberta	
405	33	27	32	6.5	Tonquin vy	✗
404	42	31.5	33	7.5		✗
406	38.5	34	31	8.5	Mt Park	
156	41	35	30	7.5	Laggan	
Ph 698	41.5	31.5	31.5	7	L. Louise	
632	43	34	31	7.5	Mt O'Brien	
mean	38	31	31.5	7.5		
range	10	8	3	2		

No. 2056, Ptarmigan val., L. Louise, a queer melissa like specimen, but with an argyrognomon aspect of the falx has F 46, H. 32, U. 37 (NB!) and E 7

exclude this and No 405 discuss them separately.

Color Plates

Except as noted, captions are by Kurt Johnson
(following the current international taxonomic code)
and edited by Stephen H. Blackwell.

Old World

Color Plate 1 Special taxa [*cormion, coridon*]

Nabokov named what he suspected was a new Blue from Europe in 1941, *Lysandra cormion* Nabokov. Perhaps this was drawn after the publication (cited at upper left) as a celebratory exercise. *L. cormion* turned out not to be a new species, however, but a rare hybrid among the *L. coridon* (Poda 1761) complex. The parentheses around the author of *coridon* denote that the butterfly was originally named within another genus. See also Figures 1 and 2.

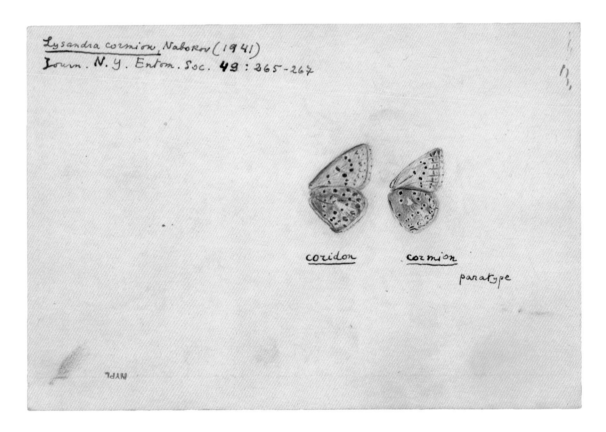

Color Plate 2 [*pheretes*]

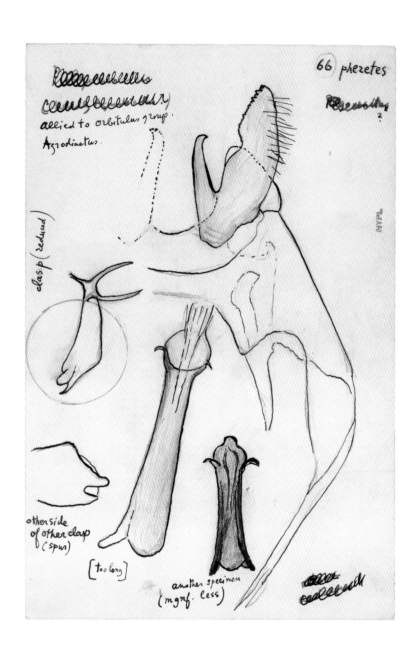

Color Plate 3 [*posthumus*]

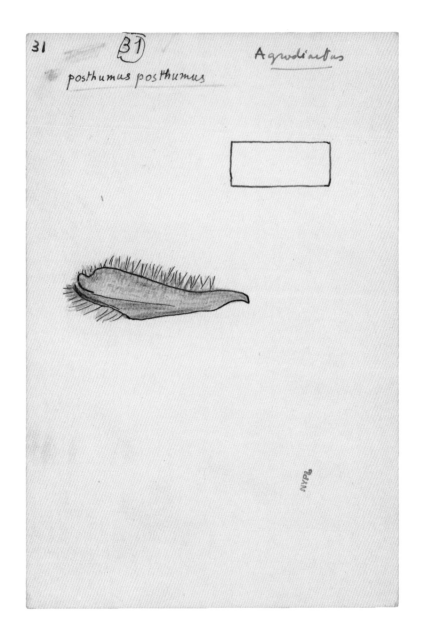

Color Plate 4 [*phyllis*]

Color Plates 2–4: Characteristics of various Blues historically associated with the species *pheretes* Huebner 1805, a brilliantly colored Blue from the Himalayan region (variously assigned in Nabokov's day to *Albulina* Tutt 1909, *Plebejus* Kluk 1780, and *Polyommatus* Latrielle 1804). Nabokov begins with numerous structures of *pheretes* (and notes its possible relationship with *orbitulus* Prunner 1798, a similar Blue occurring westward into Russia then generally associated with *Agrodiaetus* Huebner 1822). He then compares these with traits of the male valve in *posthumus* Christoph 1877 (Iranian region) and *phyllis* Christoph 1877 (Turkish region), which he denotes as members of *Agrodiaetus*. DNA analyses have since suggested that *Agrodiaetus, Lysandra,* and *Plebicula* Higgins 1969 are distinct lineages within the larger *Polyommatus* kinship group. Confirming Nabokov's suspicions about the distinctiveness of *pheretes,* it is placed today in *Albulina* whereas the other species Nabokov figures here are placed in *Agrodiaetus*.

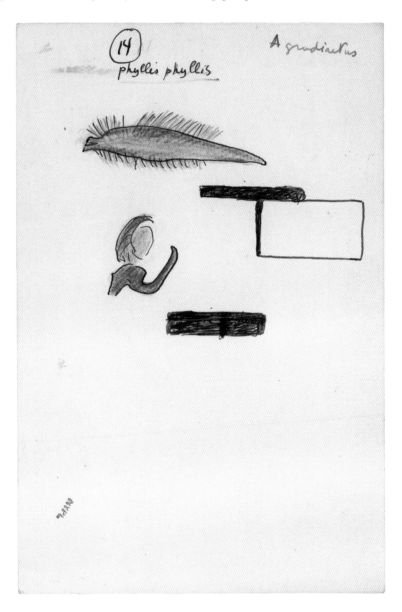

Color Plate 5 [*iphigenides*]

In Nabokov's time butterflies from the Black and Caspian Sea regions eastward into the mountains of Central Asia were far more poorly known than they are today. Here he compares Blues from the region that today are generally assigned to *Polyommatus* Latrielle 1804: *Polyommatus menalcas* (Freyer 1837), a bright silvery Blue with contrasting dark veins from the Turkish region (with the common name Turkish Furry Blue) and *Polyommatus iphigenides* (Staudinger 1886), a Blue from the Trans-Altai Mountains where the Pamirs and the more eastern Tian Shan ranges come together. *Polyommatus iphigenides* shows less silver on the forewing, and there are races that are darker (for which Nabokov here uses the name *melania*, meaning "dark"). Historical uses of *melania* have been confusing; different early authors applied at least three *melania* names to Eurasian Blues.

Fortunately, they are used for subspecies or races, as here, and thus don't overcomplicate the regional nomenclatures. See also Figures 4 and 5.

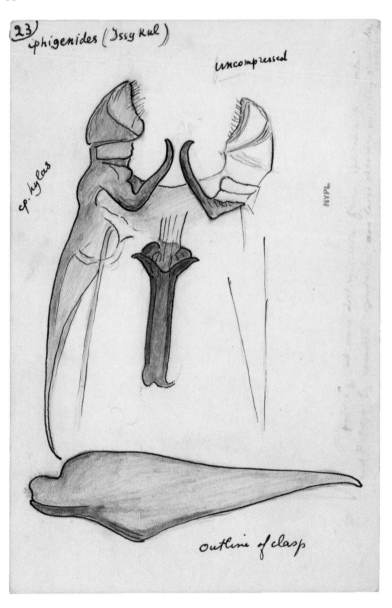

Color Plate 6 [*iphidamon*]

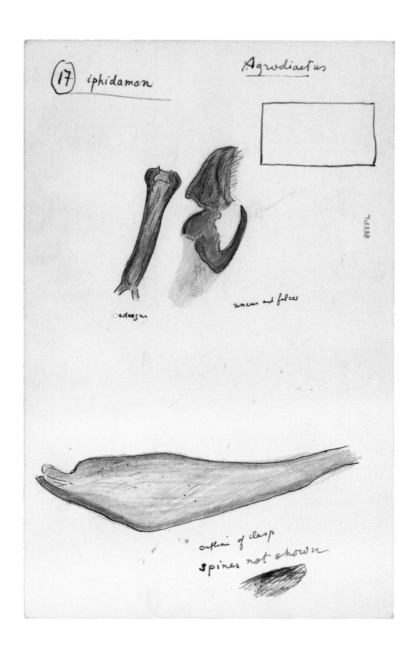

Color Plate 7 [*hoppferi*]

Color Plates 6 and 7: Here Nabokov compares confusing Blues from the Turkish region assigned in his day to either *Polyommatus* Latrielle 1804 or *Agrodiaetus* Huebner 1822. He uses the name *Agrodiaetus*. The first drawing is genitalia of *Agrodiaetus iphidamon* (Staudinger 1899), a lovely silvery Blue with darker wing margins. The second is *Agrodiaetus hopfferi* (Herrich-Schaeffer 1851). By either expertise or luck, Nabokov's generic assignment is correct; both taxa are generally placed today in *Agrodiaetus* now that DNA studies have confirmed it as a distinct generic lineage. Butterflies from this region between Europe and Asia have always befuddled Eurasian classifications because fieldwork across the region was, for political and safety reasons, not always possible.

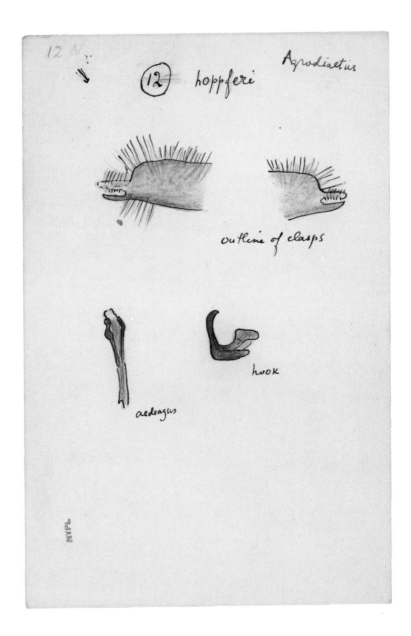

Color Plate 8 [*glaucias*]

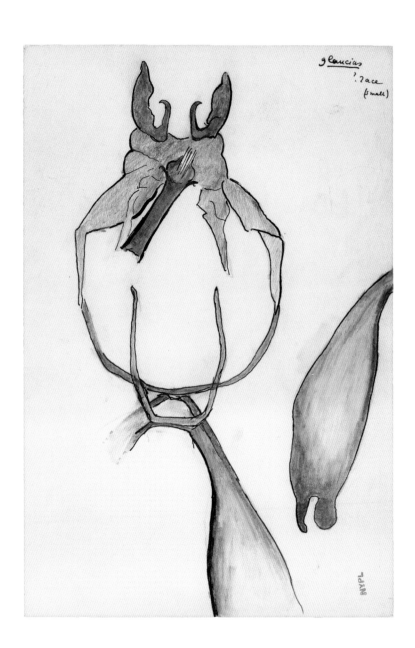

Color Plate 9 [*nivescens*]

Color Plates 8 and 9: The taxa discussed on these cards are Blues from farther west in Europe and in North Africa that Nabokov had to skillfully study in relation to the many scattered specimens he had available to him from the Middle East and Asian regions. Color Plate 8 shows the entire genitalia of *Polyommatus glaucias* (Lederer 1870), known as the Catalonia Furry Blue, an excellent drawing indicating the relation of functional parts of the male genitalia in the Old World Blues Nabokov was studying. In Color Plate 9 Nabokov draws genital elements of the European Blue *Polyommatus nivescens* (Keferstein 1851) and notes below a suspected relationship to *Polyommatus hylas* (Esper [1793]), named from Austria but also reported in North Africa. Now that more is known of biological distributions, both are usually considered today as subspecies or races of the more broadly ranging *Polyommatus dorylas* (Denis & Schiffermueller 1775).

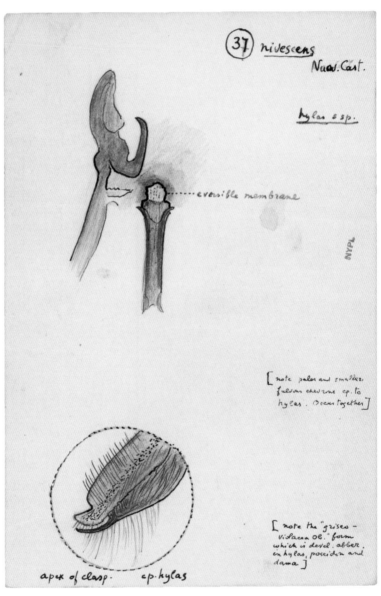

Color Plate 10 [*bellargus*]

Terminus of the male valve shown at top (with terminus down); for an Old World Blue, *Agrodiaetus bellargus* (Rottemberg 1775). Widely distributed in Europe, it is known as the Adonis Blue. Details of such structures helped Nabokov distinguish new genera among the Latin American Blues. He also spent significant time trying to assess the efficacy of characteristics in the complex dorsal terminus of the male genitalia, drawn in color for a species with a multifarious structure. It is interesting that Nabokov assigns the species here to *Agrodiaetus* Huebner 1822, since many other lepidopterists have associated it with *Lysandra* Hemming 1933. Modern DNA studies support the assignment to *Agrodiaetus*.

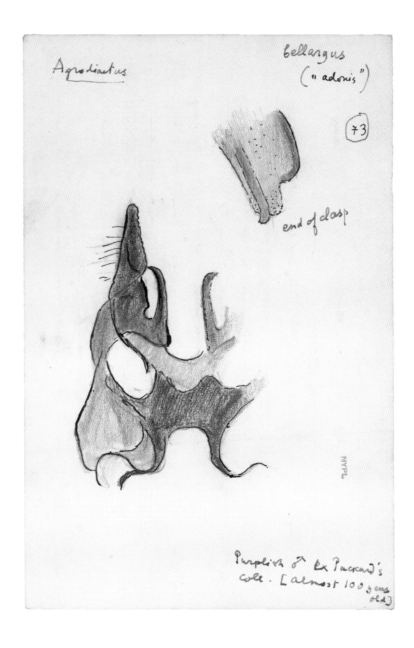

Color Plate 11 [*admetus*]

Here Nabokov evaluates the validity of subspecies names used for Blues at the same geographic locality. In these two drawings Nabokov compares examples from the "*Agrodiaetus admetus* complex." The term "complex" denotes confusing clusters of similar-looking entities for which taxonomists disagree about what names apply. Nabokov compares two specimens identifiable as perhaps the same taxon or two closely related taxa reported from the same geographic locality ("Budapest"). Technically, subspecies of a species should not be able to exist at the same precise geographic locality because interbreeding would erase their anatomical distinctions.

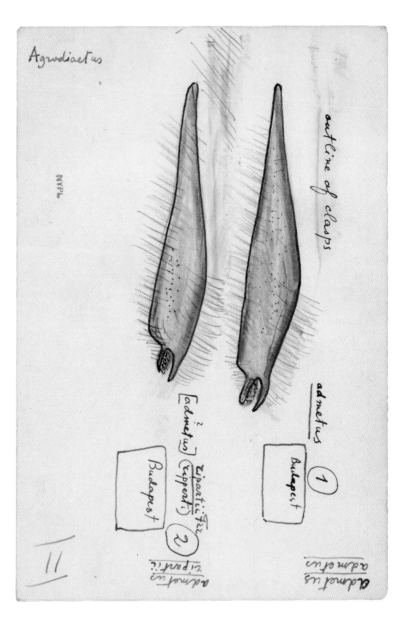

Color Plate 12 [*argus*]

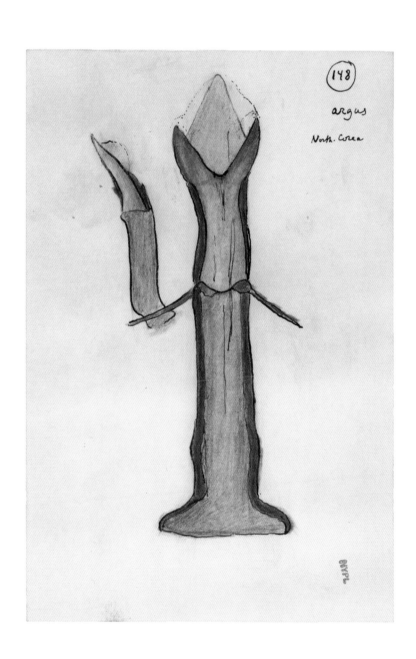

Color Plate 13 [*argus*]

Color Plates 12 and 13: Nabokov was particularly concerned about the identity and characteristics of Blues with distributions near the Bering Strait as he suspected, and later hypothesized, that New World Blues had evolved by Beringian migrations over millennia. In Color Plate 12 he draws the aedeagus (left in lateral view, right in larger ventral view) of a specimen of *Plebejus argus* (Linnaeus 1758) from North Korea. This species, often called the Silver-studded Blue, occurs widely across the Old World, so Nabokov was certainly curious about its traits near the Bering Strait. In Color Plate 13 he adds views of the coxcomblike terminus of the valve and the dorsal terminus, showing the sharply bent falx/humerulus in lateral views.

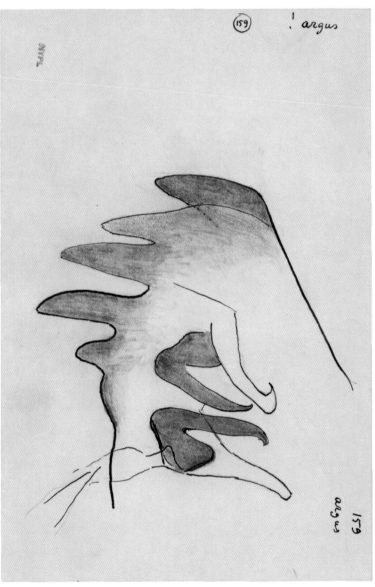

Color Plate 14 [*chinensis*]

Nabokov was fascinated by the confusing convergence of characteristics in two widely distributed Blues genera, *Aricia* Reichenbach 1817 and *Plebejus* Kluk 1780. Many of the Blues he studied had a history of being reassigned back and forth between these two generic names, and Nabokov suspected an evolutionary relationship—possibly as diverging and then parallel lineages. Concentrating on specimens from near the Bering Straits, Nabokov here records characteristics of a specimen identified as "*chinensis*" ("from China"). He comments on peculiar features along the dorsum where the clasper meets the genital ring. The species *chinensis* Murray 1874 is today usually placed in *Aricia* Reichenbach 1817, where it occupies its own unique species group (one Russian scientist suggests a separate genus *Umpria* Zhdanko 1994). It is truly a Beringian region butterfly, ranging across the Transbaikal to central and northeast China, Mongolia, Korea, and northeast Siberia.

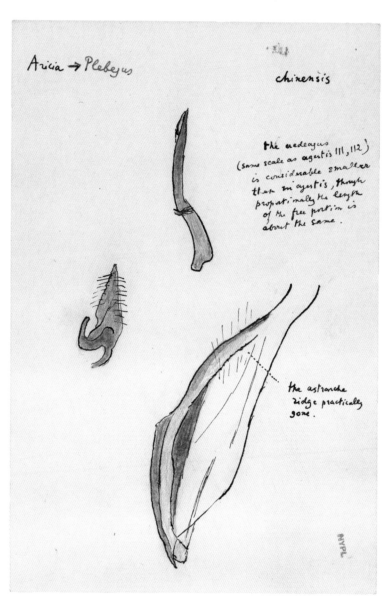

Color Plate 15 [*chinensis*]

For Blues with distributions abutting the Bering Strait region, Nabokov was careful to study and record the wing patterns. Here he identifies wing characteristics for "chinensis," referring to *chinensis* Murray 1874, which is today usually placed in *Aricia* Reichenbach 1817. He draws the undersurface, showing the distinctive red-orange distal margins on both fore- and hind wings. Conscious of other Blues that might be confused with it or need further study, he mentions *pylaon* Fischer de Waldeheim 1832 (another unique trans-Siberian Blue sometimes placed with *Plebejus* Kluk 1780, *Aricia*, or even genus *Kretania* Beuret 1959) and the Blue *eurypilus* Freyer 1852, named from Azerbaijan and also sometimes placed by modern lepidopterists in *Kretania*. The use of a separate genus—*Kretania*—again reflects the unique traits of Siberian Blues, which are also, in local and regional lists, most often recorded as "rare" and "local."

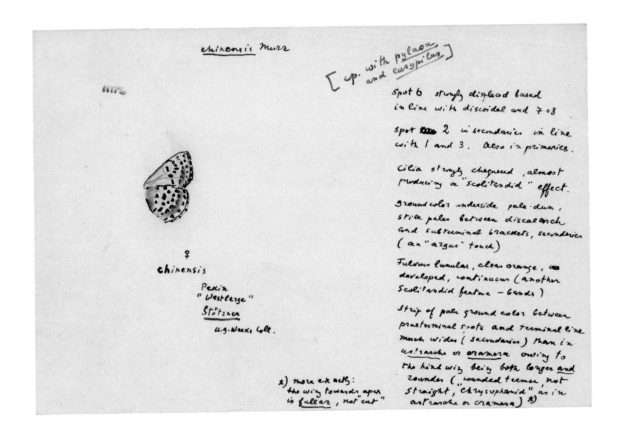

Color Plate 16 [*miris*]

Another unusual Blue, the Miris Blue (*miris* Staudinger 1881), again a species known from Central Asia and northwest China and relevant to Nabokov's musings about Beringia as an ancient migration corridor. He places it in *Agrodiaetus* Huebner 1822, a group of Blues that conventional modern taxonomists have until recently considered a subgroup (or subgenus) of *Polyommatus* Latrielle 1804. But the species *miris* Staudinger 1881 today is considered one of two members of the genus *Rimisia* Zhdanko 1994 distinguished by Russian scientists, a uniqueness that may well have drawn Nabokov's attention to its anatomy. Little is known of the other species, *Rimisia avinovi* Stshetkin 1980, named from Uzbekistan.

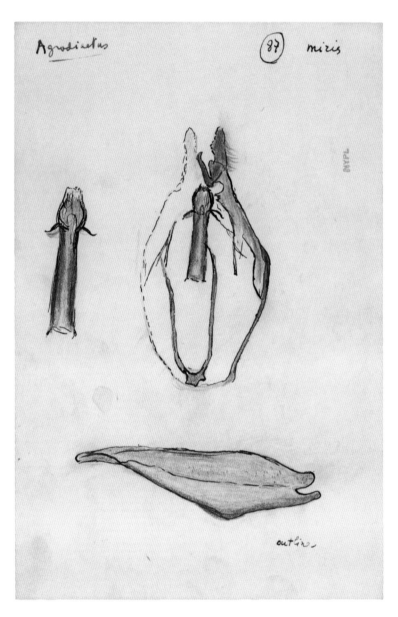

Color Plate 17 [*argyrognomon, insularis*]

Always attentive to Blues whose ranges abutted on the Bering Strait, Nabokov here draws detailed renderings of the undersurface of Blues from the *argyrognomon* Bergstraesser 1779 and *insularis* Leech 1893 complexes. In his day, determination of the evolutionary relationship of *argyrognomon* and *idas* Linnaeus 1761, two very old names with possibly broad Old and New World distributions, was still a priority. Today, these butterflies' taxonomic placement is well established from DNA studies, which often show that trying to determine their identities from wing pattern, or even genitalia, is futile. On the right Nabokov figures the geographically restricted entity "*insularis*" then known from China, Korea, and Japan—again reflecting his interest in the Asian origins question. Many of these were known only from J. H. Leech's classic 1892 book *Butterflies from China, Japan, and Corea.*

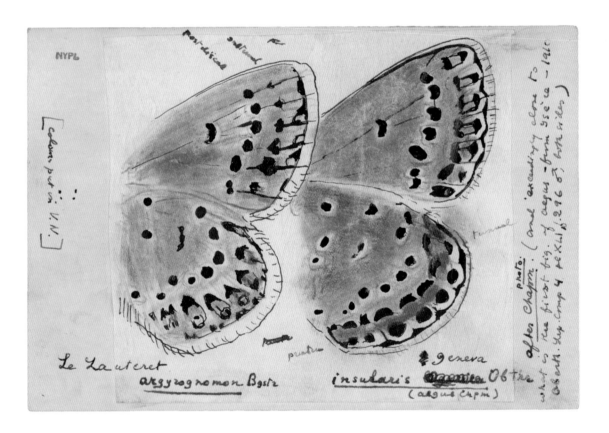

Color Plate 18 [*ismenias* subspecies]

Here Nabokov draws the wing pattern of *ismenias* Meigen 1829, noting the Altai Mountains, a pivotal transition point between West and East Asia. He appears to be suggesting the need for a new subspecies name for this entity "pseudocleobis." The addition of "pseudo" to well-known species names (as here *cleobis* Bremer 1861) was often done in Nabokov's day when the examination of genitalia showed that a Blue resembling a well-known species was actually something else.

Color Plate 19 [thius]

For comparative purposes owing to his interest in the dorsal terminal structures of the genitalia of Blues (the famous uncus and falx/humerulus combination of his magic triangles), here Nabokov draws what those structures look like in their companion subfamily, the Theclinae, or Hairstreaks. He uses the well-known species *thius* Geyer (the name is today attributed to Geyer not Huebner) today placed in the genus *Strymon* Huebner 1818 (arguably the "flagship" genus of the Theclinae). Certainly, when European lepidopterists thought of Hairstreaks, a species of the well-known genus *Strymon* would usually come to mind.

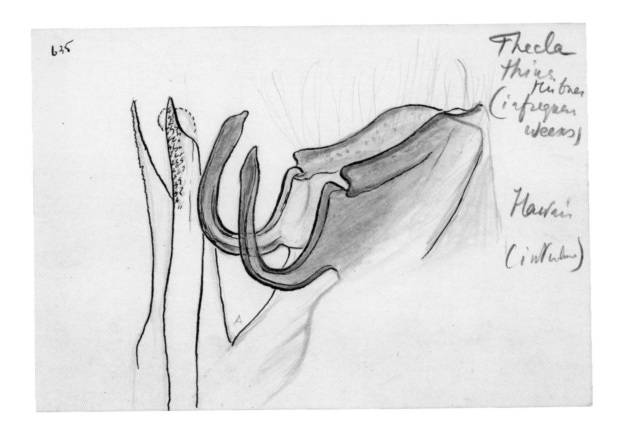

Color Plate 20 [*Lucia*]

One of Nabokov's new Latin American genera was *Pseudolucia*. So it is understandable here that he would figure the standard genitalic characteristics of the Old World genus *Lucia* Swainson 1833. He chose the name "*Pseudolucia*" because the South American species comprising it show yellow-orange "copper"-colored patches on their wings reminiscent of the Old World Coppers of *Lucia*.

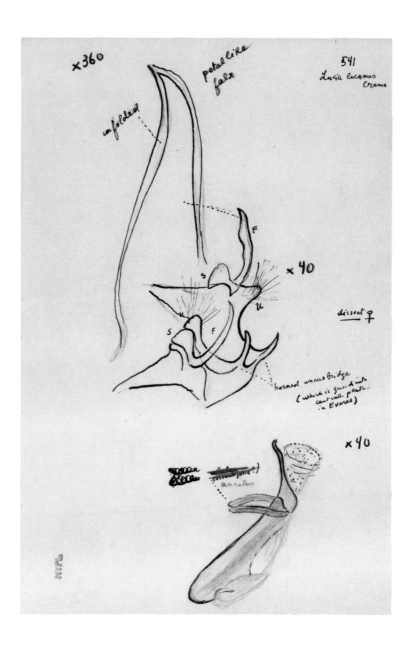

Color Plate 21 [*Scolitantides*]

Trying to decipher the meaning of Old World names *Scolitantides* Huebner 1819 and *Glaucopsyche* Scudder 1872 with regard to the New World, Nabokov here figures a lateral view of the genital apparatus of the taxon *fatma* Oberthuer 1890 known from North Africa. Today *fatma* is considered a subspecies of the more widely distributed Eurasian *Pseudophilotes bavius* (Eversmann 1832). *Pseudophilotes* Beuret 1958 is a relatively recent generic name used to further distinguish some members of the burgeoning complex called either *Scolitantides* or *Glaucopsyche* in Nabokov's day.

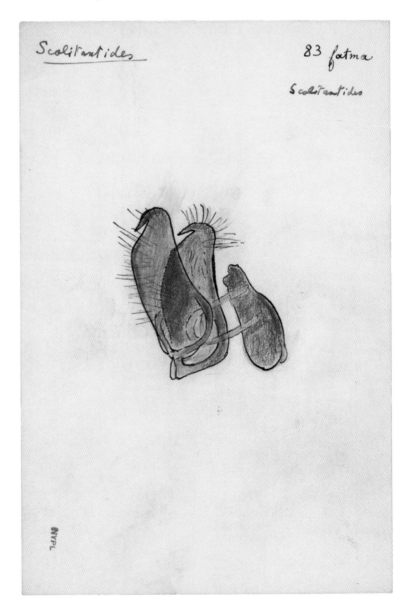

Color Plate 22 [*divina*]

Again interested in a Blue whose range is near the Bering Strait, Nabokov here questions the placement of the Blue *divina* Fixsen 1887 in *Scolitantides* Huebner 1819 or *Glaucopsyche* Scudder 1872. The Large Japanese Blue is found on the Korean peninsula and Japan and has often been placed in the genus *Shijimiaeoides* Beuret 1958, named for a group of unusual northeast Asian Blues.

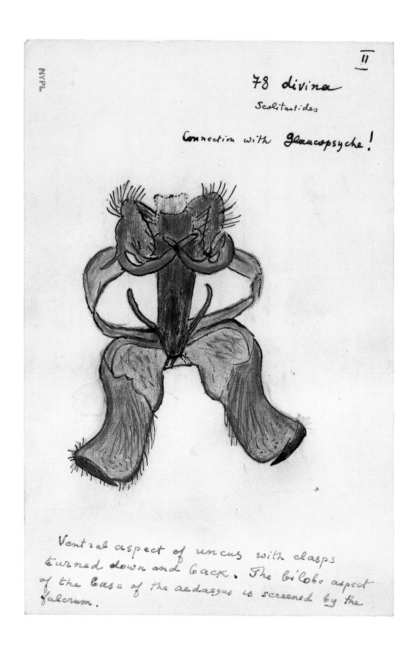

Color Plate 23 [*Brephidium*]

Brephidium Scudder 1876 is a genus of worldwide Blues that, although not part of Nabokov's polyommatines, had relevance to Nabokov's understanding of the relationships of global Blues. Although *Brephidium*s are called Pigmy Blues and were considered the smallest-sized species of Blues in the world, Nabokov's Blues (like *vera* Balint and Johnson 1993) are as small or smaller than the famed Western Pigmy Blue (*exilis* Boisduval 1852, shown above).

Color Plate 24 [*Agriades*]

Agriades Huebner [1819], the so-called Arctic Blues, were of special interest to Nabokov because of their members' modern-day trans-Beringian distributions. Nabokov here draws the male valve of *felicis* Oberthuer 1886, a Tibetan *Agriades*.

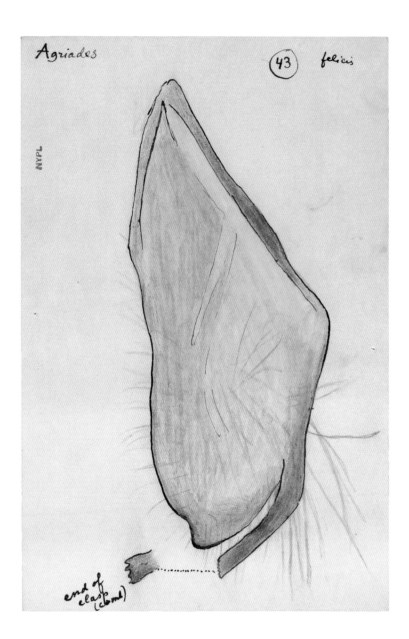

Color Plate 25 [*Agrodiaetus*]

A Blue central to sorting out which Old World Blues belonged in the available genera *Agrodiaetus* Huebner 1822, *Plebejus* Kluk 1780, and *Polyommatus* Latrielle 1804 was the taxon *elvira* Eversmann 1854, known from Uzbekistan and southern Kazakhstan. In fact, historically, nearly twenty genera have been used to contain Blues Nabokov sought to narrow to generally these three on his laboratory cards. Only in the last half-decade have DNA studies of the Asian Blues of these groups begun to really sort them out. Elvira's Blue is a unique Blue, even elevated to its own genus *Elviria* Zhdanko 1994 by a Russian taxonomist. This may be an early baseline drawing for Nabokov because he not only uses some terms seldom seen in his publications (penis, gnathos) but also, when using genitalic terms, cites specific authors.

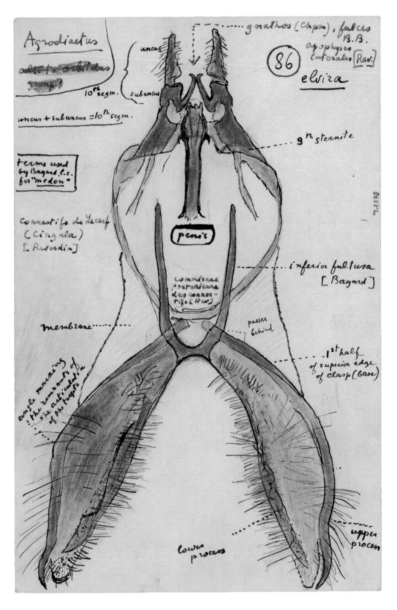

Color Plate 26 [*felicis*]

Continuing to draw details of the anatomy of *Agriades* Huebner [1819], the so-called pan-Beringian Arctic Blues, Nabokov here renders color-shaded details for the taxon *felicis* Oberthuer 1886. Again, Nabokov chooses a Tibetan *Agriades* relevant to his investigations of Asian Blues and their possible evidence for the Asian origins of the New World Blues.

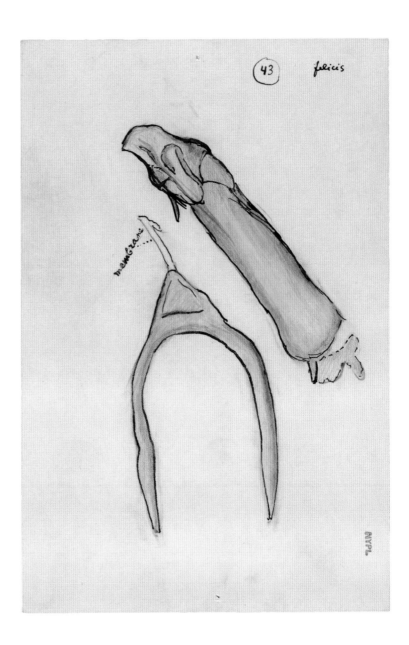

Color Plate 27 [*sonorensis*]

We see from this card that Nabokov did examine a true *Philotes: sonorensis* C. & R. Felder [1865] (see also Figure 35). That he drew this *colored* wing image suggests that he noted this species' unique qualities: spots displaced outside the general parameters for Blues' wing patterns (noted on the card) and the red highlights across the outside of the forewings. Because this species was rare in collections during Nabokov's time, his rendering in color (perhaps taken from a book, since no specimen data are noted) was intended for his own diagnostic work.

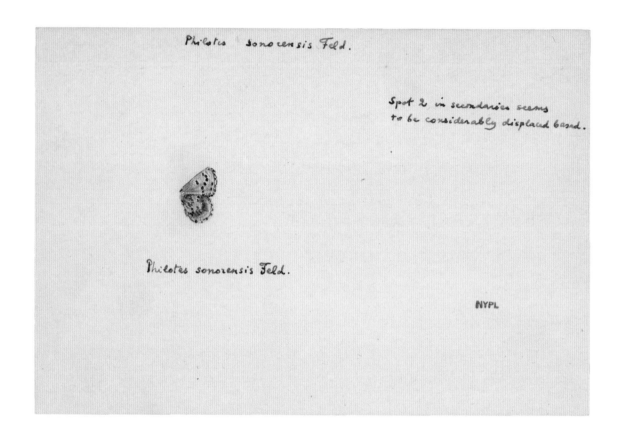

North America

Color Plate 28 [*ferniensis*]

Nabokov renders the entire male genital apparatus for *Lycaeides idas ferniensis* (Chermock 1945), the Fernie Blue of western Canada.

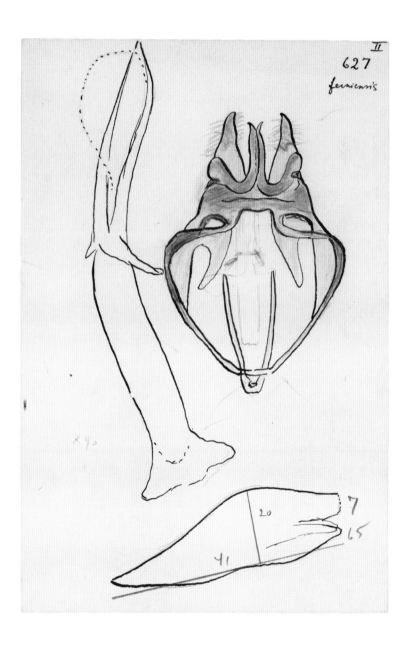

Color Plate 29 [*idas, argyrognomon*]

Here Nabokov depicts the male genitalia of a specimen of uncertain identification from the *Plebejus idas* (Linnaeus 1761) or *P. argyrognomon* (Bergstraesser 1779) complexes from Labrador.

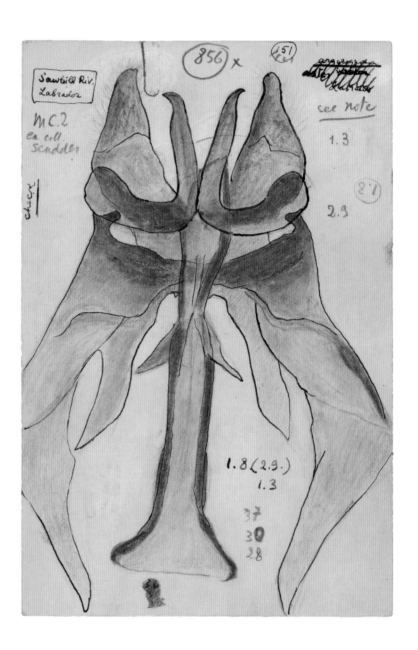

Color Plate 30 [*anna*]

This is an infrequent color sketch of the female genitalia representing *Plebejus anna* (W. H. Edwards 1861). The female genitalia in Nabokov's Blues are very simple, consisting mostly (as Nabokov shows) of layers of membranes (indicated in yellow) overlaying a lightly sclerotized orifice and inwardly directed tube (shaded gray in pencil).

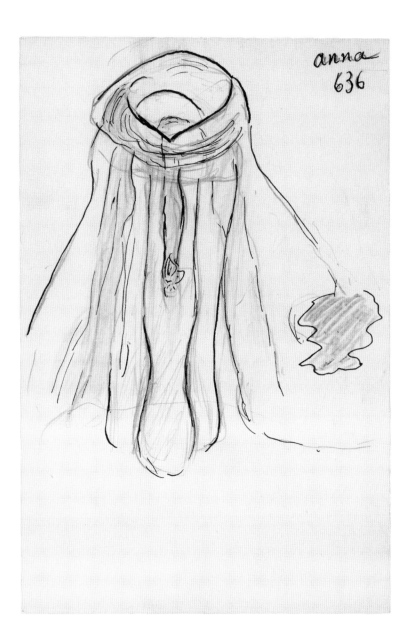

Color Plate 31 [*samuelis*]

In this sketch of *Lycaeides melissa samuelis* (now *Lycaeides samuelis*), the endangered Karner Blue, Nabokov renders the wings' undersurface in color. It is interesting that he refers to the specimen as "*samuelis*"; though he always suspected that it was a full species, he was reluctant to name it as such, perhaps from a fear of being called a splitter (at the time, lepidopterists identified the Karner Blue as the eastern population of a much more widespread assemblage). Modern DNA studies have proven Nabokov correct.—Lauren K. Lucas, Matthew L. Forister, James A. Fordyce, Chris C. Nice, and Kurt Johnson

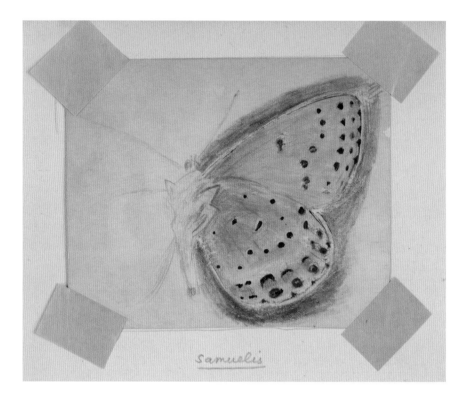

Color Plate 32 [*thomasi*]

Nabokov recognized the distinct Caribbean genus *Cyclargus* Nabokov 1945, which, although previously considered the same as Huebner's old name *Hemiargus* (1818), differed drastically in anatomy. For accuracy, Nabokov dissected the definitive "holotype" specimen of the species defining *Cyclargus*—its "type species," *thomasi* Clench 1941. What is particularly of interest is that Clench himself did not dissect his own definitive specimen of *Hemiargus thomasi* Clench. If he had, he might have noticed that its genitalia were extremely different from other members of the *Hemiargus* lineage. Like most lepidopterists of the day, Clench relied on wing pattern characteristics for classifying Blues.

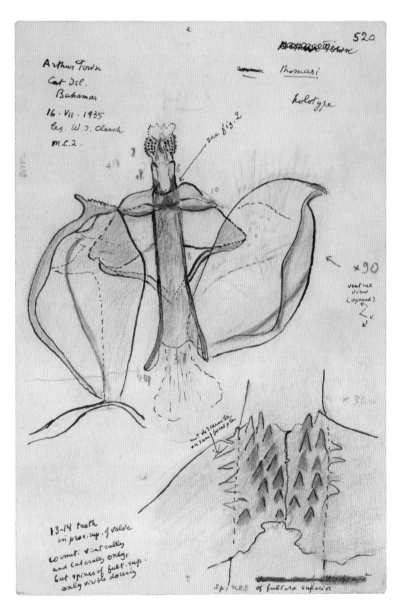

Latin America

Color Plate 33 [*martha*]

Here Nabokov draws an expansive spinate sagum (see Figure 44 and Color Plate E2) as seen on the South American species *martha* Dognin 1887, which became a member of his new genus *Echinargus* Nabokov 1945. Lepidopterists were reluctant to accept Nabokov's new genus because, unfamiliar with species like *martha* from South America, they knew only the common American species *isola* Reakirt [1867]. Further, many had not read Nabokov's 1945 paper, so they assumed he was making a new genus just for *isola*, which they opposed. The sagum is, in fact, so large on *isola* and *martha* that some readers of his 1945 paper might have thought it was the male valve. In any event, it is difficult to comprehend how other lepidopterists summarily ignored such a prominent structure, shown here, that was used to distinguish an obviously distinct genus. Most likely it was because they were used to the old nomenclature based on superficially similar wing patterns.

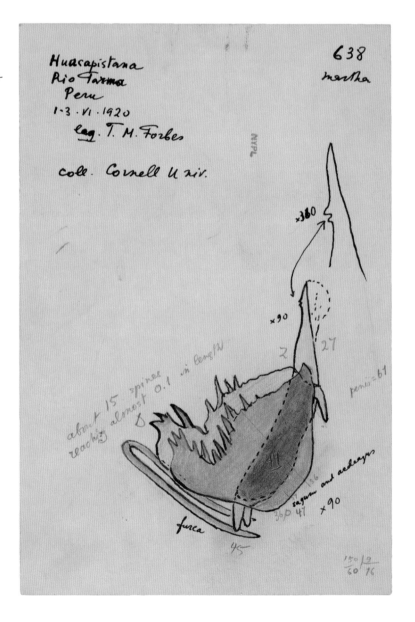

Color Plate 34 [*gyas*]

Continuing his survey of *Hemiargus* Huebner 1818, Nabokov records characteristics of a specimen from California (*gyas* W. H. Edwards 1871). He emphasizes the lack, or diminutive occurrence, of the prominent sagum structure he has found in other Latin American Blues (like his new genera *Echinargus* Nabokov 1945 and *Cyclargus* Nabokov 1945). Since, by taxonomic rules, he could not reliably restrict the definition of *Hemiargus* without having a definitive specimen by which to characterize that name, Nabokov decided to make a "neotype" for the defining, or type, species of *Hemiargus: hanno* Stoll 1790. He published that "neotype," or new type, in his 1945 paper, illustrated on its plate 4. See also the discussion at Figures 60 and 61 in this volume.

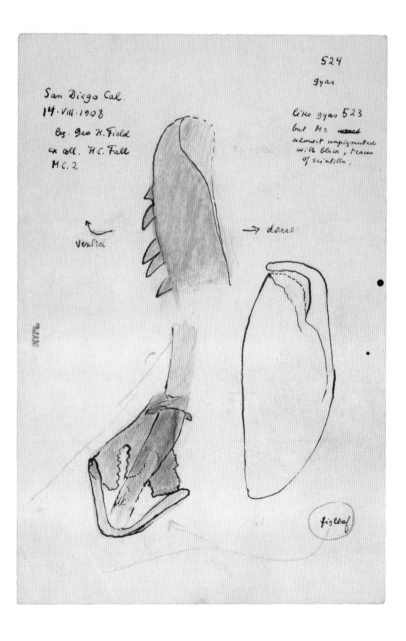

Pseudolucia

Color Plate 35 [*chilensis*]

Nabokov devotes this card to the long filament, noted in Figure 65, that appears to emanate from the juncture of two forks of the furca. He illustrates the filament and the spinate structures at its terminus. Nothing further seems to be known about this structure. Nabokov did not illustrate it in his 1945 publication on the group. See also Figure 66.

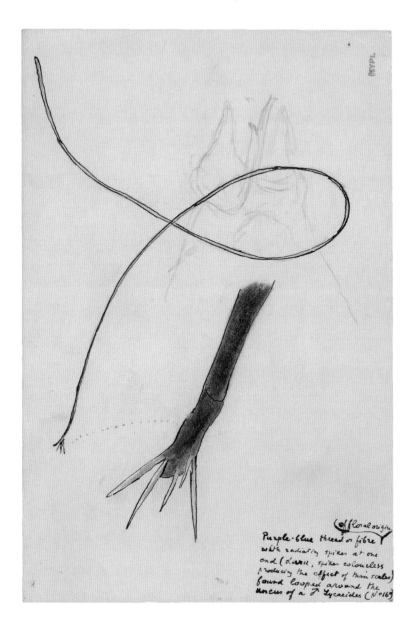

Parachilades (now *Itylos*)

Color Plate 36 [*titicaca*]

One of the best-known South American Blues in nineteenth- and early twentieth-century collections is the Titicaca Blue from around Lake Titicaca, Bolivia. Distinct in its small size and oblongate (and brilliantly blue) wings, it was certainly a Blue of great interest regarding the evolution of the butterflies of South America's Andean region.

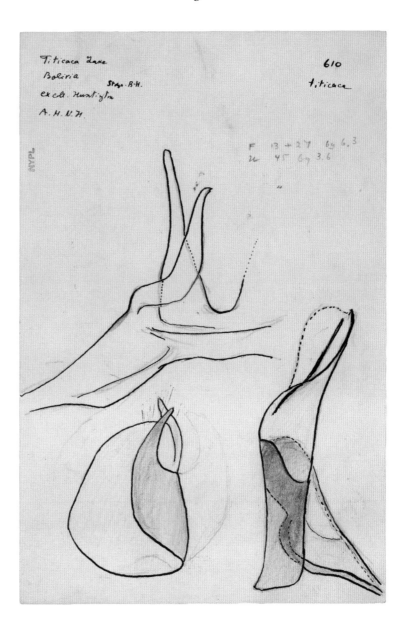

Color Plate 37 [magic triangles]

In this card Nabokov shows details of his magic triangles method for using quantitative measure comparisons to characterize qualitative differences in general shapes of both the dorsoterminal elements of the male genitalia (uncus and falx/humerulus) and the male clasper (valve). Here, using a specimen of *scudderi* W. H. Edwards 1861, he depicts in color (at center right) a lateral view of the uncus ("32"), falx ("41")/ humerulus ("32") with the magic triangle formed by his measurement lines.

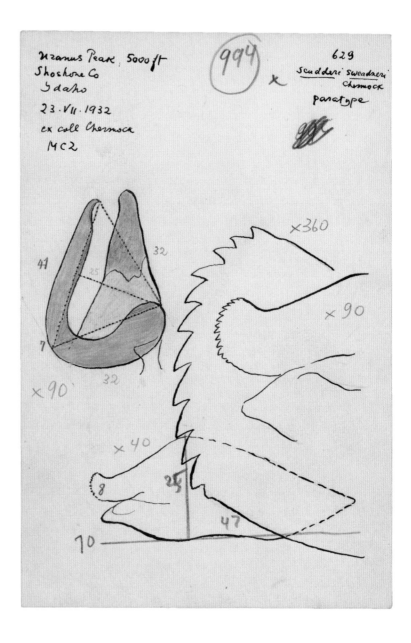

Wings

Color Plate 38 [various wings, butterflies, and schematics]

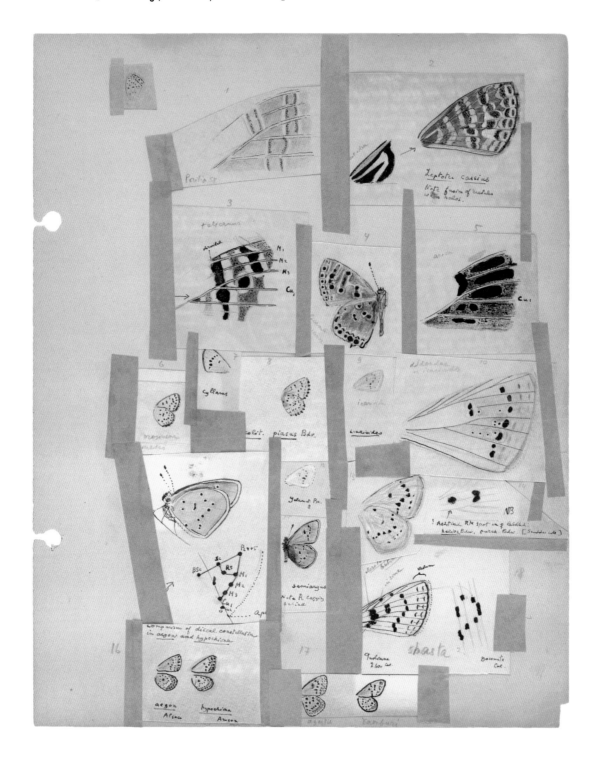

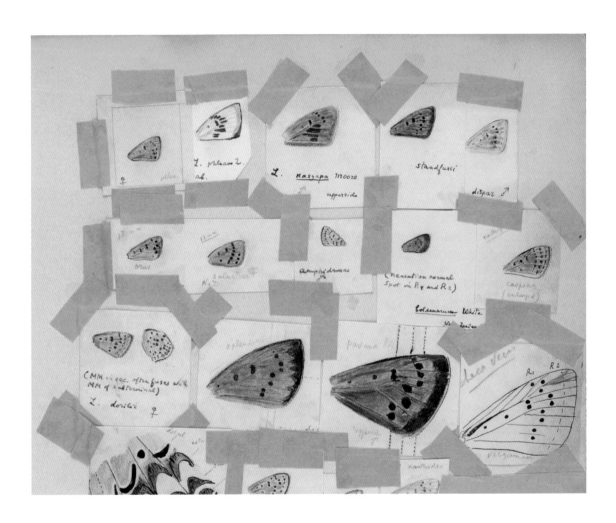

Color Plate 40 [various wings, butterflies, and schematics]

Color Plates 38–40: These cards display a panoply of concise and detailed wing pattern drawings combined in a collage of many species for simultaneous viewing and comparison. They also show examples of his correlations of detailed full wing drawings with more rudimentary schematic drawings of spots and macular patterns, with wing cells labeled at bottom right. Such schematics are found throughout Nabokov's cards and often portray extensive series of specimens. See also Figures 87 and 89.

Color Plate 41 [various wings, butterflies, and schematics]

These detailed color renderings of the undersurfaces of wings across a diversity of butterflies demonstrate Nabokov's exceptional skill as a scientific illustrator. He appears to be showing the progression from wing patterns dominated by macules over a monochrome ground pattern (as at the top) to increasingly complex patterns where macules and ground color combine to produce complexly mottled camouflage-like patterns. At bottom he details a butterfly in "perching" position; it is curiously reminiscent of many drawings of butterflies that Nabokov included in dedications of books to friends and family.

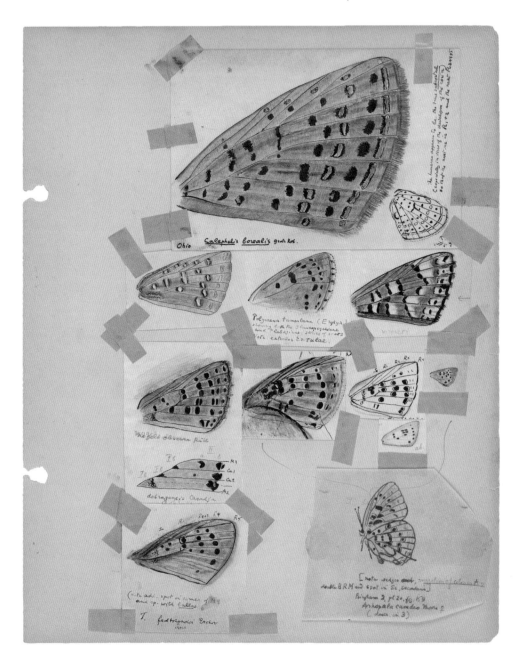

Color Plate 42 [wing scales, *samuelis*]

Top: From 1869 to 1872, Victorian microscopists John Watson and John Anthony studied strange, beautiful scales—scattered among the usual fluted scales of male Blues—that looked like battledores (antique badminton rackets). These inflatable structures bore longitudinal ribs that were studded with miniscule, mushroomlike "elevations." They are androconial scales that waft pheromones over females during courtship. Nabokov's battledore scale of the Northern Blue (*Plebejus idas,* top center) resembles that of the Karner Blue (*P. samuelis*) but is more intricately textured.—Robert Dirig

Left, middle: This is a color rendering of the wing pattern of Nabokov's Karner Blue, *Lycaeides melissa samuelis* Nabokov 1944, now shown by DNA studies to be a full species, *Lycaeides samuelis.* Nabokov rotates the forewing into alignment with the hind wing so he can study the relative position of spots and macules typifying the various veins and cells of both wings.—Kurt Johnson

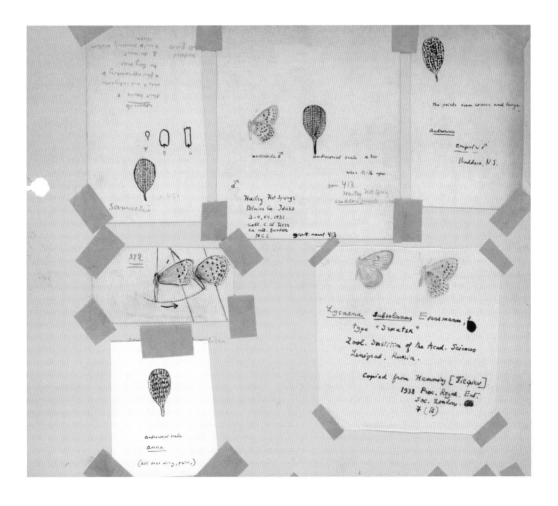

Color Plate 43 [*cleobis*]

Nabokov's particular interest in East Asian Blues, pertinent to his Beringian origins hypothesis, is reflected here in his rendering of Japanese Blues named by Shōnen Matsumura, Japan's preeminent postwar entomologist. In the upper right, for comparative purposes he draws *cleobis* Bremer 1861 (using the classical generic name *Lycaena*, in which many Blues were originally situated). Matsumura, a contemporary of Nabokov, was also interested in Blues.

Color Plate 44 [*boeticus*]

A fine example of Nabokov's skill as a color renderer is this highly detailed drawing of the hind wing undersurface of a female specimen of *Cosmolyce* Toxpeus 1927 (now *Lampides* Huebner 1819) *boeticus* (Linnaeus 1767), including even the fine fringes along the wing margins. Nabokov notes the "intensified pigmentation of macules." He discusses the maculation of this Old World species in "Notes on the Morphology of the Genus *Lycaeides* (Lycaenidae, Lepidoptera)."

Color Plate 45 [maculation comparative]

Before experts like Nabokov mastered the knowledge of genitalia, wing patterns were the means of identifying most butterflies. The Russian lepidopterist Boris Schwanwitsch (1889–1957) pioneered the detailed study of wing pattern "maculation," as it was called, hoping to find keys the intricacies of evolution. Nabokov here divides wing undersurfaces into concentric regions, with numbered scale rows, as a baseline for plotting data he is accumulating on maculation patterns. Comparing scores of such plots, he was determined to discover historical patterns that might decipher the path of butterfly evolution. Schwanwitsch's tombstone, in Saint Petersburg, Russia, is engraved with a detailed wing pattern frieze.

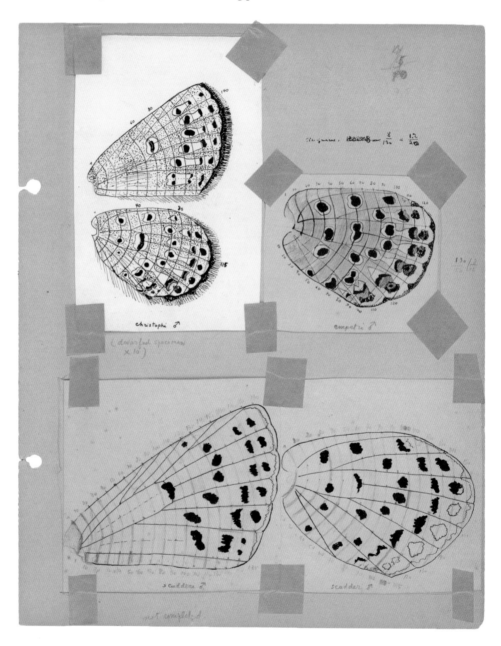

Color Plate 46 [wing cell comparative]

In his scrapbook, Nabokov assembled these individually detailed drawings of the marginal macule in Cell CuA1 of the undersurface hind wing from specimens across his *Lycaeides* Huebner 1819 study group, including *agnata* Staudinger 1889, *argyrognomon* Bergstraesser 1779, and *melissa* W. H. Edwards 1873. He prepared drawings of these comparative pie slices of the hind wing for scores of Blues as he sought to ascertain the pattern of evolution in this group.

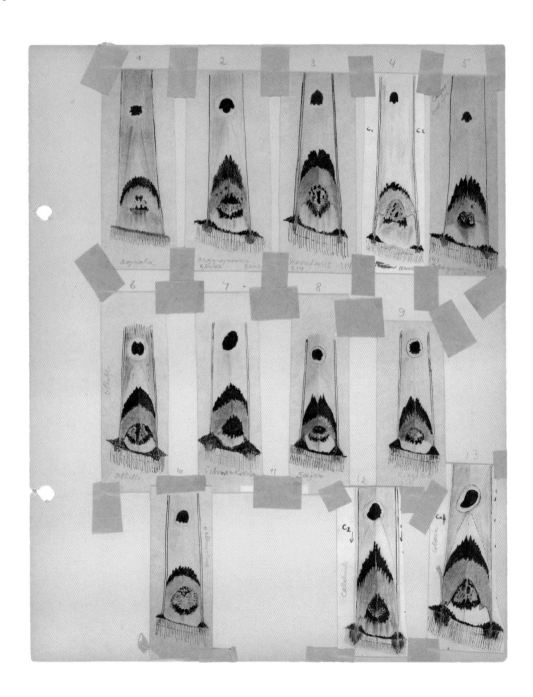

Color Plate 47 [wing sector, *Lycaeides*]

In this schematic of a wing sector for *Lycaeides*, Nabokov labels all the physical structures and components of the "marginal ornamentation of the underside of the secondary" with particular clarity.

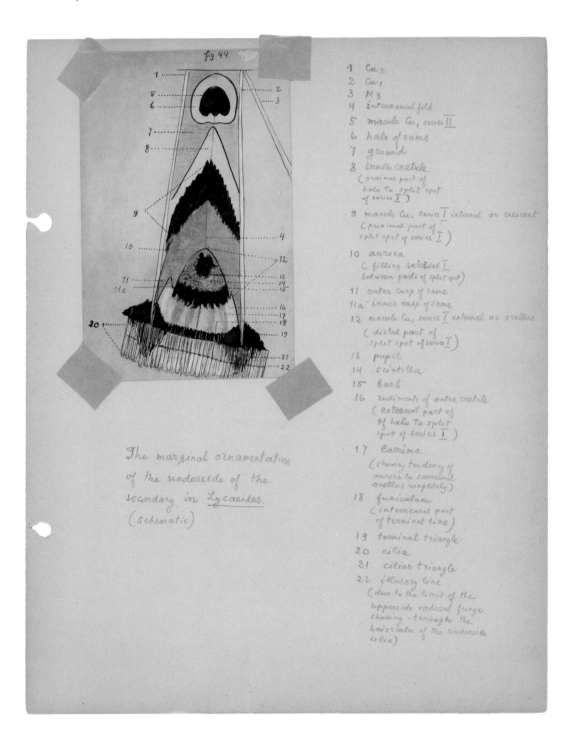

Color Plate 48 [wing sector, *samuelis*]

This detailed drawing of a hind wing cell (the space between wing veins) illustrates the position of a black spot, Cell CuA1, and the colored, iridescent pattern element at the wing margin called an aurora, in *Lycaeides melissa samuelis* (Nabokov 1944). Nabokov found the position of this particular black spot on the hind wing to be the most variable character across *Lycaeides* wings. Early lepidopterists were particularly interested in this macule because, when further developed and especially when accompanied by a marginal tail, it was suspected to create a false head image that could distract predators into attacking the butterfly's wings instead of its head.—Lauren K. Lucas, Matthew L. Forister, James A. Fordyce, Chris C. Nice, and Kurt Johnson

Color Plate 49 [wing sectors, *melissa*]

On this card Nabokov compares the wing sectors of two female and one male Melissa Blue butterflies, *Plebejus* (formerly *Lycaeides*) *melissa*. Each drawing (with added ladder marks for precise measuring) shows one hind wing sector, Cell CuA1, radiating out from the insect body like the ribs of a fan between wing veins (or nervules) CuA2 and CuA1. See Figures 90 and 91 for a schematic view of relative wing cell positions and Color Plate 47 for a fully labeled sector. These illustrations show how the spots move from the proximal end of the cell to their final positions. Nabokov imagined that during development, the haloed spots, in effect, move (in reaction-diffusion waves) toward the peacock tail markings at the distal end of the wing. He also thought that the expanding peacock markings prevented the haloed spots from advancing further. Understanding the direction of pigmentation diffusion and how the different elements within the cell constrain each other allowed him to imagine how a change in one element could affect other elements. Nabokov realized that the interactions occurring *within* the cell were more relevant than interactions *between* cells and thus that even if spots sometimes diffuse laterally and connect across veins, forming a line, they should still be understood as independent marks and not as parts of a broken line.

—Victoria N. Alexander

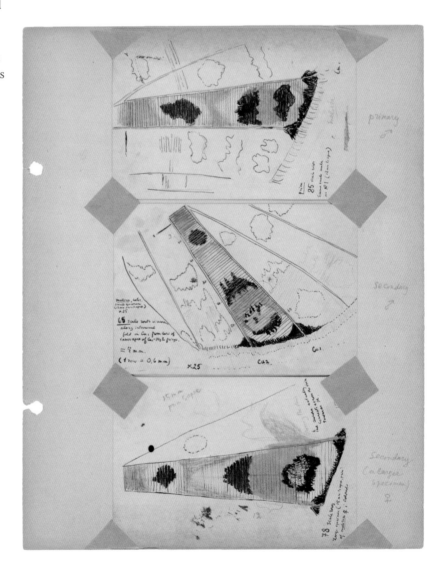

Color Plate 50 [wing sector comparative]

Nabokov prepared detailed illustrations of many of these pie-slice sectors through the undersurface of the hind wing, especially illustrating the large marginal macule in Cell CuA1. The larger figures at right depict two Old World species, *Cosmolyce boeticus* (Linnaeus 1767) and *Agrodiaetus elvira* (Eversmann 1854), which Nabokov was most likely using as an Old World comparison to his New World *melissa samuelis* complex. The left side of this drawing appears to show a cruder prototype series Nabokov prepared as part of his ongoing wing pattern maculation studies.

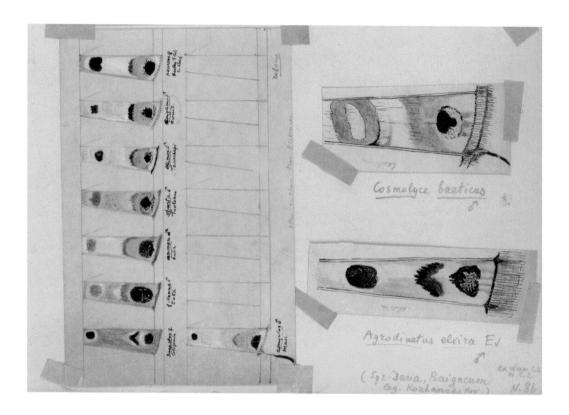

Color Plate 51 [maculation study, *icarioides*]

The right side of this drawing shows a prototype comparison drawing of an *Icaricia icarioides* (Boisduval 1852) female wing pattern, underside left and upper side right, with Nabokov noting that "very rarely there is a formation of aurora in *icarioides* ♀ underside, more often in upperside, prob. Colorado." At lower left, he appears to have crafted a schematic composite drawing concerning degrees of expression of wing macules—ranging from a "double macule" (two spheres side by side or adjoined) to a "composite macule" (two spheres overlapping), to a whole macule (one sphere), and then to various reductions seen in differing expressions of crescentlike markings (from wider to successively narrower). These would all figure in his understanding of wing pattern variation and, he hoped, the evolution of the wing patterns in the Blues.

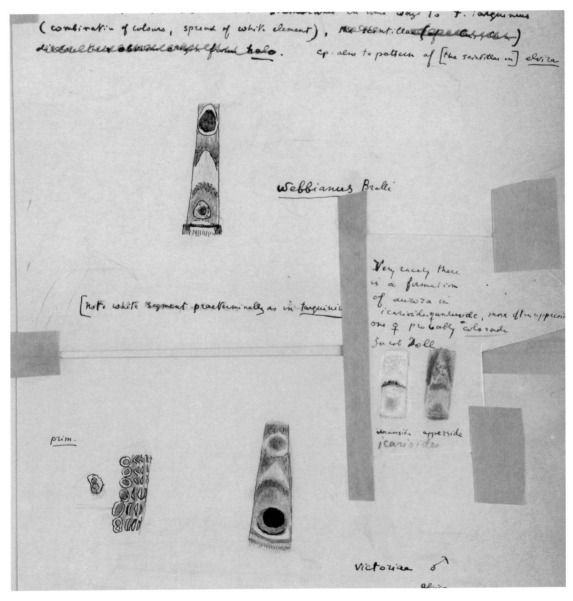

Color Plate 52 [wing cross section]

Nabokov used this illustration, representing a hypothetical folding of the wing, to compare the upper surface pattern, cell by cell, with the corresponding undersurface pattern. Having prepared the drawing this way, he could visually fold the wing over on itself and see detailed correlations of the upper and undersurface macular patterns.

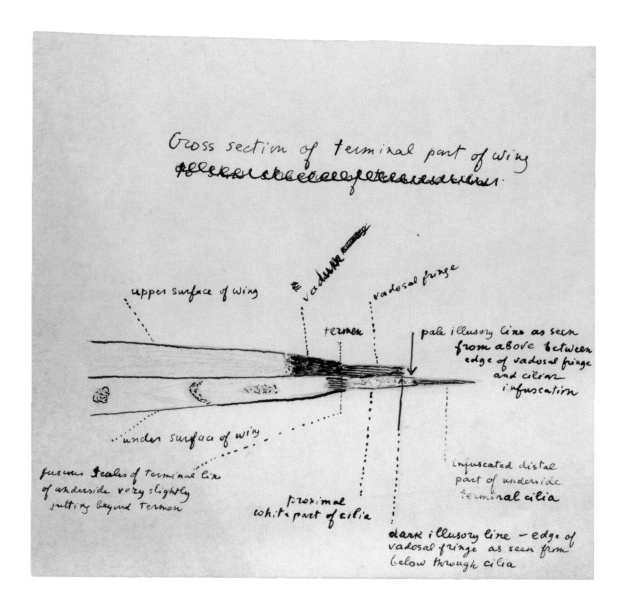

Color Plate 53 [wing scale comparative]

At the top and lower left of this drawing, Nabokov shows how stems of cover scales and darker basal scales of the European Common Blue (*Polyommatus icarus*) fit into regularly spaced sockets, each formed by a single cell. At the bottom center, he draws a cerulean scale from the hind wing spots of a European nymphalid butterfly, the Small Tortoiseshell (*Aglais urticae*). In the trio at the right, he depicts refractive scales from the blue hind wing band of a Neotropical moth, *Coronidia orithea* (Sematuridae). All may produce shimmering structural colors when sunlight strikes finely textured elaborations of the scale's ribs or interior.—Robert Dirig

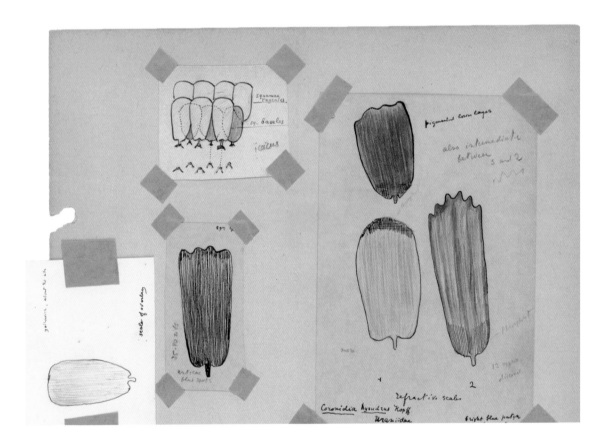

Color Plate 54 [wing scale comparative]

Each wing scale is a nonliving cuticular product of one epithelial cell. Scales are formed of two flattened cuticle layers, with longitudinal ribs and numerous microscopic cross-ribs that form "windows" into the hollow interior. These scales of the Melissa Blue, *Plebejus* (formerly *Lycaeides*) *melissa*, include, from left to right: (4), a "scintillating" scale from the hind wing venter; (1), violet-blue basal(?) scale from dorsal side; (2–2a), turquoise basal scales on the venter; and (3), perhaps the same as (4).—Robert Dirig

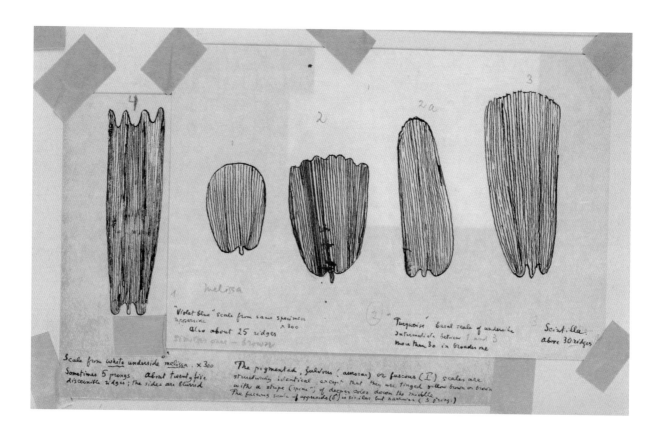

Color Plate 55 [*argyrognomon, ismenias*]

Nabokov here presents detailed renderings of the undersurface wing pattern from the genus *Lycaeides*-related complex, including the complexes related to *argyrognomon* Bergstraesser 1779 and *ismenias* Meigen 1829. Below left are upper and undersurface drawings of *bergi* Kuznetsov with an accompanying rendering of the male genital valve; below right, he shows a male and female wings of *Plebejus idas* (Linnaeus 1761) subspecies *mira* Verity 1913.

Color Plate 56 [maculation comparative]

This card depicts comparative maculation schematics of Cells M3 and CuA1 on the undersurface hind wing of three taxa, including the well-known Hairstreak *Satyrium titus* (Fabricius 1793), the Coral Hairstreak. Nabokov's interest here is probably in comparing the maculation of butterflies that have two prominent marginal hind wing macules with that of most Blues, which usually have one, in Cell CuA1.

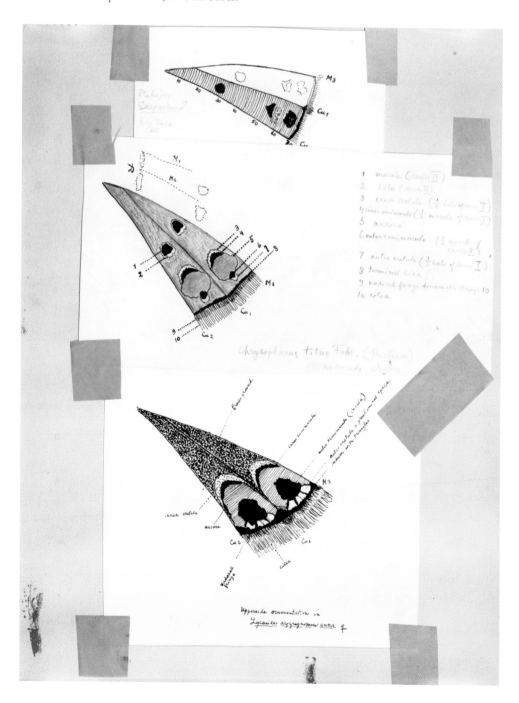

Inscriptions to Véra

Color Plate 57 *"Vanessa verae"*

As readers of *Pale Fire* know and those of Brian Boyd's *Nabokov's "Pale Fire"* know better, *Vanessa* is the genus of the most prominent butterfly in that novel (but see Robert Dirig's essay in this volume placing the Toothwort—or West Virginia—White in close contention for that honor). The appearance of this drawing of a fanciful blue-barred *Vanessa verae,* in the copy of *Pnin* Nabokov inscribed to Véra in 1957, when *Pale Fire* was only beginning to take shape in his mind, serves as a moving example of butterfly as artistic inspiration. As Kurt Johnson told Sarah Funke in 1999 for *Véra's Butterflies,* this specimen resembles the genus *Araschnia* and "may represent what Nabokov would have liked to have seen in *Vanessa.*"
—Stephen H. Blackwell

Color Plate 58 *"Vanessa incognita"*

From the inscription copy of *Pnin* to a birthday inscription copy of just-released *Pale Fire*, the inspiring butterfly has transformed into *incognita*—the unknown one, unknown woman—perhaps invoking Alexander Blok's poem cycle of that title ("*Neznakomka*"). This motif is rich with connections to Russian Symbolism, and one wonders here if Nabokov implies a specific bond with Blok or with Symbolism—or if perhaps he expresses something about the mysterious essence of his just-published work that eludes commentators to this day. Kurt Johnson told Sarah Funke in 1999 for *Véra's Butterflies* that this was the last of four drawings, beginning with *Vanessa verae* (above), in which "Nabokov adopts blue colors which would not appear in these groups in nature." The first of these was, as noted above, inscribed in *Pnin*—a novel in which Blues, more specifically some of Nabokov's own Karner Blues, make a stealthy appearance. Yet, as Johnson notes below (see Color Plate 65), Nabokov inscribed few Blues in his gift copies. Their color's unlikely presence in these wings blends Nabokov's artistic precision and scientific passion in a single imaginary taxon.—Stephen H. Blackwell

Color Plate 59 "*Colias lolita* Nab."

Among the minor characters in *Lolita* is Edusa Gold, the director of Quilty's play *The Enchanted Hunters,* in which Lolita was to have starred at Beardsley. Her strange name derives from the long-superseded species name *Colias edusa* (now *Colias crucea*), the Clouded Yellow.

Nabokov met his first love on a Biarritz beach in 1909, when he was ten and she was nine, and he drew on their summer romance as the model for Humbert's first love with Annabel Leigh on a Mediterranean beach. By 1909 Nabokov was old enough to identify and recollect butterflies. In his account of his "first love," whose real identity he protects in his autobiography under the name "Colette," he could therefore recall: "At a tremendous pace a stray Clouded Yellow came dashing across the palpitating plage" (*Speak, Memory,* 147). The "Colette" name he bestowed on his Biarritz love and the genus name of *Colias edusa* he saw there appear to combine here with Lolita's link (since to Humbert, Dolores Haze seems the very image of his Riviera love) to "Colette."

The specimen shown here of the new species, *Colias lolita*—a species named by Nabokov, according to this mock captor's label—is identified (of course!) as a female. Nabokov's description of Humbert's famous epiphany as he looks down from a mountain road at a town in the valley below, at the end of *Lolita,* depicts and commemorates Telluride, where in 1948 he had traveled to capture the hitherto unknown female of the subspecies he had named *Lycaeides argyrognomon sublivens* (see *Lolita,* 309–10, and Nabokov's afterword, 318). Feeling very much an enchanted hunter, he caught twenty of these hitherto unknown females on the slopes above Tomboy Road.

Sublivens means "livid," referring to the unusual deep bluish coloring of the butterfly's wing. The color of Nabokov's invented *Colias lolita* echoes the theme of yellow (the characteristic color of the genus *Colias*) being transformed into the complementary color blue (here, a deepish blue), as in other butterflies drawn for Véra, like *Colias verae.* Perhaps Nabokov has in mind his own particular area of lepidopterological expertise, the Blues, as if he were putting his stamp on butterflies of other kinds and colors, like these *Colias*es and the *Vanessas,* never blue until his imaginary *Vanessa verae* and *Vanessa incognita.* Or perhaps in this case he also has in mind the unnaturalness of a blue *Colias* as a parallel to the unnaturalness of the thoroughly American *Lolita* appearing in a translation—even if it is his own—into Russian. Here, after the Russian inscription, "To my Verochke, October 1967," Nabokov has written the place-name "Montreux" as a comical combination of Roman and Cyrillic letters.—Brian Boyd

Color Plate 60 "*Paradisia radugaleta*"

This butterfly is named after "Ardis," Ada Veen's paradisal family home in *Ada*, and "Radugalet," the "other Ardis," where Van's father and uncle share an estate. *Raduga* means "rainbow" in Russian, and "-let" is the English diminutive ("-let" itself becomes a minor motif in *Ada*). At the same time *raduga leta* means "rainbow of summer," appropriate to *Ada*, most of whose action focuses intently on two summers at Ardis. Nabokov had finished composing *Ada* just ten weeks before drawing this butterfly and would have been reading Véra his daily output on the novel since 1966.

Rainbows were a personal motif for Vladimir and Véra, strongly associated with their first meeting, and partly for that reason they form a motif in *Speak, Memory* as well as in the writer's butterfly inscriptions for his wife. The eyespot here is highly like a female human eye, which the artist accentuates further by means of the lashes at the edge of the hind wing. Véra first appeared before Vladimir wearing a mask, which framed and highlighted the eyes peering at him when much of the rest of her face was obscured. Compare this eyespot with the even more masklike eyes of the forewings in the butterfly drawn on the *Lolita* "duplicate for Vera," duplicated or mirrored in the eyes of the hind wing.

Nabokov rarely drew a butterfly ovipositing. *Paradisia radugaleta*'s egg-laying may be ironic here in conjunction with *Ada*, since "acarpous" Ada cannot have children. Stephen Jan Parker, a former student of Nabokov's and the founding editor of the *Vladimir Nabokov Research Newsletter* (since 1984, *The Nabokovian*) chose a monochrome reproduction of this image (butterfly, eggs and stalk only, without inscription), as the cover emblem of the journal from number 5, Fall 1980. The emblem remains on the *Nabokovian* even now.—Brian Boyd

Color Plate 61 "*Maculinea aurora* Nab." (male)

It is curious that among all the butterfly drawings and paintings dedicated to family and friends on various first editions and other books given by Nabokov as gifts, he didn't include many Blues. This color rendering of "*Maculinea aurora* Nab." is an exception. In this fictitious Blue, Nabokov exaggerates the extent of orange marginal coloration on the hind wing upper surface of this male specimen, certainly a trait that would have excited him if he had ever seen such a Blue in nature. He would have known it was unknown to science and pursued it with his net with vigor! On the undersurface, a number of Blues rendered in color in his laboratory scrapbooks approach this density of marginal orange (see Color Plates 17 and 44). Nabokov was obviously fascinated by what a Blue might look like with that amount of orange on top! The genus *Maculinea* van Eecke 1915 is a real genus of Blues, today usually considered the same as *Phengaris* Doherty 1891. In terms of their genitalia, these Blues fall into the groups that Nabokov rendered often on his laboratory cards concerning the genus *Glaucopsyche* Scudder 1872. *Phengaris* is today considered one of the genera in the larger Glaucopsyche Section. As the captions to many genitalia drawings indicate, Nabokov spent a lot of time trying to sort out the Old and New World nomenclature for *Glaucopsyche*.—Kurt Johnson

Color Plate 62 "*Polygonia thaisoides* Nab."

The large numbers of butterfly renderings in Nabokov's laboratory scrapbooks show his fascination with the intricately reticulate and cryptic colorations of many butterflies, especially butterfly wing mimicry (see Color Plates 39 and 59). And so it does not surprise that in the fictitious entity *Polygonia thaisoides* Nab., all these elements are at play, along with patterns that invoke stained glass (especially on the upper surface of the hind wing). Interestingly, if the distinct scallops and dentations of the wing margins were removed (if the margins were what taxonomists call "entire"), *P. thaisoides* could easily be taken for a Metalmark (Riodinidae), not a Nymphalid (Nymphalidae). *Polygonia* Huebner 1918 is a real Nymphalid genus, and any lepidopterist viewing *thaisoides* would likely place it there because of the wing shape; otherwise, *thaisoides* is an odd *Polygonia* indeed. Here are two more titillating facts to add to the speculation about why Nabokov coined this name: in Thai, "thai-so" is slang for "ha ha ha" or "laughing out loud," and taxonomists use the suffix "-oides" to mean "looks like."—Kurt Johnson

Plates to Essays

Captions are by the authors of the respective essays.

Color Plate E1 *Pseudolucia* butterflies [Introduction]

Nine photographs of Nabokov's Blues by pioneering modern fieldworker Dubi Benyamini. Benyamini, Zsolt Bálint, and Kurt Johnson's South American collecting expeditions are recounted in the book *Nabokov's Blues*. (1) *Pseudolucia vera* (Bálint & K. Johnson 1993), perching, showing brilliant orange-yellow upper surface patches. (2) *P. charlotte* (Bálint & K. Johnson 1993), feeding on yellow flower. (3) *P. charlotte* (Bálint & K. Johnson 1993), perched with wings folded up, showing undersurface. (4) *P. lanin* (Bálint & K. Johnson 1993), perching. (5) *P. zina* Benyamini (Bálint & K. Johnson 1995), perching with wings folded, showing mottled undersurface. (6) *P. chilensis* (Blanchard 1852), perching. (7) *Nabokovia faga* (Dognin 1895), mounted specimen. (8) *P. lanin* (Bálint & K. Johnson 1993), showing incredible camouflage of the upper wing surface among pebbles. (9) *P. lanin* (Bálint & K. Johnson 1993), mounted specimen. Yellow-orange patches on the wings of many of these Blues apparently mimic the yellow-orange colors of *P. chilensis*, which feeds on a toxic plant. Yellow and orange are common warning colors to potential predators among the Lepidoptera. Mimicry in Nabokov's South America Blues is described in *Nabokov's Blues*.

Color Plate E2 Blacktail Butte, Wyoming [Lucas et al.]

This trail leads to an admixed population of *Lycaeides* first described
by Nabokov on Blacktail Butte in Wyoming.

Color Plate E3 Dissected male *Lycaeides melissa* genitalia [Lucas et al.]

Nabokov made four measurements of an uncus for *Lycaeides* species
delineation: F, H, E, and U.

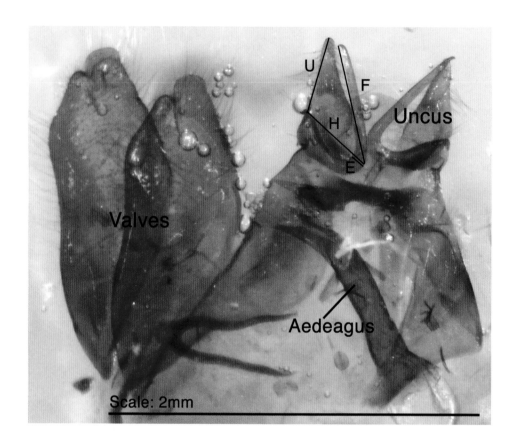

Color Plate E4 Genetic composition of butterflies [Lucas et al.]

Genetic composition of butterflies sampled from a *Lycaeides melissa* population, a hybrid population first described by Nabokov (Blacktail Butte), and a *L. idas* population. Each row is a unique locus, or location in the genome. Each column is a unique individual. Each square is colored according to an individual's genotype at a locus. Dark blue indicates an individual is homozygous for the allele more common in *L. melissa*, intermediate blue indicates an individual is heterozygous, and light blue indicates an individual is homozygous for the allele more common in *L. idas*. Hybrid individuals are homozygous for *L. melissa* alleles at some loci, homozygous for *L. idas* alleles at a greater number of loci, and heterozygous for many loci. We used these DNA sequence data to test the hypothesis that *Lycaeides* at Blacktail Butte are hybrids and to study the role of natural selection in shaping the composition of their genomes. Illustration created by Zachariah Gompert with the software environment R.

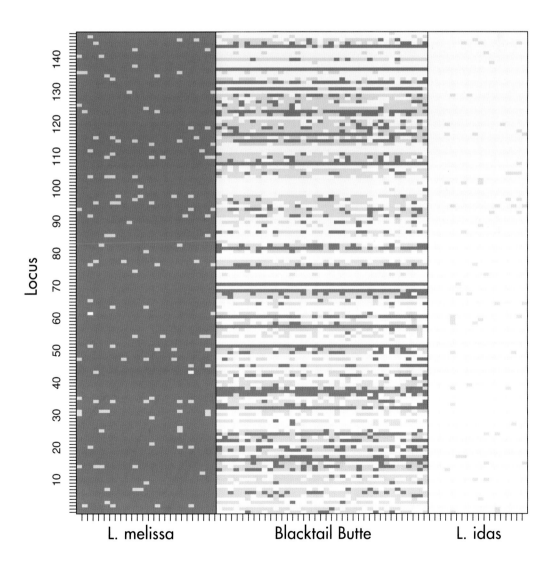

Color Plate E5 Leafwing butterfly [Mallet]

The Leafwing butterfly, *Zaretis isidora*, photographed in Corcovado National Park, Costa Rica, 1981, reared from caterpillar by James Mallet.

Color Plate E6 Nabokov's "genitalia cabinet" [Pierce et al.]

Nabokov's "genitalia cabinet" in the Museum of Comparative Zoology, showing corked vials of butterfly genitalia in glycerin, exactly as Nabokov left them. Inset: The tin of sherry pralines containing forty-nine papered specimens from Ashland, Oregon, collected in 1953.

Color Plate E7 [Pierce et al.]

Top: Part of Nabokov's rearrangement of "Plebejinae" recorded in the marginalia of Barnes and McDunnough's 1917 *Check List of the Lepidoptera of Boreal America*.

Bottom: Nabokov reprints from the Museum of Comparative Zoology collection. Nabokov decorated these with a favorite *Lycaeides* butterfly and presented them to Nathaniel Banks, a coworker in the Entomology Department.

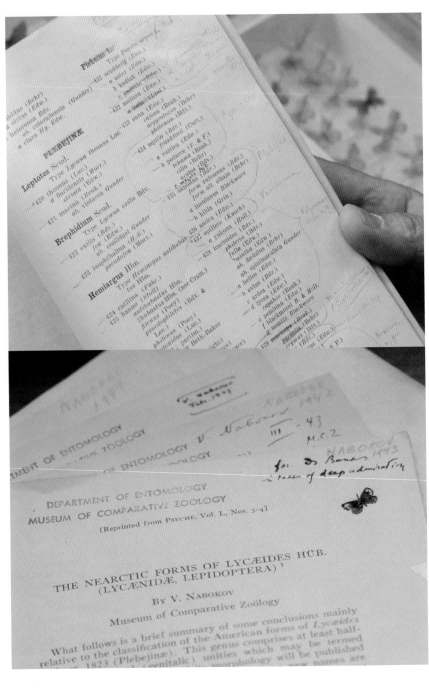

Color Plate E8 [Pierce et al.]

Top: From the Museum of Comparative Zoology (left), Nabokov's type specimens of the Karner Blue, *Lycaeides melissa samuelis*, in the Museum of Comparative Zoology collection and (right) polyommatine lycaenids collected and curated by Nabokov.

Bottom: Map of Nabokov's field trips and specimens collected while he was based at the Museum of Comparative Zoology. Nabokov's journeys were re-created using digitized information from specimens collected by Nabokov deposited in the museum (Table 1, see text) and Dieter Zimmer's "Chronology" and "Whereabouts" and drafted in ArcGIS using known points as fixed locations. The routes drafted between these locations were inferred using the 1940s and 1950s highway system, but whether Nabokov traveled these routes is unknown.

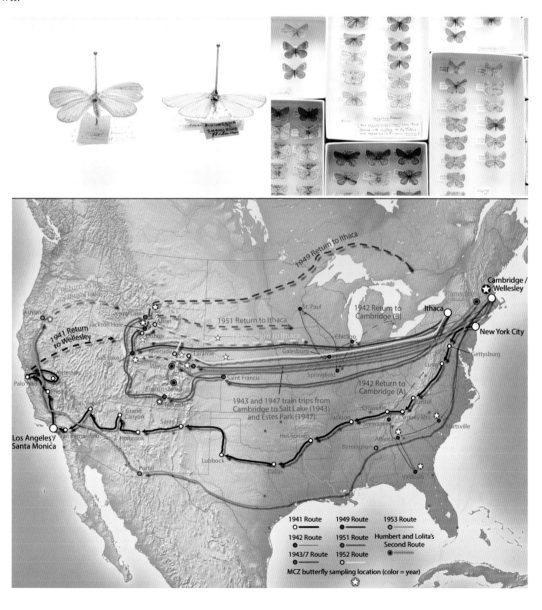

Color Plate E9 Butterflies and wildflowers at McLean [Dirig]

Dirig Figure 1. (A) Scenic with White Trilliums. (B) Gray Comma. (C) Mourning Cloak. (D) Male Cherry Gall Azure. (E) Two-leaved Toothwort, larval host of Toothwort Whites. (F) Carolina Spring Beauty, their favorite nectar flower. (G–L) Toothwort White behavior and life history. See text for details. All except A are at approximately the same scale.

Dirig Figure 2. (A) Outer envelope with his field notes and food plant sketch (*right edge*). (B) Inner glassine envelope housing an individual specimen. (C) Drawer of envelopes with butterfly specimens from the 1959 trip. (D, E) Male (*top*) and female specimens from Gatlinburg, with set wings and labels.

Color Plate E11 Toothwort Whites in the Great Smokies, April 23–24, 2013 [Dirig]

Dirig Figure 3. (A) Scenic. (B, C) Fresh female. (D) Male Early Hairstreak nectaring at Fringed Phacelia. (E) Fringed Phacelia, the local favorite flower of Toothwort Whites. (F, G) Female Early Hairstreak, open (*right*) and closed, photographed on Mount Greylock, Massachusetts. (H) Four male Toothwort Whites communally puddling while reflectance-basking, a rarely seen behavior. (I, J) Fresh male. (K, L) Slender Toothwort, flowering stems (*left*) and basal leaves (*right*). All are at approximately the same scale, except A, H, and L.

Color Plate E12 Cayuga Heights Natural History [Dirig]

Dirig Figure 4. (A) "Hazel Shade's Shagbark" in leaf, casting its dense shadow, where her swing hung in 1957 (June 2013). (B) The same tree without leaves, viewed from the lawn (January 2013). (C) Freshly emerged male Luna Moth on hickory trunk (April 29, 2004, wing expanse 4 1/2 inches). (D) Cut-leaved Toothworts in the woods, 500 feet from the Nabokovs' 1957 house in Cayuga Heights (May 5, 2014). (E) Flowering Dogwood (May 20, 1997). (F) Male Red Admiral perched in evening light (June 20, 2004, wingspread 2 inches).

Color Plate E13 William Henry Edwards's three "criterial" butterflies, at and near Coalburg, West Virginia [Dirig]

Dirig Figure 5. (A) Historical marker honoring Edwards. (B) Type series of *Pieris virginiensis* from Coalburg vicinity, 1871. (C, D) Closeups of type specimens, showing labels; *Kan.* = Kanawha River valley (*top*), *Coalb.* = Coalburg (*bottom*). (E, F) Diana Fritillaries, orange male above (left label); larger black female below; her label (on right) reads "bred." [reared], "em[erged from pupa] 5/30/82" [May 30, 1882]. (G) Atlantis Fritillary collected by Theodore L. Mead, with label, venters (*left*), dorsals (*right*). Figures E–G are at the same scale.

Drawing with Words

The Toothwort White and Related Natural History Motifs in *Pale Fire*

ROBERT DIRIG

On May 31, 1959, Vladimir Nabokov sent a letter from Flagstaff, Arizona, to his friend and entomological colleague John G. Franclemont, a Lepidoptera specialist at Cornell University in Ithaca, New York. In it he mentioned that in April of that year, he and Véra had "stayed for a week in the Great Smokies, Tenn., where we had a delightful time with *Pieris virginiensis*."[1] This sentence may puzzle many readers, but between two lepidopterists familiar with eastern North American habitats, it conjured images of a vernal woodland filled with early wildflowers, tender leaves, birdsong, warm sunshine, and a gentle white butterfly. Franclemont might have smiled as he read Nabokov's understated testimonial of triumph that marked the end of a long-standing quest in which he had played a part.

Immersed in lepidoptery from the age of seven, Nabokov grew up surrounded by the best paintings, literature, and music of Russia and Europe.[2] He learned to draw and paint but also *to see beyond the visual:* to capture every nuance in lines on paper, in words on a page, and in his memory. A charismatic interaction with a Swallowtail butterfly (*Papilio machaon*) in June 1906 at Vyra, a small town about fifty miles south of Saint Petersburg, mesmerized him with its beauty but also prompted scientific questions. Thus began the interplay of art and science in his approach to Lepidoptera—a recurring theme of his lifelong passion for these insects, which appear in many of his literary works. His parents encouraged their son's developing interest in butterflies and moths, which was enhanced by a well-stocked family library of splendid tomes that illustrated species from several continents. Among them was Samuel Hubbard Scudder's classic work on New England butterflies, in which the boy might have found mention of *Pieris virginiensis*, an endemic eastern North American "White" that had been reported from West Virginia by someone named Edwards.[3]

The years from 1906 to 1940 brought dramatic changes to Vladimir and his family—leaving Saint Petersburg (1917) and then Russia (1919), attending Cambridge University (1919–22), his father's death in Berlin (1922), marrying Véra Slonim (1925), the birth of their son, Dmitri (1934), and his mother's passing (1939). As Russian expatriates in Berlin, Vladimir and Véra earned their livelihood by tutoring, writing, typing, and translating, but their life was stressed and insecure. Hitler's growing power drove them from Germany to France in 1937. In 1940, they sailed for America, hoping for a better life.[4]

Nabokov's collecting skills and scientific knowledge of Lepidoptera had become sufficiently sophisticated for him to publish detailed papers on butterflies of the Crimea in 1920, and of the French Pyrenees in 1931. As an affiliate in entomology at New York's American Museum of Natural History (AMNH), and later as acting curator of Lepidoptera at Harvard's Museum of Comparative Zoology (MCZ), he published a dozen formal scientific papers on butterflies in a flourish of activity that lasted from 1941 to 1949.[5] These presented descriptions of several new species, including the Karner Blue (*Plebejus samuelis*). At the same time, he taught Russian language and literature at Wellesley College.

In July 1948, the Nabokovs moved to Ithaca, where Vladimir began his position as professor of Russian literature at Cornell. This job left less time for butterflies, but he was acquainted, through correspondence, with W. T. M. Forbes (Cornell's venerable "Dean of American Lepidopterists"), and when Franclemont came to Cornell as associate professor of entomology in 1953, they formed a congenial trio.[6] Nabokov wanted to see living *Pieris virginiensis* but had not found the butterfly in either the New York City area or near Boston, since it reaches the limit of its range off the coastal plain in the uplands of eastern New York and western New England. It was also known to fly around Ithaca, and Nabokov asked Franclemont to help him locate adults there.

When Robert Michael Pyle and I shared a day exploring Cornell's Nabokoviana on October 6, 1994, we had the privilege of hearing eighty-two-year-old Franclemont recount his acquaintance with Nabokov. They were cordial when they met in the Lepidoptera collection, and Franclemont, always the quintessential gentleman, twice drove Vladimir to Cornell's celebrated McLean Bogs Reserve, a pristine eighty-four-acre woodland-wetland complex, to seek the long-desired pierid.[7] These excursions were made on warm, sunny days in the second week of May, probably on May 8 and 9, 1957.[8] Forbes and Franclemont had collected the Whites at McLean between 1916 and 1940, but Nabokov and Franclemont were puzzled and disappointed by not finding them there on either visit.[9] Afterward, Franclemont drove Vladimir back to his house in Cayuga Heights and was invited in for a drink of scotch. Franclemont considered Nabokov to be "one of the really good field collectors and observers, among the very best."[10]

About a decade after Nabokov's and Franclemont's excursions, Arthur M. Shapiro, author of a comprehensive 1974 update on New York State butterflies, also unsuccessfully sought the Whites at McLean, but David P. Shaw and I, while visiting the reserve on May 4, 1974, rejoiced over finding them reestablished there, flying in all their pristine glory![11] They have appeared in spring ever since at McLean and were filmed at the reserve in May 1996 for a French documentary on Nabokov.[12]

The seasonal atmosphere, habitat, larval and nectar plants, and other butterflies that Nabokov and Franclemont, Shapiro, Shaw and I, and others who have sought the Whites would experience during visits to this famous reserve are distilled in Color Plate E9 (Figure 1). Between late March and mid-May, multitudes of wildflowers thrust through the beige leaf carpet, come into bloom, and set seed in this early sunny season, reaching their zenith just as the trees begin to leaf out (1-A). At the same time, the first butterflies appear, lending their special grace. Overwintered adult Gray Commas (*Polygonia progne*, 1-B) and Mourning Cloaks (*Nymphalis antiopa*, 1-C) are the earliest, followed by resplendent Cherry Gall Azures (*Celastrina serotina*, 1-D), which emerge from chrysalides that hibernated on the forest floor. As Two-leaved Toothworts (*Cardamine diphylla*, 1-E) and Carolina Spring Beauties (*Claytonia caroliniana*, 1-F) reach the perfect stage of development, the Whites (*Pieris virginiensis*) appear, resting on the leaf carpet (1-G) and drinking nectar from the Spring Beauties (1-H). They fly for about three weeks at any location. When they are ready, these butterflies engage in a mating dance (1-I), the male closely attending the female, which flies, then rests on a flower in the warm afternoon sun. She raises her abdomen to indicate lack of interest or holds it level when receptive. Mating (1-J) lasts for about an hour, revealing the wing undersides, which are striped with pale gray along the veins. After separating, the female lays small, yellow eggs on Toothwort leaves, the Whites' only plant host. These hatch into downy, green caterpillars that grow rapidly, exactly matching the color of the food plant's leaves (1-K). After three weeks of feeding, they crawl off the plant in early June and find a leaf or stick on which to pupate (1-L)—there to rest until next May's sunshine draws them forth as the centerpiece of another spring.

Individual Whites spend the few days of their adult existence in this heaven of perpetual spring, reveling in the sunshine, siphoning nectar from delicate, pink-lined, white flowers, and sleeping on a violet leaf or tree trunk. Males are often pure white (Color Plate E9 [Figure 1-I, left]), the females usually having a dusting of gray scales above (1-G). Males weave over the wildflowers all day in search of fresh females, occasionally stopping to warm themselves by adjusting their wing angle to bombard the thorax with reflected sunbeams. The Whites live mostly in steeply sloped forests but occasionally fly along woodland edges or cross open spaces to seek nectar, especially as growing leaves begin to close the tree canopy. They are subtle, insouciant denizens and occur so locally that they may be missed in the rush of spring, without a deliberate search, soon disappearing after their courting and egg-laying are done. Although Nabokov did not have the good fortune to see them at McLean, they would remain a butterfly grail.

WHEN THE AMERICAN EDITION of *Lolita* was published in 1958, it sold so well that Nabokov could retire from teaching early in 1959. Having delivered his last two Cornell lectures on January 19, he left Ithaca with Véra on February 24, heading for New York City to transact some literary business. From there they set out, on April 18, on an extended vacation trip through the southern United States. The culmination of his quest for *Pieris virginiensis* came during the first week of their trip, when "they stopped for four days among the flowering dogwood and . . . alluring lepidoptera of Gatlinburg, Tennessee."[13]

Archived weather reports for Gatlinburg between April 21 and 24, 1959, indicate daily high temperatures of 77°F, 55°F, 64°F, and 71°F, with fog and drizzle (the "smoke" of the Great Smokies) on the first two days, some morning fog on April 23, and clear skies on April 24.[14] Vladimir caught his first *Pieris virginiensis* on April 21,[15] then hit the jackpot two days later, collecting fourteen specimens and writing extensive field notes on the tan envelope in which he stored them (Color Plate E10 [Figure 2-A]). He gave the locality as up to three thousand feet in elevation, about five miles south of Gatlinburg, Sevier County, Tennessee (on a seven-mile stretch between Gatlinburg and Chimney Tops) on Newfound Gap Road (US Route 441), along the West Fork of the Little Pigeon River—in the Great Smoky Mountains National Park, just north of the North Carolina border. The butterflies were fairly common, the females ovipositing on Toothwort ("*Dentaria*," now called *Cardamine* spp.), their hind wing undersides varying from heavy gray on the veins to almost unmarked. He also sketched the food plant.[16] A second envelope contained eleven additional specimens collected "up to 10 miles S on U.S. 441" the next day. They continued their journey on April 25, following Route 73 toward Maryville, Tennessee, finding seven more *P. virginiensis*, six Falcate Orange Tips (*Anthocharis midea*), and a fresh Gemmed Satyr (*Cyllopsis gemma*) during a good-bye stop in Blount County, fifteen miles west of Gatlinburg.[17]

In order to understand the Nabokovs' field experience in the Smokies, I made a pilgrimage to the same location in April 2013, arriving April 22, then seeking the Whites and scrutinizing their habitat along Route 441 on April 23 and 24, exactly fifty-four years after their visit.

Perfect weather (blue sky, bright sun, 72°F, light breeze) heightened my excitement as I entered the Great Smoky Mountains National Park on the south edge of Gatlinburg—a lovely, wild enclave, with a green haze of newly unfolding leaves on trees and shrubs, over-arching a rich carpet of wildflowers. Route 441 gradually climbs south through thick forest on the west side of the Little Pigeon River's West Prong, occasionally bordered by Flower-ing Dogwoods (*Cornus florida*) over the 6.2 miles to the Chimneys Picnic Area and Cove Hardwood Nature Trail at about three thousand feet elevation. I spotted two Whites flying across the road as I drove up, so knew they were out. After parking among the picnic tables, I proceeded uphill to the trail, immediately flushing a pristine female White on a sunlit, flower-covered knoll (Color Plate E11 [Figure 3-A]) that settled and opened her wings for photographs. She was unusually dark (almost "smoky") on top and below (3-B, C). While targeting the White, I was aware of a Spring Azure (*Celastrina ladon*) flashing like a chip of the sky behind me.[18] Another tiny, dark butterfly in my peripheral view proved to be a fresh male Early Hairstreak (*Erora laeta*), a rarely encountered eastern butterfly, which was nec-taring at the intricate white flowers of Fringed Phacelia (*Phacelia fimbriata*, 3-D, E). (I later learned that this was only the third sighting of this butterfly in the park.)[19] Female Early Hairstreaks have more blue above (3-F), and both sexes are green with scarlet spots beneath (3-G), making them our most colorful spring butterfly.[20] A worn Green Comma (*Polygonia faunus smithi*) sprang up from his perch a few steps farther up the trail but did not linger for photographs. This is a boreal butterfly that I usually see in northern New England, but it also resides at high elevations in the southern Appalachians. What a welcoming committee in my first few minutes at the site!

The nature trail climbed through a joyous exaltation of wildflowers, intermingled in a solid patchwork over several acres, between the massive, mossy-trunked trees of this cove hardwood forest. Here were many familiar flowers, which also occur at McLean—White Trillium, Smooth Yellow Violet, Canada Violet, Squirrel Corn, Dutchman's Breeches, Bish-op's Cap, Foamflower, Smooth Solomon's Seal, Carolina Spring Beauty, Mayapple, Dwarf Ginseng, and Sharp-lobed Hepatica—among others that are more southern, including Yellow Trillium, Sweet White Trillium, and the Fringed Phacelias.[21] Finding glossy oval leaves and splendid lavender-and-white blooms of Showy Orchis was a welcome surprise, as I had not seen this plant since 1980. Trees included Yellow Birch, American Beech, Yellow Buckeye, Silverbell, Tuliptree, Sugar Maple, and Eastern Hemlock; with shrubby Alternate-leaved Dogwood, American Hornbeam, Wild Hydrangea, Toothache Tree, and Maple-leaved Viburnum in the understory.[22] Abundant White-tailed Deer have so heavily browsed woodland herbage in the Northeast that I often see only a few stems or sparse single plants of many of these flowers and shrubs, which here grew luxuriantly all around me. While eating lunch, back at the car, I noticed a few Whites flying through the area, and shortly after I found several males that were probing a roadside seep for minerals while reflectance-basking in unison, behaviors I had not seen before in fifty years of field expe-rience with this butterfly (Color Plate E11 [Figure 3-H]). Adults also repeatedly nectared at Fringed Phacelias (3-E) here, in the same way they feed all day at Carolina Spring Beauties (Color Plate E9 [Figure 1-F]) in New York. I looked carefully for toothworts, finding a few leaves of two species, but no flowers on April 23.

Returning the next day, I roused a fresh male White from the same spot where the female had flushed the day before. He briefly opened for photos (Color Plate E11 [Figures 3-I, J]), but it was cool and clouding over, so he soon rested quietly on the ground. I explored a hollow behind this knoll that was filled with tumbled boulders on a thirty-degree slope and finally found abundant toothworts! I had been puzzled by Nabokov's drawing of the plant (Color Plate E10 [Figure 2-A]) but now understood, for the leaves on these flowering stems were very small with narrow leaflets, in contrast to those of the Two-leaved Toothworts in Ithaca (Color Plate E9 [Figure 1-E]). The basal leaves of these closely resembled those from New York but were darker, pine-needle green with paler veins (recalling the color of English Ivy, *Hedera helix*). This was a different species, Slender Toothwort (*Cardamine angustata*, 3-K, L). The plants were abundant, growing in duff on mossy boulders, with as many as eighty-five basal leaves and forty-two flowering stems in one small patch. It is the White's presumed larval food plant at this location. Spring was late in 2013 in the Smokies, according to local wildflower enthusiasts. This was confirmed by Slender Toothwort plants that were only budded or in early bloom during my visit, in contrast to the drooping seedpods in Nabokov's field sketch from the same date and place in 1959.

The beautiful weather, wildflower peak, and abundance of cars on the road had led me to expect throngs of chatty pedestrians on the trails. Instead, I encountered absorbed, mostly silent people, quietly strolling with cameras and reverent attitudes throughout this outdoor temple. I noticed that most of the wildflowers were white (with a few accents of lavender and yellow), while here and there, the glorious blooms of Flowering Dogwood (Color Plate E12 [Figure 4-E]) soared overhead, like a flock of squarish white butterflies that had settled all along the branches, gently rustling their translucent, delicately veined wings. In combination with a real white pierid fluttering at ground level among the old growth trees, they strongly reinforced a pure and sacred theme.[23]

In 1959, Vladimir and Véra may also have seen white butterflies crossing the road as they drove slowly uphill through the full splendor of the season. After parking, we can envision them grabbing gauzy nets and his Band-Aid box of glassine envelopes (Color Plate E10 [Figure 2-B]) out of the back seat of their black 1956 Buick and hurrying to follow the Whites that were weaving above the flower carpet.[24] The butterflies would have been patrolling, basking, nectaring, courting, and egg-laying with their usual unhurried grace, accompanied by the Ovenbird's ringing crescendo and the Black-throated Green Warbler's wispy, five-note cadence, among much other birdsong, and surrounded everywhere by the striking landscapes, designs, and palette of an old forest. The model for butterfly study in the 1950s was *making a collection*, and the Nabokovs took a good series of specimens of the Whites during their sojourn. These unique physical souvenirs still carry the magic and excitement of their revel with a new, long-sought butterfly in a beautiful place on a perfect spring day, which happened also to be his sixtieth birthday! One now needs a research permit to collect anything in the park; perhaps policies were less formal, half a century ago. Without the vital clue of Nabokov's *specimens*, still in their original packaging, which I chanced on at the Cornell University Insect Collection (Color Plate E10 [Figure 2]), this story could not have been written.

Traveling to mountainous regions, seeking new butterflies, had become a recurring recreation for the Nabokovs. Their experience in the Smokies likely set off many echoes of past joys with Lepidoptera. It was the first time since Vladimir's youth in Russia and their shared trip to the Pyrenees in 1929 that they had enjoyed such freedom and security. All the early horrors, privations, dangers, and stresses were gone, recent pedagogical duties were in their past, and a fresh future of writing and butterflying beckoned. When the fictional poet John Shade wrote, "I love great mountains," in *Pale Fire* (l. 510), perhaps Nabokov was commemorating their few idyllic days in the Great Smokies.

AFTER SEVERAL YEARS OF SIMMERING, Nabokov's *Pale Fire* finally appeared (finished in December 1961, published in 1962).[25] This odd, fascinating novel is presented as 999 lines of iambic pentameter, with lengthy prose commentary. In the pages that follow, I examine *Pieris virginiensis* and several related North American lepidopteral, ornithological, and botanical motifs found in *Pale Fire*, explain their scientific nuances, and suggest the special roles of this White, historical butterfly literature, associated place-names, and natural history characteristics of other plants and animals in the genesis of the novel.

The only concrete appearance of *Pieris virginiensis* in the poem is "The Toothwort White haunted our woods in May" (l. 316), where it serves as an icon of spring in a three-line sequence about passing of the seasons—the sole indicator with a natural history context.[26] The first common name of this butterfly was West Virginia White, coined by Alexander B. Klots in 1951 and widely used ever since.[27] Nabokov's Toothwort White was a new name that sparkles with his talent for choosing the perfect words, equally vested in art and science; it nicely parallels the vernacular designations of *Pieris oleracea* (the Mustard White) and *Pieris rapae* (the Cabbage White) in referencing the larval host.[28]

John Shade's line 316 referred to his house and yard in New Wye, which was inspired by a dwelling the Nabokovs rented for a year in Cayuga Heights, starting in February 1957.[29] This community abuts the north edge of the Cornell campus, on a steep slope of 650–850 feet elevation that overlooks Cayuga Lake, having many splendid residences sited on expansive lots. Brian Boyd described the Nabokovs' ranch-style house, which was situated at the north end of a large meadowlike lawn, and had a picture window into which Cedar Waxwings (*Bombycilla cedrorum*) occasionally crashed after feeding on fermenting rose hips that grew in the yard.[30] A large Shagbark Hickory (*Carya ovata*), with a horizontal limb supporting a tire swing, dominated the east side, rising on a mossy hummock. Explorations of this property in January and June 2013 showed the house surrounded by a thin woods of oak (*Quercus* spp.), Pitch Pine (*Pinus rigida*), more Shagbark, Sugar Maple (*Acer saccharum*), Hop Hornbeam (*Ostrya virginiana*), Hemlock (*Tsuga canadensis*), and Eastern Redcedar (*Juniperus virginiana*). The stream bed and ravine of Renwick Brook pass on the northeast and north sides, and the now-huge Shagbark still reigns on its knoll, minus the swing (Color Plate E12 [Figures 4-A, B]).[31] The meadow has shrunk to half its original size, after two new houses were built on the south side in the 1960s and 1970s, while the famous picture window was removed during a 1996 renovation. A wildflower enthusiast who lives about five hundred feet downhill from the Nabokov's former residence shared observations of native Cut-leaved Toothworts (*Cardamine concatenata*, 4-D), which occur in masses in the woods, meadow, and lawn surrounding her house. Historical specimens of these wildflowers from Cayuga Heights (pressed between 1915 and 1932) corroborate the abundant presence of this plant there for at least a century.[32] If Nabokov found Toothworts on the edges of his lawn or in nearby treed corners of Cayuga Heights, they would have reminded him of *Pieris virginiensis* and prompted the trips with Franclemont to McLean.

I was originally puzzled by Nabokov's use of "haunted" in line 316—it seems to darken the bright, flowery enclave this butterfly occupies—but the Toothwort Whites' habit of gently floating in and out of forest shadows could qualify them as friendly ghosts. Another interpretation might be *their absence where they were expected* in the 1950s, as at McLean and places with Toothwort in Cayuga Heights.[33] A third sense of this might be *an enduring aura of Hazel Shade's presence made more poignant by her absence*—recalling a similar scene at the end of *Lolita* (chapter 36), where Humbert looked down on a schoolyard and noted "the absence of her voice from that concord" as a painful remnant of Dolores Hayes's presence in his associations.[34]

Lines 55–57 in *Pale Fire* may also subtly refer to Toothwort Whites (and metaphorically to Hazel, in the shadow of the Shagbark):

> White butterflies turn lavender as they
> Pass through its shade where gently seems to sway
> The phantom of my little daughter's swing.

If toothwort plants grew in Paul Hentzner's "Dulwich Forest" (a short distance uphill, where Flowering Dogwoods bloomed) in the novel, or in the real treed surrounds of the Nabokovs' 1957 Ithaca yard and on wooded downhill slopes—and had Toothwort Whites been present in any numbers—they would have flown across their lawn and around its edges.[35] This butterfly might also be the familiar Cabbage White, which frequents sunlit yards and gardens; one can imagine Vladimir racing after every White that flew over his lawn that spring, hoping for *Pieris virginiensis*, when it was only *P. rapae*. The wings' color change from sunlit white to shaded lavender resembles the bluish purple shadows that ornament sailing rafts of cumulus or hide in hollows behind sparkling snowdrifts.[36]

The Shagbark Hickory (Color Plate E12 [Figures 4-A, B]) was John Shade's favorite tree (*PF*, ll. 49–54, pp. 89, 93). It is a solid, brawny species with strong wood, huge, vigorous buds, and tough, curling slabs of armor on its dark trunk. The large, compound leaves cast a heavy shadow that shelters and protects but also masks the area around it.[37] The Shagbark's presence implies the co-occurrence of Luna moths (*Actias luna*, 4-C): a "leaf sarcophagus / (A Luna's dead and shriveled-up cocoon)" was kept by John Shade's Aunt Maud (*PF*, p. 114). This was likely found in the shaded area beneath the Shagbark, hickories being a favorite larval food plant of Lunas. Their caterpillars feed for several weeks on the leaves, then descend and spin thin cocoons of white or beige silk in debris beneath the tree, remaining there through the winter. The gorgeous, long-tailed, green moths that emerge the following May are among the most dramatic incarnations of spring. Lunas appear when the Shagbark's leaves are unfolding, and Toothwort Whites are flying, and migrant Red Admirals are arriving from the south.

Hazel Shade is given much attention in her father's poem. Her name has botanical implications, being tied to three shrubs of eastern North American forests—American Hazelnut (*Corylus americana*), Beaked Hazelnut (*C. cornuta*), and Witch Hazel (*Hamamelis virginiana*). The American Hazelnut's plump, oval nut is enclosed in rough leafy wrappings, while the Beaked Hazelnut's kernel is borne inside curved bracts that resemble a long horn, nose, or elephant's trunk. Both might have suggested this unfortunate girl's appearance. Witch Hazel's leaves have been used medicinally, and its forked branches for dowsing ("water-witching"), indicating a magical quality.[38] In the poem, Hazel seeks communication with spirits and becomes one of *Pale Fire*'s ghosts—while it is not surprising that she, as a water-seeker, would drown herself![39] Thickets of Beaked Hazelnut often grow in deep shade, while American Hazelnut and Witch Hazel occupy partially shaded fringes of hedge-rows and forests; both situations imply Hazel's surname. Beaked Hazelnut and Witch Hazel also frequently share the Toothwort White's habitat in New York.

Characteristics of the Toothwort White's early stages further suggest Hazel: the beige pupa (Color Plate E9 [Figure 1-L]) has a long "nose," awkwardly humped profile, and wall-flower obscurity that fades into its background, while the sedentary larva's camouflaging leaf-green color and softening hairs help it to "hide in plain sight" (1-K)—as does shy, so-cially invisible Hazel. But whereas her suicide occurs in an atmosphere of male rejection, the opposite happens in the Toothwort White's universe, where each female is immediately desired and pursued by every male she encounters (1-I). These are vivid examples of how Nabokov may have used a suite of scientific details to derive more facets of a fictional char-acter's name, appearance, and behavior.

Might Toothwort Whites ultimately represent the Nabokovs' own release from the de-manding routine of teaching into their quiet, happy, productive, and secure seclusion at Montreux?

ONE OF THE MOST FASCINATING ASPECTS of *Pale Fire* is its setting, which was deliberately unspecified but cleverly coded by Nabokov. *New Wye, Appalachia, U.S.A.* (p. 13) is (1) at the *latitude of Palermo* (assumed to be the one in Italy by most readers; p. 19); (2) about *400 miles by rail from NYC* (p. 277); (3) at an *elevation of about 1500 ft.* (p. 169); (4) a locality for the *Toothwort White* (l. 316, p. 183); and (5) a place where two other butterflies, *the southern Appalachian Diana Fritillary and boreal Atlantis Fritillary, unexpectedly flew together* (p. 169).[40] The last condition is the most difficult to satisfy.

In 1991, Brian Boyd likened New Wye to Ithaca and Cornell.[41] Having lived and worked in Ithaca for more than forty years, I was constantly reminded of real scenes and buildings in this small city and at the university as I reread *Pale Fire* while preparing this essay. De-spite its abundant suburban and academic similarities, Ithaca does not fit any of the criteria except (4). In his later analysis of this novel, Boyd urged readers to "trust Nabokov" because he "always rewards the curious."[42] Following this advice, participants in a NABOKV-L list-serv discussion between April 2007 and July 2008 (and likely many other readers before and since) have considered a number of Appalachian locations in Virginia, Maryland, and West Virginia that are near the proper latitude and distance from New York City but didn't perfectly match the elevation or co-occurrence of the three butterflies. To make the prob-lem more interesting, Nabokov skillfully inserted subtle locational decoys, including one for Boston, one for "New Wye, New England," and two for Virginia.[43] He even stated that "many of . . . [the poem's discarded original lines preserved in Charles Kinbote's commen-tary] are more valuable artistically and historically than some of the best passages in the final text" (*PF*, p. 16, and a similar comment on p. 297)—presumably as pointers to his hints. With deep respect for Boyd and all others who have grappled with this puzzle, and after an intricate and exciting detective exercise of my own, I offer a solution, feeling "as certain as anyone can about anything Nabokov does" that it is the right one.[44]

The location is *Coalburg* (Victorian spelling "Coalburgh"), at the southeastern edge of Charleston, in *Kanawha County, West Virginia* (Color Plate E13 [Figure 5-A]). It satisfies the five criteria thus: (1) *Palermo*, in adjacent Lincoln County, West Virginia, is at approximately the same latitude (38°10'N), twenty-four miles west of Coalburg (38°12'N)! Lacking knowledge of this obscure Appalachian hamlet, the Italian Palermo also works, at 38°7'N. (2) Coalburg is about *400 miles from New York City*. (3) Elevational summits reach about *1500 ft.* at Coalburg, and in other highlands surrounding Charleston. (4) The *Toothwort White* (5-B, C, D) has been recorded in Coalburg, the adjacent Kanawha River valley, and surrounding hills (see next paragraph). And the clincher, (5) the *Diana Fritillary* (*Speyeria diana*, 5-E, F) historically occurred along the Kanawha River valley at Coalburg and widely throughout southern West Virginia; while the *Atlantis Fritillary* (*S. atlantis*, 5-G) was collected about nineteen miles upstream near McDougal in western Fayette County, within the range of Diana, by sixteen-year-old Theodore L. Mead, who worked as an apprentice with William Henry Edwards, a well-known regional butterfly expert, in the summer of 1868![45] Edwards was "much surprised at finding several *Atlantis* among Mr. Mead's collections. They were seen near Turkey Creek."[46] Nabokov would easily have discovered this information in the first volume of Edwards's *Butterflies of North America* at various university or museum libraries. This beautifully illustrates how meticulously he adhered to natural history facts in allusions that may seem casual, or involve minor irrelevant details (as Kinbote supposed), but actually point to authentic original sources.

Charleston, the state capital of West Virginia, is in the heart of Appalachia and, with nine colleges and universities and the West Virginia State Archives, is an important center of learning, history, and culture. The region's rich coal deposits drew Edwards to the area in 1864, where he helped develop coal mining and railroads along the Kanawha River while intensively studying butterflies on the side. In 1871, on the south side of the river at Coalburg, he built a large, Italianate house (later called Bellefleur), which was placed on the National Register of Historic Places in 1990. In scientific circles, Coalburg is internationally known as the locus of Edwards's pioneer butterfly work and as *the most important geographic site for the Toothwort White—its type locality*, the place from which it was described as new to science by Edwards in 1870 (Color Plate E13 [Figures 5-B, C, D]).[47] Edwards also extensively studied Diana, the most dramatic North American fritillary (5-E, F), at and near Coalburg, and was the first person to associate its unique, large, bluish black females with the orange-and-black males and to work out its life history.[48] Even if Nabokov never traveled to Charleston or Coalburg, he certainly would have seen a fascinating article about visiting Bellefleur in an early journal of the Lepidopterists' Society, written by Cyril F. dos Passos (a contemporary authority on North American butterfly nomenclature, with whom Nabokov talked and corresponded, and uncle of the writer John Dos Passos, with whom Nabokov was also acquainted).[49] That account and Edwards's original butterfly literature would have given him enough details to fix on the area surrounding this famous natural history landmark as an appropriate setting for *Pale Fire.*

Place-names that occur within ten miles of Coalburg may have prompted Nabokov's creation of the Zemblan universe.[50] *Charleston* (originally *Charles Town*), *Charlton Heights*, *Kingston*, *Monarch*, *Crown Hill*, and *Gaymont* evoke Charles Xavier (*Monarch* also has a lepidopteral meaning), while *Cedar Grove* and *Cabin Creek* suggest the cabin at *Cedarn*, Utana, the site of Kinbote's final exile from Wordsmith University. *Big Chimney* and *Blount, West Virginia*, coincidentally echo *Chimney Tops* and *Blount County, Tennessee*, where the Nabokovs found *Pieris virginiensis* in 1959.

In many ways, Nabokov's earlier Karner Blue saga resonates with the Toothwort White story presented here.[51] Both butterflies appear in May at the height of spring, were first seen alive by Nabokov after a focused search of several years, and were described as new to science from "legendary lands."[52] Klots coined a common name for both in 1951.[53] Each has a life cycle entwined with a beautiful, color-coordinated wildflower that blooms when the insect flies; was separated as a new species from a complex of closely related taxa (*Lycaeides, Pieris*), with which it was long combined; and has been considered rare and conservation worthy. These ephemeral butterflies are delicate centerpieces of gorgeous habitats—Nabokov described the Karner Blue's as "a sandy and flowery little paradise," similar to my characterization of the Toothwort White's haunts as a heaven of perpetual spring.[54] Just as the Karner Blue's story is tied to Samuel Hubbard Scudder, so the Toothwort White's tale is associated with William Henry Edwards, the other great Victorian student of our butterflies, who, like Scudder, produced a three-volume, exquisitely illustrated treatise on American species. Nabokov's hidden tribute to Edwards in *Pale Fire* has the same level of subtlety he employed when naming the Karner Blue "*samuelis*" to honor Scudder. Nabokov's "delightful" experience with Toothwort Whites near Gatlinburg obviously impressed him at a level that greatly influenced the genesis of *Pale Fire*, however minimally this butterfly appears in Shade's poem. *Pieris virginiensis* ultimately serves as a natural history link between the outdoor universe in Ithaca (which inspired New Wye) and Edwards's genuine butterfly haven in Appalachia.

THE RED ADMIRAL (*Vanessa atalanta*) is the other, much more visible lepidopteral protagonist in *Pale Fire* (Color Plate E12 [Figure 4-F]). Brian Boyd and Dieter Zimmer have thoroughly reviewed the real and mystical appearances of this magnificent nymphalid butterfly in Shade's poem and Kinbote's commentary.[55] Here are a few supplemental details.

Boyd discussed parallel natural history metaphors that describe Hazel Shade's transcendence after her death.[56] The terrestrial, lepidopteral model has the dingy, sedentary, shy, and pure female Toothwort White (Color Plates E9, E11 [Figures 1-G, 3-B, C]) become her diametrical opposite, a flamboyant, confident, interactive male Red Admiral (Color Plate E12 [Figure 4-F]). The aquatic, ornithological variation (ll. 318–19) features a gray cygnet (Nabokov charitably avoided "ugly duckling") transforming into a male Wood Duck (*Aix sponsa*) in resplendent breeding plumage. The sex reversal from feminine to the more colorful masculine state in both examples is interesting.[57] The rich dark hues, bold patterns, and splendid iridescence of a male Wood Duck are subtly duplicated in the Red Admiral's ventral hind wing pattern, which under magnification looks like an intricate "oriental rug," woven in scales of lavender, mahogany, iridescent teal green, plum, brown, and black. The red bill and eye of the duck contrast sharply with surrounding black and white markings of its head and neck, in the same way that the Red Admiral's scarlet wing bars smolder in their dark, white-spotted setting.

The jaunty Red Admiral enters the story repeatedly, but the Toothwort White appears only as a shadowed presence, paralleling their contrasting "species personalities." The *Vanessa* actually functions as another decoy, drawing attention away from the hidden but pivotal role of *Pieris virginiensis* in decoding the novel's locational puzzle.

Male Red Admirals alertly perch (Color Plate E12 [Figure 4-F]) in the same elevated, tree-bordered, sunlit lanes for several days or even weeks, from late afternoon to twilight, nimbly darting at any movement—other butterflies passing, birds flying, or people walking by—and may even land briefly on a person's head or arm.[58] The function of this ritual is finding a mate (when ready, a female will enter a perching arena to seek a male). Nabokov's rendering of this behavior and the Red Admiral's beauty (ll. 993–95 and especially Kinbote's commentary to these) resembles his "blue snowflakes" passage (a hidden tribute to Karner Blues) in his novel *Pnin*.[59] These figure among his most sublime verbal depictions of American butterflies. Vested in highest art but also perfectly true to science, his words draw living, moving, color-saturated images in the reader's mind.

Nabokov was among the first to recognize migrant butterflies (other than Monarchs) in North America, after many years of observing them in Russia and Europe. The Red Admirals that Charles Xavier/Kinbote noticed in the Palace Gardens (*PF*, p. 172) were autumn migrants passing south through Zembla. Fall migrants of this butterfly in eastern North America are visible along the Atlantic coast but are usually more subtle inland. In recent springs (especially 2001, 2010, and 2012), millions have flown into the Northeast in April and May in a crowded, manic, headlong rush north, which has been very exciting to witness. Autumn migrants overwinter in Florida and other southern parts of North America. Adults return north at the same time that Toothwort Whites fly.[60]

In Shade's poem, Red Admirals associate with his wife's beauty (ll. 269–71), and in Boyd's reading they figure as an exalted reincarnation of their daughter.[61] I suggest an additional interpretation below.

At the end of the poem in *Pale Fire, eight* people were present when Jack Grey's bullet killed John Shade: the poet; Charles Kinbote (Shade's next-door neighbor and "best friend," the Waxwing of his alter ego's crest, and the disguised real target of Gradus), who gripped Shade's outstretched fingers at the very end (*PF*, 177, 294); "Balshazar" (Kinbote's gardener and lodger, approaching his nine o'clock bedtime, who disarmed Grey after two gunshots and assisted following the murder); and Jack Grey/Jakob Gradus (the assassin), as physical beings.[62] But also Sybil Shade (who had just driven away in her car, yet remained, touching his sleeve, as his lovely "dark Vanessa"); Hazel Shade (the sometime Toothwort White, here gripping her father's arm, in Boyd's metamorphosed sense, as the Red Admiral); King Charles Xavier (Kinbote's grandiose alternate personality, the ultimate target of Gradus, embodying the emblematic Zemblan Silktail, a Waxwing)—as the perched, randy male Red Admiral, with four glowing red bars and lines of four white stars on his splendid Admiral's uniform, glorying in the lingering rays of dusk, while waiting to spar with other males or engage in a tryst—also laying his hand on Shade's sleeve; and Samuel Shade (John's deceased father, an ornithologist, as the namesake of another Waxwing, *Bombycilla shadei*, and the source of John's self-assumed "waxwing" identity)![63] Thus the people closest to John Shade—his wife, daughter, father, and "best friend"—were also at his side when he died through the error of a psychopath, lending a surprisingly warm, comforting, and positive touch at the end of this strange, melancholy ghost story, which commentator Kinbote now can conclude with a line conveying many layers of subtlety:

> A man, unheedful of the butterfly—
> Some neighbor's gardener, I guess—goes by
> Trundling an empty barrow up the lane.
> I was the shadow of the waxwing, slain.[64]

Acknowledgments

This essay is dedicated to the late John G. Franclemont, my earliest mentor in lepidopteral studies, who had an important role in this story, and to Brian Boyd for his inspirational treatise on *Pale Fire*. An initial catalytic discussion with Boyd at the Cornell Nabokov Centenary Festival in 1998 prompted this examination of the Toothwort White and related natural history motifs in *Pale Fire*.

Lee B. Kass, Torben Russo, Scott LaGreca, Carolyn Klass, Matthew Dirig, David Werier, Thelma Turner, Gavriel Shapiro, Brian Boyd, Stephen Blackwell, Kurt Johnson, and other coauthors reviewed early drafts, offered support, and provided helpful comments. Blackwell kindly shared an archived NABOKV-L dialogue from 2007–8 on possible locations of New Wye and offered a local perspective on the natural history of the Great Smoky Mountains National Park. Indispensable help with natural history details in Cayuga Heights (near Ithaca) was offered by R. Parker, John B. Whitman, and Charles R. Smith. Toothwort specimens from Ithaca and Tennessee were made available at the Bailey Hortorium Herbarium at Cornell University by Anna M. Stalter, Peter Fraissinet, and Kevin C. Nixon. James K. Liebherr, E. Richard Hoebeke, and Jason Dombroskie allowed me to access Franclemont's and Nabokov's butterfly specimens at the Cornell University Insect Collection. William Henry Edwards's and Theodore L. Mead's historical butterfly specimens from West Virginia were examined at the Carnegie Museum of Natural History in Pittsburgh, through the kindness of John Rawlins and the hospitality of Lee Kass and Robert Hunt. Peter Fraissinet assisted with the preparation of Figures 1–5. Photographs are by the author.

Notes

1. *NB*, 528; *VNAY*, 381.

2. See Dirig, "Nabokov's Rainbow," and Shapiro, *Sublime Artist's Studio*.

3. Scudder, *Butterflies*: for *Pieris virginiensis*, see 1191, 1193–94, 1198, 1204. "Whites" is a generic common name for mostly white pierid butterflies (Pieridae), used in the same way as "Blues" for the lycaenid group to which Nabokov's famous Karner Blue belongs.

4. *VNAY*; Dirig, "Nabokov's Rainbow."

5. D. Zimmer, *Guide*, "Nabokov's Non-Fictional Writings on Lepidoptera," 365–69.

6. Hoebeke, Root, and Liebherr, "John George Franclemont," 12–15.

7. Dirig, "Definitive Destination," 4–16.

8. Franclemont told Pyle and me that their trips to McLean happened "just as *Lolita* was starting to take off." *VNAY*, 314, describes this as beginning in spring 1957. Ithaca temperatures reached 75°F and 80°F, respectively, on May 8 and 9, 1957, both sunny days, according to archived weather reports at http://www.farmersalmanac.com/weather-history/search-results/.

9. In this episode, the Whites had also disappeared from other places in central New York and Ontario, for unknown reasons, and were an early conservation concern of the Xerces Society. They rebounded nicely in New York's Finger Lakes Region between 1974 and 1990 and were withdrawn from Canada's endangered species list in 1990.

10. Franclemont, "Remembering Nabokov."

11. A. M. Shapiro, "Butterflies and Skippers," 21; A. M. Shapiro, "Butterfly Mysteries."

12. Desmortiers, *Vladimir Nabokov*. A brief sequence in this documentary shows Robert Dirig netting a Toothwort White at McLean on May 7, 1996. Kurt Johnson also has a cameo in this film.

13. *VNAY*, 376, 380–81, 386–87; Boyd, *Nabokov's "Pale Fire,"* 277n8.

14. See http://www.farmersalmanac.com/weather-history/37738/1959/05/21, /22, /23, /24.

15. Boyd, *Nabokov's "Pale Fire,"* 277.

16. Unfortunately, Nabokov made no drawings of *Pieris virginiensis,* having finally found this butterfly on his next-to-last North American field trip (April to August 1959), long after he had left his microscopic studies behind at the MCZ. But he did quickly sketch their associated toothwort on

his field envelope (Figure 2-A). At first glance, this rendering seemed so abstract as to be of little use for identification purposes, but it closely matched plants of the Slender Toothwort (*Cardamine angustata*) that I found growing at the site in 2013.

17. The rest of this trip can be partially reconstructed from dates on butterfly specimens at the Cornell University Insect Collection that were collected en route: After leaving the Smokies, they traveled southwest through Alabama, Mississippi, and Louisiana (April 26 and following days) into southern Texas (May 2–11), then on to Arizona (May 17–July 15) and California (July 30–August 12), tiptoeing into Nevada (August 3) while visiting the Lake Tahoe area, before returning east via Wyoming (August 21) and South Dakota (August 22). See Boyle, "Absence of Wood Nymphs," for a detailed description of the Nabokovs' daily collecting routine in Arizona during this trip.

18. Nichols, *Butterflies and Skippers*.

19. See Jeffrey Glassberg's photograph of a female Early Hairstreak, taken on April 25, 2009, in the Great Smoky Mountains National Park: Glassberg, "Photograph of Early Hairstreak."

20. Figures 3-F, G photographed on Mount Greylock, Massachusetts. Early Hairstreaks are also rarely sighted at McLean.

21. Figures 1 and 3 compare the habitats, larval food plants, nectar flowers, and behavior of *Pieris virginiensis* near the northern (McLean) and southern (Great Smokies) limits of its range. There is a substantial overlap in the landforms, flora, associated butterflies and birds, and even a small crustose lichen (*Porpidia albocaerulescens*) that grows on boulders, over this wide separation of 650 miles.

See White et al., *Wildflowers*, for color photographs, scientific names, and descriptions of these plants.

22. See *Vascular Plants of the Great Smoky Mountains* for scientific names and other details.

23. Although ostensibly describing a television advertisement for feminine beauty aids, lines 413–15 in *Pale Fire* ("A nymph came pirouetting, under white / Rotating petals, in a vernal rite / To kneel before an altar in a wood") might also describe this Great Smokies habitat of *Pieris virginiensis*.

24. Boyle, "Absence of Wood Nymphs," 127.

25. *VNAY*, 464.

26. Kinbote appended a discarded marginal variant of this line, "In woods Virginia Whites occurred in May," in his comments (*PF*, 183–84).

See later in the text and n. 43, below, for the hidden significance of this additional new common name.

27. Klots, *Field Guide*, 201, 208.

28. Boyd, *Nabokov's "Pale Fire,"* 276n5; D. Zimmer, *Guide*, 228–29.

29. *VNAY*, 303, 313–15, and three photos on both sides of the leaf preceding p. 227; Boyd, *Nabokov's "Pale Fire,"* 282n3.

30. *VNAY*, 315; Cedar Waxwings (ll. 1–4, 131, commentary to l. 1000 [*PF*, 292]) are serene and elegant residents in Ithaca, where they also feed on fruits of Eastern Redcedar (ll. 181–82), Flowering Dogwood, and other woody plants. Their gorgeous, black-masked face may have suggested a disguise, and the high crest a crown for Nabokov's characterization of Charles Xavier in *Pale Fire*. Nabokov probably derived the Zemblan Silktail, another waxwing used as Charles Xavier's heraldic emblem, from the larger but very similar Bohemian Waxwing (*Bombycilla garrulus*) that occurs across northern Europe, Asia, and North America.

31. In January 2013, the tree's trunk was three feet in diameter (at the base), thirty inches wide (at breast height), with the top gone, perhaps damaged by lightning.

32. Courtesy of the Bailey Hortorium Herbarium at Cornell University.

33. Perhaps the west-facing, semiopen, sun-baked slopes of Cayuga Heights were too hot and dry for the pupae to survive there; they typically rest under heavy shade in damp forests from mid-May to October. See n. 9, above.

34. Boyd, *Nabokov's "Pale Fire,"* chap. 9.

35. *PF*, 185; Boyd, *Nabokov's "Pale Fire,"* 87–88, 135, 142.

36. G. Shapiro, *Sublime Artist's Studio*, 102–5, discussed Nabokov's use of related snow motifs. See also Boyd, *Nabokov's "Pale Fire,"* 288n17.

37. Kinbote laments that their obstructive quality prevents his regular spying on the Shades (*PF*, 23, 86).

38. Brooks, *Catskill Flora*, 38–41.

39. She died about midnight (l. 483), perhaps on the spring equinox ("a night of thaw, a night of blow, / With great excitement in the air," ll. 494–95)—both faerie junctures, mystical edges of time and space that are imbued with magic. Hazel's last moments are recounted in italics by her father after "time forked" (l. 404), suggesting the Y-shaped dowsing rod drawing her toward

her "crossing" (l. 488) at the lakeshore (two more faerie junctures).

40. *PF*, 169: "The extraordinary blend of Canadian Zone and Austral Zone that 'obtained' . . . in that particular spot of Appalachia where at our altitude of about 1,500 feet northern species of birds, insects, and plants commingled with southern representatives." See also Boyd, *Nabokov's "Pale Fire,"* 136.

41. *VNAY*, 427.

42. Boyd, *Nabokov's "Pale Fire,"* 22, 43.

43. For Boston, see *PF*, 22: "Parthenocissus Hall" on the Wordsmith University campus might have been covered with *Parthenocissus tricuspidata* (Boston Ivy). For "New Wye, New England," see *PF*, 139: Odon's mother was an American, from this fictitious place. And finally, for Virginia: "Parthenocissus Hall" could also have been covered with native Virginia Creeper (*Parthenocissus quinquefolia*) or Woodbine (*P. inserta*) vines, which are sometimes grown ornamentally. The other clue is the common name "Virginia White" for *Pieris virginiensis*, which Nabokov coined in this novel as an inept variant to line 316, reported in Kinbote's commentary. This is a humdrum literal translation of the species epithet *virginiensis*. Nabokov also coined "Toothwort White" in line 316, which far surpasses Klots's common name "West Virginia White." See Boyd, *Nabokov's "Pale Fire,"* 276n5, and Klots, *Field Guide*, 201, 208. Creating new common names for a butterfly was an odd thing for Nabokov to do, even in a novel, but it was necessary here to avoid revealing too much information about the novel's setting.

44. Borrowing a phrase from Alexander B. Klots's letter to me of March 31, 1975.

45. *Guide to the Theodore L. Mead Collection.*

46. Occurrence of these two fritillaries was reported in William Henry Edwards's "Supplementary Notes" (six unnumbered pages just before the "Systematic Index") at the end of volume 1 of his sumptuously illustrated *Butterflies of North America*, a rare and indispensable classic, serially published in 1868–72, that is now accessible online as a free Google ebook. Specimens documenting these written records were found among Mead's and Edwards's butterfly collections, which now rest in the Section of Insects and Spiders at the Carnegie Museum of Natural History in Pittsburgh (Figures 5-E, F, G). Mead's records of Atlantis Fritillary may have been the first from West Virginia, where this butterfly reaches its southern limit, and Diana its northern limit; Allen, *Butterflies of West Virginia*, 121, 126. Historical records in New York indicate that Atlantis has disappeared from former sites since the 1920s. Dirig, "Definitive Destination," 14; Cech and Tudor, *Butterflies of the East Coast*, 158–59, 165.

47. Edwards, "Description of New Species," 13–14; Edwards, "On *Pieris*," 95–99, pl. 3; Brown, "Types of Pierid Butterflies."

48. Edwards, "Description of the Female"; Edwards, *Butterflies*, vol. 1, *Argynnis* I, *Argynnis Diana* (n.p.), vol. 2, *Argynnis* VII, *Argynnis Diana* (n.p.); Dirig and Kawahara, "Moonbeams." Edwards, "Description of the Female," observed Dianas frequently taking nectar at Ironweeds (*Vernonia* spp.), and they also like Common Milkweeds (*Asclepias syriaca*) in nearby Lafayette County, West Virginia (pers. obs.); both are noted as part of their habitat in *PF*, 186.

49. Dos Passos, "Visit," 61–62. See Calhoun, "Extraordinary Story" for further information about Edwards's butterfly work.

50. Gleaned from *West Virginia Atlas and Gazetteer*, 43, 52–53.

51. Dirig, "Theme in Blue."

52. From the first line of Nabokov's poem "On Discovering a Butterfly." Karner Blues were described from the world-famous Karner Pine Bush near Albany, New York; Bailey, "Center, N.Y."; Rittner, *Pine Bush*. Joseph Albert Lintner, "Calendar," first found Karner Blues in the United States at Center, New York (later renamed Karner), in 1869, working at the same time that Edwards was describing the Toothwort White in Coalburg. Edwards had erroneously included a Karner Blue specimen in his type series of *Lycaena scudderii* in 1861; Karners were later recognized as distinct and described by Nabokov in 1943 (see his "Nearctic Members," 537–38, for historical details of the scientific synonymy). Nabokov surely knew a great deal about Edwards's butterfly literature, and Coalburg is an equally legendary locus in the history of American butterfly scholarship.

53. Klots, *Field Guide*, 165–66, 176; Dirig, "Theme in Blue," 208n14.

54. See Dirig, "Theme in Blue," fig. 19.6.

55. Boyd, *Nabokov's "Pale Fire"*; D. Zimmer, *Guide*, 275–78, pls. C-3, O-1.

56. Boyd, *Nabokov's "Pale Fire."*

57. Female Wood Ducks are dingy, like the cygnet, as are males when not courting females. Only male Red Admirals behave as Nabokov described;

females are somewhat less brightly colored and much less brazen, being most often found slowly fluttering around their nettle food plants (Urticaceae). However, female Red Admirals and Toothwort Whites retain the power to seek or refuse male attention.

58. In the Ithaca area, I watched the same male Red Admiral (identifiable by damaged wings) bask in sunny spots along a spruce lane from 7:30 to 8:15 p.m. on several evenings between June 21 and July 10, 1997.

59. Dirig, "Theme in Blue," 207. Nabokov hinted at the Karner Blue's appearance in a puddling assembly in *Pnin* (*Pnin,* 127) by mentioning Wild Lupines bordering a sandy road with pines (120), the presence of scrub oaks (139), and "Vladimir Vladimirovich [Nabokov]," who knew "all about these enchanting insects" (127). Nabokov later declared this identity in "Novelist as Lepidopterist," 46. The New England setting of "Cook's Castle"/"The Pines," where the butterflies were found in *Pnin,* was probably somewhere in the Merrimack River valley between Nashua and Concord, New Hampshire. Nabokov did not find them there before 1949, but Karners lingered at Concord until the 1990s.

60. Swanson, *20 Years of Butterfly Revelations,* recorded daily afternoon antics of perched male Red Admirals over two decades in his Florida backyard near Orlando. He noted a distinct gap in residence in summer and autumn, when adults migrate north and reproduce before their progeny return south for the winter.

61. Boyd, *Nabokov's "Pale Fire."*

62. In *PF,* 292, Kinbote wrote that he and his gardener were "the last two people who saw Shade alive," but there was also the assassin and, by implication, four others, as "shades."

63. Charles Xavier: although the associations with Sybil and Hazel are apt, Xavier seems the ideal match for the Red Admiral in this scenario, based on the scientific facts that surround male perching behavior in *Vanessa atalanta.* Samuel Shade: *PF,* 100.

64. John Shade was walking in the "shadow" of Kinbote/Charles Xavier, target of Gradus, in the Zemblan reality; became the effigy of Judge Hugh Warren Goldsmith (*PF,* 82–83) to the vengeful madman Jack Grey; was the scion of his father's namesake Waxwing; and was the human metaphor of a real Cedar Waxwing, killed by a deceptive reflection in the glass.

A Few Notes on Nabokov's Childhood Entomology

VICTOR FET

The marvelous compendia by Brian Boyd and Robert M. Pyle, Kurt Johnson and Steve Coates, and Dieter Zimmer present exciting reading to anyone interested in Nabokov's butterflies.[1] The main emphasis in these volumes, however, is on the double passion of Nabokov in his adult age. Nabokov's childhood activities in lepidoptery were so brilliantly described by the writer himself (chapter 6 of *Speak, Memory* / *Drugie berega* [Other shores]) that one finds it hard to add anything to his own account. Here I sketch possible lines of inquiry that surround childhood involvement in natural science—an issue of a great importance in Nabokov's case—that interested scholars could pursue.

A lay reader, I suspect, still readily conjures an image of a Victorian child with a butterfly net and perceives lepidoptery to be a trivial, childish activity—a less serious form of child's play than that of more technically inclined, adult-imitating children who build engine models and computers. In modern Western culture, a boy with a butterfly net is perceived as engaging in an old-fashioned, though excusable, activity. Steve Coates, who cowrote *Nabokov's Blues* with Kurt Johnson, offers a perspective from his own childhood: "I grew up in rural western North Carolina, and a lot of the boys in the neighborhood had fabulous, well-organized insect collections and knew a great deal about entomology. As I grew older and came to think of myself as more 'sophisticated,' I dismissed the whole thing as an unhip, rustic pursuit, but this of course was exactly what Nabokov was doing at the turn of the century."[2]

Nabokov's lepidoptery long posed a question: Was he an amateur or a professional entomologist? Today, it has been amply demonstrated that he *was* a professional. Kurt Johnson says, "For Nabokov, as with many, fascination with the big picture books of butterflies as a young child grew to concerted collecting as a youngster. As with many scientists, these impressions of youth become a driving life force."[3] Nabokov started collecting butterflies in 1906, at age seven, and never ceased; he published his first book of poems ten years later, at age seventeen; his first research paper on butterflies, at age twenty; and his first novel, *Mashenka* [Mary], at age twenty-six. To quote Dieter Zimmer, "For Nabokov lepidoptery was not a mere hobby. It was a lifelong passionate interest that began when he just turned seven, eight years before he began to compose his first poems, with his first Old World Swallowtail in Vyra."[4]

Entomological work for Nabokov started very early and included not only self-training but also the careful guidance of his polymath father, who was also a butterfly collector—in this case, a well-informed amateur. Precocious Nabokov, with his early English and French, could read serious scientific volumes (such as the *Entomologist*) in those languages; his childhood notes on butterflies (which do not survive) were written in English.[5] We witness the early "imprinting" that those voluminous books had on his visual and linguistic memory by finding lepidopterological names, allusions, and puns scattered throughout his ouevre in both Russian and English. As Brian Boyd relates, "Even before he read and reread all of Tolstoy, Flaubert, and Shakespeare in the original languages as he entered his teens, he had mastered the known butterflies of Europe and [by 1910] 'dreamed his way through' the volumes so far published of Adalbert Seitz's *Die Gross-Schmetterlinge der Erde*."[6] Johnson and Coates comment further on the classic foreign entomology books Nabokov had close at hand and on the beauty and importance of Seitz's monumental work.[7] Although Nabokov studied German at the Tenishev School, he enrolled only in January 1911; therefore he was evidently self-trained in technical German of the *Schmetterlingenbüche* (having had no early tutoring in German).

Dieter Zimmer reminds us that most of the basic knowledge in entomology (as well as other areas of zoology and botany, I should add) until recently "was collected by amateurs who either possessed the means to devote themselves to a consuming hobby or who earned their living in some other way."[8] This is still the case in the twenty-first century: as in Nabokov's time, quite a lot of descriptive work is done, reasonably well, by self-trained zoologists who do not earn a living from this activity. Collecting, moreover, is commonly done by amateurs: there is simply not enough funding to support such extensive fieldwork.

IN AFFLUENT FAMILIES OF THE GENTRY in Europe, including imperial Russia, children could spend their time and allowance on collecting. Expensive foreign butterfly books were readily available to young Nabokov; his own collections of Russian fauna were augmented by exotic specimens purchased through mail-order catalogs.[9] Of course, money always mattered for funding zoological research, collecting, and travel. The largest museums of the European empires—British, German, French, Austrian, Russian—were founded and supported by the royal dynasties, as was the case with the famed Imperial Zoological Museum in Saint Petersburg (now the Zoological Institute of the Russian Academy of Sciences, just across the Neva River from the Winter Palace). Nabokov's favorite imagery of minor, fictional European royalty (see *Pale Fire*) includes references to a few historical figures who were naturalistically inclined, not always just as amateurs. The foremost figure in this regard was the Grand Duke Nikolai Mikhailovich Romanov, one of the great Russian lepidopterists fondly mentioned in *The Gift*, "Father's Butterflies," and elsewhere throughout Nabokov's works. The grand duke was murdered in 1919 by the Bolsheviks, along with many other Romanovs.

Another curious personage appears in *Pnin*, where we read that "the figure of the great Timofey Pnin, scholar and gentleman, . . . acquired in Victor's hospitable mind a curious charm, a family resemblance to those Bulgarian kings or Mediterranean princes who used to be world-famous experts in butterflies or sea shells."[10] Similarly, in *Pale Fire*: "How often is it that kings engage in some special research? Conchologists among them can be counted on one maimed hand."[11] Brian Boyd explains that both Emperor Hirohito of Japan and Prince Albert I of Monaco were marine biologists.[12] But Bulgarian "kings," technically speaking, never existed (except in Voltaire's *Candide*), and Nabokov surely meant here the first Bulgarian tsar of the twentieth century who was also an avid amateur naturalist—Ferdinand I of Bulgaria, aka Prince Ferdinand of Saxe-Coburg-Gotha (1861–1948).

I am not sure what Nabokov knew of this truly Ruritanian ruler, but Ferdinand was a very visible figure on the European scene before World War I. He became the first ruler of independent Bulgaria, first as prince [knyaz] beginning in 1887, and then, from 1908, as tsar. On Ferdinand's ascent to the Bulgarian throne, Queen Victoria (his father's first cousin), stated to her prime minister, "He is totally unfit . . . delicate, eccentric and effeminate . . . . Should be stopped at once."[13] Ferdinand was a keen lepidopterist and botanist, and in his youth organized an expedition to South America. Alas, Ferdinand's flamboyant politics were less successful than his natural science: he was an active but often unsuccessful participant in all of the Balkan wars he could find a way into and was forced to abdicate in 1918; his son Boris became the next tsar.

Another noteworthy fact possibly linking Ferdinand to *Pale Fire* is that he was the first head of state ever to fly in an airplane—with the Belgian pilot Jules de Laminne, on 15 July 1910. It is highly possible that Nabokov was thinking of this Bulgarian royal lepidopterist-aviator when he invented King Alfin, who crashed his Blenda IV aircraft in 1918 (many European monarchies crash-landed that year). Later in 1910, Ferdinand and his children Kirill and Boris flew several times in Sofia with the famous Russian pilot Boris Maslennikov (one of the prototypes of Colonel Gusev in *Pale Fire*?) who in 1910 founded the first aviation club in Bulgaria, and then the first Russian aviation school, Oryol (The eagle), in Moscow.[14] Maslennikov flew in the first, disastrous, Saint Petersburg–Moscow flight contest by nine pilots on July 10 (23), 1911, widely covered in the journal *Niva* (of nine pilots, only one reached Moscow; three, including Maslennikov, crash-landed; one passenger died). The twelve-year-old Nabokov would have known about these important technological events. (Under the Bolsheviks, Maslennikov was exiled to Siberia and spent eight years in Stalin's gulags.)[15]

There is one significant historical episode involving Ferdinand of Bulgaria that to my knowledge has never been published in English. The episode most likely remained unknown to Nabokov but it originates from the same epoch and subculture of royal lepidoptery—and reads like a *Pale Fire* scene. My friend and colleague Alexi Popov, the former director of the National Museum of Natural History in Sofia, tells this story about his grandfather, zoologist Ivan Buresch (1885–1980), son of a Czech immigrant. In 1903, seventeen-year-old Buresch collected butterflies in the highest Bulgarian summit, Musala (elevation 9,596 feet [2,925 m]), where he came across the future tsar, then Prince Ferdinand. "Why do you collect my butterflies?" exclaimed the prince in anger, but then softened as he recognized in young Buresch a fellow entomologist. The prince invited Buresch to climb the ridge together and talk about butterflies, and he was so impressed with the young biologist that he gave him his royal cape as a gift. The very next year, Ferdinand appointed Buresch as a technician in his Natural History Museum that occupied one of the royal palace buildings (it is still there today). Ivan Buresch traveled with Ferdinand on his many expeditions, survived both world wars in Sofia, and continued as a director of the same museum under the Communists until his peaceful retirement in 1959. One fancies that a similar fate, under slightly different circumstances, could have been Nabokov's own.

Such "kingly" naturalists as Tsar Ferdinand or Grand Duke Romanov cut mildly Quixotic, often tragic, figures. There were other images of naturalists found in Nabokov's childhood reading in Russia. From Jules Verne, one recalls the absent-minded but heroic geographer Jacques Paganel from *Les Enfants du capitaine Grant* (The children of Captain Grant), and also the absent-minded but comical entomologist Cousin Benedict from *Un Capitaine de quinze ans* (A captain at fifteen)—a thoroughly ridiculed and pathetic figure. Alas, insect collection in European cultural and literary tradition was an oddity even in the enlightened nineteenth century. The public perception of an entomologist as a nut with a net (bordering on the more familiar modern cliché of the mad scientist) has hardly changed since

Nabokov's *Lolita: A Screenplay*. Still, even Cousin Benedict stands among Jules Verne's many immortal scientists with a selfless passion for knowledge.

THE IDEA OF *naming a new species* "in that incompletely named world in which at every step he named the nameless" has been made famous through Nabokovian writings.[16] In *Speak, Memory*, Nabokov relates how, at the age of nine (!) he wrote to the great lepidopterist Nikolai Kuznetsov (1873–1948), proposing a new Latin name for a distinct form of Poplar Admirable he found. Kuznetsov, then already a mature researcher, "snubbed" the young entomologist.[17] This did not mean, however, that Nabokov's conclusions were *wrong*! "Proposing a new name" means that the nine-year-old Nabokov simply did not know all the existing research literature—this happens to mature taxonomists as well. In this case, Nabokov did not recognize that the subspecies in question was already described from Bucovina (now in western Ukraine, then in the Austro-Hungarian Empire) as *Limenitis populi bucovinensis* Hormuzaki, 1897. "How I hated Hormuzaki! And how hurt I was when in one of Kuznetsov's later papers I found a gruff reference to 'schoolboys who keep naming minute varieties of Poplar Nymph!'"[18] It is important to note that Kuznetsov did not reject the fact that the form Nabokov identified exists in reality—he just pointed out that it was already described by another researcher, in this case Constantine von Hormuzaki—an Austrian professor at Czernowitz University. Thus, at age nine, Nabokov already could, and did, observe the minute diagnostic features of butterfly varieties (subspecies) correctly.

I have not found the "gruff reference to schoolboys," but among Nikolai Kuznetsov's papers published within the same period is one that is indeed very gruff and quite relevant to the issue.[19] This "methodological" paper has no research content; it consists only of lengthy complaints against aimless Latin naming of varieties of butterflies by amateurs and irresponsible scientists due to high commercial interest and sheer vanity. It reads much like many similar statements today, in which authors lament the "taxonomic vandalism" of irresponsible namers and self-published journals. Clearly, young Nabokov had read this paper, as a lot of Kuznetsov's "gruff" comments are recognizable in "Father's Butterflies" and *Speak, Memory*. It is one of the sources for some of Nabokov's (and K. K. Godunov-Cherdyntsev's) opinions, incorporated in the same way as Central Asian explorers' texts are in *The Gift*, as Dieter Zimmer has shown.[20]

The issue of Poplar Admirable varieties appears in Kuznetsov, in a paragraph that translates: "The overproduction business has reached the point where not only among serious opponents, but also among the admirers of the nomenclatural enrichment of entomology, some already are perplexed about where their further activity in this direction will lead, as these authors no longer know what to do with the names and 'established' forms of their favorite *Parnassius apollo* L. or *Limenitis populi* L."[21] A reference follows to a paper by a splitter, A. A. Yakhontov, who in turn discusses butterflies described by another fellow splitter, Leonid Krulikowsky. Among those varieties we find a Siberian form *Limenitis populi fruhstorferi* (Krulikowsky 1909), which appears to linger in the background of *Ada*.

Brian Boyd has suggested that the name of Krulikowsky, a prominent Russian lepidopterist, was well known to Nabokov, and much later became a source of the "leporine" Dr. Krolik in *Ada*.[22] We see now that Krulikowsky's name could have been even more important to Nabokov at a very early period, for both fell under the same criticism from Kuznetsov as they tried to establish new "minute varieties of Poplar Nymph" at about the same time. It was Dr. Krolik who christened a butterfly species, *Antocharis ada* Krolik (1884)—"as it was known until changed to *A. prittwitzi* Stumper (1883) by the inexorable law of taxonomic priority."[23]

The "passion for naming," what Kuznetsov terms "German" *Namengeberei*, is still a great force that drives and plagues taxonomic research. Criteria by which a species is defined are constantly in flux—many different species concepts have been proposed in the hundred years since Kuznetsov's gruff remarks. The subspecies concept also continues to be murky; many modern taxonomists see no value in giving names to geographic varieties and want to operate only at species rank. (In fact, the jury is still out on the validity of the many *Limenitis populi* forms referred to above.) While experts may not agree on criteria of taxonomic delineation, they all rely on primary data, based on meticulous documentation of morphology—as well as on DNA marker data available today. Much has been said about Nabokov's keen attention to taxonomic delineations, many of which proved to be spectacularly true. Further, Nabokov appears to be the only trained zoologist who also carried this intuitive skill, honed in his formative years, into the highest ranks of literary art.

SERGEI AKSAKOV (1791–1859) was the first and only professional writer in Russia to describe butterfly collection by children of the gentry (*Sobiranie babochek* [Collecting butterflies], 1858). We know that Nabokov was highly critical of Aksakov's essay: in *Drugie berega* (chapter 6) he called it "extremely dull" (*bezdarneyshee*) (the passage is absent in *Speak, Memory*). Fyodor in *The Gift* dismisses Aksakov's nature writings in his imaginary dialogue with Koncheev: "My father used to find all kinds of howlers in Turgenev's and Tolstoy's hunting scenes and descriptions of nature, and as for the wretched Aksakov, let's not even discuss his disgraceful blunders in that field."[24]

Was it the genuine disdain of a professional toward a hopeless amateur? Probably. In his commentary on *Eugene Onegin*, Nabokov called Aksakov "a very minor writer, tremendously puffed up by Slavophile groups."[25] But then we know how caustic Nabokov often was toward many literary luminaries, most famously Fyodor Dostoevsky. In the sentence quoted above he did not spare Ivan Turgenev or even his beloved Lev Tolstoy, albeit via double-proxy opinion (Fyodor repeating his father's words). Maybe we should not judge Aksakov's earnest accounts of natural history as harshly as Nabokov did. Recently, I came across a note by the prominent Russian lepidopterologist Yuri Korshunov (1933–2002), who thought that Nabokov was completely unfair to Aksakov. Korshunov insists that Aksakov committed no "disgraceful blunders" in his texts addressing butterflies, contrary to Fyodor's claim. Perhaps the issue requires an impartial look by an expert on Russian butterflies into Aksakov's pages.[26]

In all candor, one just cannot compare Aksakov to Nabokov: for Vladimir, lepidoptery was not a mere collecting pastime but natural science, in which from the very beginning he followed the highest standards of the field as it was in the 1900s. Aksakov, on the other hand, was a true amateur who wrote his butterfly notes as an old man, reminiscing about his golden childhood in central Russia during a very different epoch. *Sobiranie babochek* was written a year before Aksakov died and addresses events that happened more than sixty years earlier. Aksakov was born in 1791, which means he was hunting and rearing butterflies in the end of eighteenth century, even before Alexander Pushkin was born—more than a hundred years before Nabokov! Upper-class children in Russia, like those in England, were trained in the sportsmen's pursuits of hunting and fishing. It is fitting that Aksakov wrote enormously detailed treatises on both activities, and he is generally considered a great authority on Russian game hunting and serious fishing—both pursuits now largely extinct in central Russia, along with forest and river habitats.

For his time and milieu, Aksakov and his schoolmates were rather advanced in natural science training. At age fifteen, Aksakov was a student in the newly opened (1805) Kazan University. He learned natural history from Carl Fuchs (1776–1846), a medical doctor, ethnographer, and one of those German polymaths who moved to the vast imperial countryside of Russia. Fuchs's house in Kazan was an intellectual center that attracted visitors ranging from Alexander von Humboldt to Pushkin. A Göttingen alumnus, like Pushkin's Lensky, Carl Fuchs was the rector (president) of Kazan University until 1827, succeeded by the famous mathematician Nikolai Lobachevsky—whom Nabokov *did* admire!

Nabokov was not aware of another interesting point: for many Russian children of later (Soviet) generations, it was "wretched" Aksakov who introduced them to lepidoptery. In 1938, Aksakov's ancient butterfly essay was reworked for children into a small book by the inveterate Soviet-era popularizer of zoology, the entomologist Nikolai Plavilshchikov. It was one of the most popular entomology books then, with about 150 species illustrated by G. Orlov arranged on fifteen color plates and an appendix telling how to collect and spread butterflies. I used its later 1950s edition, as well as some very good zoology books by Plavilshchikov himself.

"I reserve for myself the right to yearn after an ecological niche:
> . . . Beneath the sky
> Of my America to sigh
> For *one* locality in Russia."

THESE LINES FROM *Speak, Memory* are a revisitation of Pushkin's ironic dream "to sigh, . . . beneath the sky of my Africa, for somber Russia."[27] They point to a very specific "ecological niche" (a rather new scientific term, which was widely popularized only in the 1950s) for Nabokov, which he did not share with any other writer hailing from Saint Petersburg. His use of the word "locality" (rather than "place") in this context is another playful gesture toward the geographic precision of an entomological label. Nabokov's "*one* locality" for which he yearns is not the imperial city of Saint Petersburg itself but not far to its south, the few square miles of the Oredezh River valley around Vyra and Batovo. This is where he spent his ten formative collecting years of 1907–17.

Much has been said about the "Saint Petersburg text"—the semiotic concept developed by Vladimir Toporov and others. This "text" was generated by dozens of major Russian writers—Pushkin, Nikolai Gogol, Dostoevsky, Osip Mandelstam, Andrei Bely, Anna Akhmatova, Konstantin Vaginov, Joseph Brodsky, to name just the main ones. It was largely Saint Petersburg that defined Russian literature in the Silver Age of the early twentieth century, with its Symbolists and Acmeists. Pekka Tammi has demonstrated how this "text" influenced the "texture" of Nabokov's nostalgic poetry and prose, especially in his European years, but also later (for example, in *Look at the Harlequins!*).[28] Indeed, the Nabokovs' house was located in the heart of the imperial city, Nabokov went to school there, and he never had a chance to visit any other large Russian city south of Saint Petersburg—Moscow included. But the very personal space of Nabokov's "text," so tightly bound to butterfly pursuit, was well outside the city and its "text."

In the gallery of Saint Petersburg writers, Nabokov is marginal to the Silver Age not only by belonging to a different generation (he was nineteen years younger than Alexander Blok, ten years younger than Akhmatova) and not only because he left this "text" early, with his emigration at age eighteen. He is marginal in space, as well as in time. His nostalgic yearning was never for the "yellow government buildings" (Mandelstam) or the Bronze Horseman's empire, Westernized or Slavophile, but for the northern woods and bogs of Rozhdestveno and Vyra, the real, firmly geographic fringe of Peter the Great's ghostly capital. He is probably the only author whose work is deeply rooted in these northern countryside landscapes—and the one who undoubtedly best knows them, having traversed them for ten years, from age seven to seventeen, on foot and by bike. Tammi notes that "there is always winter in Nabokov's St. Petersburg" and that "Vadim in *Look at the Harlequins!* is obviously speaking for his creator when he says that he had 'never seen [his] native city in June or July.'"[29] Of course he had not, for he was busy in his ecological niche: June and July are the major butterfly collecting months, every sunny day being precious in a cold, northern climate, with dozens of species collected every summer, hundreds of specimens with carefully noted localities and other data.

We can clearly see how this so-called Boreal biogeographic zone (its southern boundary lies between Saint Petersburg and Moscow) extends to the imaginary Ultima Thule and Zembla. Always a naturalist, Nabokov carried into his exile the minutest details of Russian nature, which earlier writers generally neglected. Confined within their phantasmic city, Gogol and Dostoevsky cannot be imagined outside of it or expected to know much about the natural environment surrounding the imperial capital. Others who ventured to the countryside had a generic, Rousseauian approach to local nature and its "Finnish rocks." They rarely knew their trees or flowers—recall Chernyshevsky's opinion (reported by Fyodor in *The Gift*) that the flowers of the Siberian taiga "are all just the same as those which bloom all over Russia."[30] One can occasionally find a cliché like "a spruce, this sad trademark of northern nature" (Pushkin, *Travel from Moscow to Petersburg*), but Russia's classic writers were more comfortable praising lush Mediterranean nature, which many of them observed in person in France and Italy—or at least the Crimea, in the case of tightly controlled Pushkin, who was never allowed to travel abroad.

Not so with young Nabokov. He carried with him the imprint of the Oredezh countryside, its ecological niches, with a true naturalist's passion, which was much deeper than any bond of Turgenev-style or Tolstoyan gentry sportsmen to their coveted game. In *Speak, Memory*'s famous lines, Nabokov steps directly into the American ponderosa pine forest from Vyra's *sphagnum bog*. The very use of this precise botanical term—hardly even known to most other Russian writers—gives away a scientist who had known this distinction already as a boy when he pursued his butterflies through just such a bog.

IN HIS 1946 INTRODUCTORY LECTURE on Russian literature, Nabokov explains to his Wellesley students: "Suppose a schoolchild picks up study of butterflies for a hobby. He will learn a few things about general structure. He will be able to tell you . . . that there are innumerable patterns of butterfly wings and that according to those patterns they are divided into generic and specific groups. This is a fair amount of knowledge for a schoolchild. But of course he has not even come near the fascinating and incredible intricacies invented by nature in the fashioning of this group of insects alone."[31] This passage talks about various levels of depth in knowledge. Nabokov gently but slyly depicts here, not himself, but a quite ordinary schoolchild who has not mastered his skills at identifying "innumerable patterns" and their importance in systematics. In stark contrast, Nabokov himself already at age eight or nine could skillfully use these patterns to identify and classify those "generic and specific groups" of butterflies.

Nature needs to be documented and described. Zoology, undertaken at an early age, provides an active early training of memory and attention, focused on minute detail. Such a connection, I suspect, is underappreciated by most readers and researchers since it requires a firsthand childhood experience, as well as emotional involvement, in biological systematics. After a specimen is obtained and preserved, the subsequent zoological work is not limited to using technical literature such as species keys. It always includes other, more active research components, with constant feedback and iterative actions. It combines reading, writing, drawing; it requires observational and analytical skills. Published materials (research papers, books, keys) and one's own notes allow one to compare specimens. The work goes on, and it never ends.

The sheer amount of this work is likely not appreciated by noncollectors. One collects large series of specimens of the same species to reflect ecology and observe variation. Currently 107 species of butterflies (and many more moths) are recognized in Leningrad Oblast (province), about 30 percent of the eastern European faunal list. Nabokov's collecting around Vyra over several seasons must have yielded thousands of specimens.

Nabokov's entomological training was extremely rigorous, and it produced tangible, professional results. Along with extensive field experience, it included technical reading of specialized literature, as well as technical writing—starting with primary field notes, containing data on habitat distribution, phenology, food plants, reproduction, and so on, and ending with taxonomic descriptions of species. The tremendous attention to detail in his literary work, in my opinion, derives in many ways from the fact that such attention was a required professional skill for any systematic zoologist. Nabokov's fictional Ada was not an exception as a precocious entomologist: on Antiterra, with her "larvarium" and her hybrids, she merely elaborated further on the dreams and occupations of Nabokov when he was the same age in Vyra.

What I have tried to convey here—obvious and perhaps trivial to an expert but less well known to the average Nabokov reader—is that his early concentration on entomological work provided young Nabokov with a very specific training, which other writers simply did not have. Such was, for example, his labeling activity, itself the first mark of a professional zoologist.[32] I think that Nabokov's genius was fed from an early age not only by his artistic sensitivity to the diversity and wonders of natural objects but—first and foremost—by his zoologist's need to distinguish their details in order to describe this diversity. Nabokov's case, probably unique in the modern history of both science and art, demonstrates how a childhood emotional involvement with nature's elaborate diversity and beauty may form and inform both a scientific and an artistic response.

Acknowledgments

I thank Victoria Alexander, Stephen Blackwell, Brian Boyd, Kurt Johnson, Oleg Kosterin, Lauren Lucas, Alexi Popov, and Dorion Sagan for their help and useful comments.

Notes

1. *NB*, *NBl*, and D. Zimmer, *Guide*.
2. NABOKV-L, October 26, 1999.
3. Ibid.
4. D. Zimmer, *Guide*, 4.
5. *NBl*, 115.
6. *NB*, 4.
7. *NBl*, 114.
8. D. Zimmer 2001, *Guide*, 11 ("Writer and Scientist").
9. *SM*, 127.
10. Nabokov, *Pnin*, 88.
11. *PF*, 75.
12. Boyd, *Nabokov's "Pale Fire,"* 81.
13. Groueff, *Crown of Thorns*, 26.
14. Negenblya, "Pervyi," 48.
15. For a general discussion of early aviation in Russian culture, see Leving, *Vokzal*.
16. Nabokov, *Gift*, 119.
17. *SM*, 133.
18. *SM*, 133.
19. Kuznetsov, "O stremlenii."
20. See D. Zimmer and Hartmann, "Amazing Music," and D. Zimmer, *Nabokov*.
21. Kuznetsov, "O stremlenii,"364.
22. Boyd, "Pinning down Krolik."
23. Nabokov, *Ada*, 57; dates placed in parentheses appear to be Nabokov's or his editors' error, contrary to the conventions of zoological nomenclature.
24. Nabokov, *Gift*, 73.
25. Pushkin, *Eugene Onegin*, 3:139.
26. Yuri Korshunov's note can be seen at http://jugan.narod.ru/nabokov.htm.
27. Pushkin, *Eugene Onegin*, 1:117.
28. Tammi, "St. Petersburg Text."
29. Tammi, "St. Petersburg Text," 126.
30. Nabokov, *Gift*, 244.
31. First published in *VNAY*, 110.
32. See Fet, "Zoological Label."

Chance, Nature's Practical Jokes, and the "Non-Utilitarian Delights" of Butterfly Mimicry

VICTORIA N. ALEXANDER

> *When a butterfly has to look like a leaf, not only are all the details of a leaf beautifully rendered but markings mimicking grubbored holes are generously thrown in. "Natural selection," in the Darwinian sense, could not explain the miraculous coincidence of imitative aspect and imitative behavior, nor could one appeal to the theory of "the struggle for life" when a protective device was carried to a point of mimetic subtlety, exuberance, and luxury far in excess of a predator's power of appreciation.*
>
> —Vladimir Nabokov, *Speak, Memory*

Writing about insects that play the mimic far better than is necessary to fool predators, Nabokov notes, "I found in nature the kinds of non-utilitarian delights I sought in art." With this Rosetta nugget, we better understand how Nabokov's art and science inform each other. As a scientist, Nabokov argued that natural selection, because it is not a source of variation, has no actual creative powers.[1] As an artist, he understood creative processes; he saw himself in nature and nature in himself. As he muses in "Father's Butterflies," "Certain whims of nature can be, if not appreciated, at least merely noticed only by a brain that has developed in a related manner." Nabokov rejected Darwinian "gradual accumulation of resemblance" as an explanation for mimicry.[2] Instead, he thought that some butterfly resemblances were more or less probable coincidences (as in the similarity between the Viceroy and the Monarch) while others were unlikely coincidences (as in the butterfly that looks like a dead leaf). He did not think that fitness selection was necessary or able to explain how such resemblances came to exist. In his fiction, Nabokov compared such coincidences to a funny typo that could give new meaning to a sentence: "the chance that mimics choice, the flaw that looks like a flower."[3]

ALTHOUGH, ACCORDING TO DARWINISM, mutations are said to be random with respect to the needs of the organism, chance, in fact, plays but a walk-on part in the performance of evolution by natural selection. No organism in Darwin's narrative wins a seven-figure fitness lottery. Instead, each individual toils day in and day out, tucking away a penny here and there, until a species finally becomes a very good fit in a particular niche. The gradual process of natural selection is said to be very "scientific" in the sense that with every useful, but likely, small step, chance is virtually explained away.

Many early Darwinian theorists assumed they must try to make useful coincidences—for example, mutations that happen to work well—as minor as possible to create the most naturalistic models. As Darwin was publishing in the nineteenth century, a new literary tradition arose called naturalism that is modeled on Darwin's principles. The novel whose plot turns on fantastic luck is eschewed (likewise artistic and allusive writing). Things incidental rather than coincidental are preferred—that is, the numerous mundane causal factors, which are more or less expected to occur most of the time, are favored over the unusual intersection of unrelated causal chains: no Armageddon meteorites, no cataclysmic floods, no fortuitous meetings at the crossroads. Gradualism, not catastrophe, is what geologist Charles Lyell recommended and what Darwin accepted. Likewise, according to literary naturalists, narratives should relate everyday phenomena and avoid the unusual luck one might attribute to a deity.

Nabokov may not have believed in a traditional Creator, but he relished the unlikely coincidence in narrative. He never doubted that a coincidental likeness might be noticed by natural selection once it was in existence. For example, of a butterfly with a false second head Nabokov writes, "A bird comes and wonders for a second. Is it two bugs? Where is the head? Which side is which? In that split second the butterfly is gone. That second saves that individual and that species."[4] But Nabokov rejected the idea that natural selection, a mere proofreader, could *create* something so wonderful, playful and nonutilitarian as some instances of insect mimicry. According to Nabokov's "pure chance" view of mimicry evolution, although the patterns of separate species might have similar material causes, any functionality of similar appearances would be unrelated to their separate evolution and entirely coincidental. In step with the few sympathetic theorists of the day, Nabokov cites, among other things, the "aimlessness" of some mimetic forms—which, though amazing, would not provide greater reproductive fitness—and the "absence of transitional forms" as evidence supporting his antigradualist view of mimicry.[5]

Almost all lepidopterists have assumed that there is an extremely low probability that, say, a White Admiral butterfly species might mutate, by chance, into a Viceroy species, the match of the Monarch. To my knowledge, no one has tried to precisely determine the odds of such a chance match. If it were known how many possible distinct patterns exist and which of these may be more probable than others, then one could determine how hard selection would be required to work, if at all, to find a match. Boris Schwanwitsch in 1924 and Fritz Süffert in 1927 independently realized that all butterfly wing patterns are produced by different combinations of a limited set of elements called the "ground plan." Today the ground plan can be understood in terms of constraints on pigment diffusion across a wing surface and compared to Theodor Eimer's work on "orthogenesis."[6] To Nabokov, as well as to Goethe's admirers, the ground plan produced "variations on a theme," some common, others rare. Frederick Nijhout, who has modeled the separate elements and studied how they interact, vaguely predicts only that the number of different possible combinations would be "enormous."[7]

I would like to suggest, based on a conversation I had with mathematician Persi Diaconis about this problem, that when lepidopterists guess the probability of a chance match like the Viceroy and Monarch, they may be doing the math wrong. The question does not concern the probability that a particular species will converge with another particular species. Such a question applies only to the Darwinian-style evolution in which a particular mimic evolves in appearance to be more like its particular model. Nabokov suspected that the work natural selection would have to do to accomplish this feat would be almost miraculous.[8] The correct question to ask is: What is the probability of any of the twenty- to thirty thousand species of butterfly roughly converging with any other?[9] Logically, this might be fairly high over the course of evolutionary time. The difference here is roughly analogous to the difference between the chance that *I* will win the lottery versus the chance *someone* will win the lottery, or come very close to winning the lottery.

To test this question, I have suggested that a computer model of wing pattern development might be developed that would constitute a "movie version" of Nijhout's work. It could be left to run through various parameters.[10] Lepidopterists watching the simulation run would probably see quite a few familiar butterflylike patterns produced by chance—that is, without predator selection pressure. Such a dynamic simulation might also give us a glimpse into Nabokov's visual imagination. His descriptions of wing pigmentation indicate that he seems to have intuited the self-organizing processes that control pattern formation. His practice of carefully drawing wing patterns of specimen after specimen allowed him to imagine the process as if he were watching a flip-book in his mind's eye. In his lepidoptery papers, he describes the pattern elements as moving images:

> What we see as a transverse, more or less sinuous, "line" or "row" of spots seems to me to be the outcome of two unrelated phylogenetic phenomena. The "upper" part of the "row" . . . is formed by spots having radiated fanwise . . . owing to an apicoid extension of the wing texture; the "lower" part[s] . . . have been pulled out [from the proximal end]. . . . Insofar as spots have been evolved in this family, they occupy different positions in different species or genera, and what we see is not the remnants of a definite band in a definite place, but this or that stage of a more or less coordinated longitudinal movement of spots . . . (certain comet-tail traces of this progress are sometimes caught and fixed aberrationally) [see Color Plates 40–45, 60].[11]

Nabokov could guess where a pattern had been and predict where it was going because he understood how pigment diffuses. He could guess which types of patterns were likely and which were not. In Nabokov's words from "Father's Butterflies," "Prolonged contact with animate nature placed at his mind's disposal not only a succession of scenes but the repetition of scenes and the series in which they occurred. . . . A truth had matured that he had not consciously sought but that had harmoniously grown out of an internal association of elements he had gathered."[12]

In contrast to Nabokov, a neo-Darwinian might approach the probability of a match question by first trying to determine the typical error rate for copying a specific nucleotide and then compound that number by a factor based on the fact that a typical gene codes for a protein of slightly more than three hundred amino acids. Assuming that a thousand out of a million might have a mutation in any given gene, that most will be recessive, and that less than 30 percent will likely have a nonlethal phenotypic effect, our neo-Darwinian might thus calculate the probability of getting a homozygote with two copies of the same mutation as $(3 \text{ in } 10,000)^2$ or 9 in 100 million. Further reasoning that, for example, in order for a proto-Viceroy to converge with a Monarch, one might expect a minimum of four genes (one for orange background color, two to remove the white band of a White Admiral, and one for the vein pigmentation of a Viceroy), if each gene has the same 9-in-100-million chance of producing a homozygous phenotype, then a neo-Darwinian might guess, were he or she to do such a calculation, that the combined probability of getting all four mutant genes homozygous in the same individual would be roughly 5 in 10^{29} individuals.

With such a method, the probability of a chance match would certainly seem low. But there is reason to doubt that the main source of variation is, as many early neo-Darwinists assumed, single-point replication errors (more on this below). In any case, this approach completely misses the forest for the trees. Because wing pattern genes "code for" reaction diffusion processes, one might just model those processes, as in Meinhardt and Nijhout, and ignore the details at the level of the nucleotides (see Richard Goldschmidt).[13] This is essentially what Nabokov did forming his intuitions about the likelihood of chance matches. Although Nabokov never developed a testable theory of mimicry, he did develop a theory of species relationships and the historical ordering of Blues (Neotropical *Polyommatus*) that has been tested. While formulating his theory, Nabokov did not have access to genetic information of his specimens. Instead, he observed the *actions* of the genes, as they changed from one specimen to the next, in the formation of the genitalia structures, whose formal rules he came to understand intuitively through extensive observation. In 2011, genetic sequencing verified that his "guess" was spot on.[14]

Material constraints of wing pattern formation were first described by Reginald C. Punnett and were elaborated by Goldschmidt.[15] Punnett, who was a chief critic of the gradualist explanation for mimicry, was professor of biology at Cambridge during Nabokov's time there from 1919 to 1922. We know that in 1920, someone in the biology department offered Nabokov a position as "entomologist's assistant" on a trip to Ceylon. Punnett's butterfly fieldwork was done mostly in that colony. Punnett also worked closely with William Bateson, a leading early advocate of the theory of discontinuous (nongradual) variation.

Punnett and Goldschmidt were at the very center of the debate over mimicry in the 1930s and 1940s. Although he tends to not mention the work of like-minded thinkers, Nabokov would have been familiar with research showing that variants like those found in nature can be produced under laboratory conditions in one mutational step. "Therefore nothing else is required in nature either," Goldschmidt recapitulated in later years. "The application to the mimicry case appears obvious."[16]

In 1927, Süffert deduced that only two modifications of the butterfly ground plan are needed to turn a normal-looking butterfly into one with a dead-leaf pattern: most pattern elements become dull colored and the wing is foreshortened. This shape change pulls the central pattern elements into alignment with a band of dots, forming a single line down the center of the wing (like a leaf vein), with the dots happily taking on the appearance of random grub-bored holes in their new position. Nabokov never doubted that a butterfly that looks like a dead leaf might escape predation more often than a more conspicuous insect. He doubted that such a form had been refined gradually by the poor decisions of predators. He had reason to believe that the dead-leaf pattern could have appeared suddenly, without the help of selection, in a single generation.

Sometimes a coin toss does result in ten heads in a row, and this is perfectly consistent with the laws of probability; it can happen only rarely. Even what we may consider extremely incredible coincidences—a lost wallet being found by a woman with the same name as its owner—are to be expected, every so often. Probability theory soberly advises us not to look for further explanation for predictably rare events: the woman is not a character in a romance with a supernatural author. *Improbable events are only unusual; they are not impossible.* And yet when two butterflies resemble each other or an insect resembles an object, and from the resemblance a benefit is reaped, scientists disregard the probability of such occurrences happening strictly by chance over the course of the millennia of evolutionary time, and they turn to natural selection for an explanation. How different is this from what superstitious people do, providing explanation where none is needed? Nabokov rebelled against this idea.

Insectivores, observed Nabokov, aren't particularly fond of butterflies.[17]

MIMICRY IN NATURE REFERS TO the abilities of organisms to deceive predators or prey by resembling something else—something alarming, inedible, neutral, or attractive, whatever the survival situation demands. In nature, a mimic does not actually pretend to be something else. Instead predators mistake it for something else, and the prey's goal of survival is unintentionally accomplished. Thus Darwinists use "mimicry" metaphorically to indicate an advantage to the similarity. Darwinists argue that the interpretations and responses of predators slowly create better mimics with every individual that is eaten while another is spared. Although technically natural selection does not have creative powers—it can select only among given variation—under this view, it would have de facto creative powers in its cumulative effects. Books and websites on the subject of evolution inevitably feature the remarkable images of moths with huge glaring "eye spots" or insects that look convincingly like leaves or other toxic species. Mimics are thought to be the very image of Darwinism, standing as exemplary achievements of the painstakingly gradual process of selection.

Nabokov was convinced that mimicry is the least likely adaptation to have been created gradually by natural selection for several reasons. First, although selection might gradually shape one form into some other form, it would be much more difficult, statistically speaking, for selection to gradually shape one form into another specific, predefined goal. As Nabokov puts it, evolution by random mutation and natural selection "in no way presupposes . . . a coordination of actions [that is, mutational events] between two creatures or between a creature and its environment."[18] In contrast to mimicry, camouflage is a good example of the selection of any pattern that works. A camouflage does not match a target model; it merely blends in with its surroundings, having approximately the same degree of busy-ness and similar color. A vast number of different patterns, dots and lines going any which way, might fit the bill for bark camouflage. Essentially only one pattern would be a match for mimicry. Whereas natural selection might "create" camouflage, it is far less able to "create" mimicry. The mathematics used by population geneticists would not apply unmodified to the special case of an adaptation that requires a specific predetermined goal as a mimic requires a model.

Second, if natural selection were shaping mimicry, the butterfly model would have to be in a state of evolutionary stasis while the mimic's genome continues to mutate at a normal level. Mimicry could only evolve gradually if the model, "during the full number of centuries required by the toiler at evolution toward a gradual attainment of resemblance, would remain unchanged (in the kind immobility that a painter demands of his model)."[19] If constraints on the so-called model do exist, making its pattern more robust and stable, these constraints would be available to the so-called mimic as well. In "Father's Butterflies," Nabokov explores such a theory of shared morphological potentiality among all species. Such constraints might be sufficient to cause resemblances to arise via convergent evolution without assistance from predators.

Third, according to the Batesian theory of mimicry, an unpalatable species is mimicked by a palatable one. In Nabokov's day, Viceroys were thought to be palatable mimics of unpalatable Monarchs. Nabokov, a devout empiricist, tasted them and found both bitter.[20] It took other lepidopterists (watching mockingbirds do taste tests) almost fifty years to come to the same conclusion.[21] The Viceroy and Monarch were subsequently reimagined as Müllerian mimics, which, because both are unpalatable, can share the cost of predator education. Very recently, however, the mimicry-via-gradualism hypothesis for some Müllerians has been called into question by the discovery that the similar insects are hybrids, hence the similarity.[22]

Fourth, whatever theory of similarity creation one may have (whether through hybridization or convergence), even the gradualists have long realized that a good-enough resemblance must be caused by chance first before natural selection would be able to act on it as a resemblance, at which point selection might fine-tune it.[23] Wielding Occam's razor, Nabokov supposed that an excellent resemblance could be caused by chance. A computer simulation of wing pattern formation could test such a hypothesis.

Fifth and finally, natural selection would not be able to create better mimicry than would suffice. This is what we find in the dead-leaf butterfly. Not only does the insect look a lot like a leaf, it has faux "grub-bored holes" down the center that add a degree of realism unnecessary to fool predators. An apparent leaf with apparent holes does not look significantly more like a leaf than one without apparent holes would and so would not give the more scrupulous mimic a higher reproductive fitness. The existence of such superfluous detail could not be shaped by selection, and so, however realistic it may appear, it must be a product of pure chance. Making the point clearer, Nabokov notes another instance of nonutilitarian mimicry. He describes a round marking on a wing that looks like a drop of liquid. A line that runs through the "droplet" is shifted in a perfect imitation of refraction.[24] It is difficult to imagine what advantage could be ascribed to an imitation of a droplet on an insect's wing. This is art for art's sake. If chance could create these remarkably artistic resemblances, it should have no trouble with more prosaic ones.

IN "FATHER'S BUTTERFLIES," Nabokov imagines a caterpillar that resembles a particular flower; however, the caterpillar is out of sync with the seasons, and it appears in its disguise too late after the blooms are gone:

> Among the numerous illustrations of these blatant excesses of nature let us select the following example: the caterpillar of the quite local Siberian Owlet moth (*Pseudodemas tschumarae*) is found exclusively on the chumara plant (*Tschumara vitimensis*). Its outline, its dorsal pattern, and the coloring of its fetlocks make it resemble precisely the downy, yellow, rusty-hued inflorescence of that shrub. The curious thing is that, in conformity to the rules of its family, the caterpillar appears only at summer's end, while chumara blooms only in May, so that, against the dark green of the leaves, the caterpillar, uncircled by flowers, stands out in sharp contrast.[25]

Nabokov composed this passage sometime around 1939. In 1902, Bashford Dean, a curator at the American Museum of Natural History, had published a similar amusing anecdote, "A Case of Mimicry Outmimicked?" Dean had seen a museum display of *Kallima* "dead-leaf" butterflies on a leafy branch. He noted that the butterflies were positioned on the branch exactly where a leaf would be and that they struck leaflike poses. He became suspicious when he realized that although *Kallima* live in the tropics, the matching leaves were of a species from North America. Later, when Dean found *Kallima* in their natural environment, he noticed the butterflies alighted on leaves they in no way resembled and on which their dull color was quite conspicuous. "The leaves were bright green magnolia like much larger than the butterflies." When the leaves died, they turned "bright orange yellow. . . . There were no brown leaves . . . in the neighborhood." Dean's example is just the sort of ironic joke of nature Nabokov admired. Wrote Dean, "I am not able to repress a suspicion that in *some* cases (who knows in how many, even perhaps in the case of the classic butterflies?) our idea of the mimicry may be preconceived, rather than truthful. The fact that a butterfly looks strikingly like a given dead leaf is no adequate proof that it was evolved in mimicry."[26]

Punnett's *Mimicry in Butterflies* also contains numerous anecdotal descriptions of "mimics" whose "protective" disguises provide no advantage or are nonutilitarian.

Some have supposed that Nabokov's rejection of Darwinian gradualism as an explanation for the origin of mimicry was motivated by a belief in a divine Creator or some other such nonscientific sentiment. Brian Boyd refers to Nabokov's theory of mimicry evolution as his "dearly held metaphysical speculations."[27] It is clear to me, however, that his views were based on empiricism and logic and were very like those of other reputable scientists of his day who argued against gradualism. He may have been wrong that certain butterfly-on-butterfly resemblances are caused strictly by chance tempered by pattern constraints. However, if many mimics prove to be hybrids, for example—or beneficiaries of some other form of interspecies promiscuity (more on this below) or the product of some large mutational event—Nabokov would be proved correct in his antigradualist argument, and any functionality of the similarity would still be strictly lucky.

SCIENTIFIC RESEARCH OF THE PAST twenty years indicates that Nabokov (after Punnett, Bateson, and Goldschmidt) may have been correct to suggest that nongradual mechanisms are at play in evolutionary processes. Many biologists now argue that significant, beneficial, heritable changes can occur quite suddenly. This way of looking at evolution is referred to as "saltationism." According to this view, adaptations result from various kinds of systems-level reconfigurations, not from gradual accumulations of point mutations or replication errors. Large mutational events, such as hybridization, gene duplication (segmental and whole-genome), lateral gene transfer (between different species via viruses or bacteria), transposons (genetic segments that can move from one site to another, which are specifically active in germ-line but not somatic tissue), and symbiogenesis, produce the most significant (and probably beneficial) novelties, which can then be selected or not.[28] Some evolutionary theorists have thus begun to reduce the time scale of selection narratives rather substantially and with it the de facto creative role of natural selection.

Whether all butterfly mimic forms are created by chance via such rapid mechanisms remains to be investigated, but it seems quite possible. Research has shown thus far that at least some butterfly mimic wing patterns are controlled by relatively few genes.[29] These genes, in some cases, may be inherited as a unit such that offspring do not receive a shuffled mix of genes from both parents, eliminating gradations between distinct forms.[30] Some butterflies segregate by color, possibly making mutants reproductively isolated.[31] Given these conditions, a few "hopeful monsters"—Goldschmidt's 1940 term for very different yet quite viable offspring—might proliferate rapidly, due to a reproductive fitness that is not necessarily courtesy of reluctant predators.

In 1940, leading neo-Darwinist population genetics theorist Theodosius Dobzhansky dismissed Goldschmidt's saltational theory, claiming that it concerned "catastrophism" not evolution. He further opined that Goldschmidt's dismissal of the efficacy of small uniform change as the source of variation in favor of systemic mutations (probably in chromosomal rearrangements) constituted a "belief in miracles."[32] Satisfied with the way the new synthesis had hardened against Goldschmidt's apostasy, Neo-Darwinists in the 1940s largely stopped having conversations with theorists who promoted alternatives to gradualism. Fortunately, the self-correcting nature of science eventually reopened the discussion, and today the theory of evolution includes many different mechanisms for change, speciation, and diversity.

It was shortly after this skirmish between the neo-Darwinists and Goldschmidt that Nabokov mentioned several times that he wanted to write a big book on mimicry.[33] In 1942 and 1943, he gave presentations on "The Theory and Practice of Mimicry."[34] Unfortunately, that material has been lost, and we are left to reconstruct crucial parts of his theory from other writings, mainly "Father's Butterflies," a work of fiction, similar in approach to what Nabokov-the-Real also says about mimicry in his autobiographical writings. Richard Goldschmidt's argument squares fairly well with Nabokov's, and so, to some extent, we can use Goldschmidt as a surrogate. He was *the* saltationist during the 1940s (Stephen Jay Gould calls him the neo-Darwinists' "whipping boy"), and he was the intellectual heir to Punnett and Bateson, both of whom Nabokov had very likely studied at Cambridge.[35] Although a number of Nabokov's critics speculate that the neo-Darwinist literature would have convinced Nabokov of the error of his thinking, Goldschmidt's example suggests otherwise. Goldschmidt was extremely conversant with all aspects of population genetics, was, in fact, an important figure in shaping the discussion, and was convinced that gradual selection could not create mimicry.

SOME ASPECTS OF NABOKOV'S WRITINGS on mimicry are not found in Goldschmidt, and these have to do with the deep sympathy he felt with "that other V. N., Visible Nature," as a fellow creator.[36] Nabokov saw evidence of a kind of protointelligence in evolutionary processes. He associated nature's use of chance with *interpretation*. Mimicry was even more interesting to him because it involves *mis*interpretation. If nature is intelligent, mimicry is artistic.

Today biologists use terms like "interpretation" and "sign use" to describe cell-level processes. (Although biologists do not like to ascribe human tendencies to cells, they have found that they can hardly avoid using such language.)[37] The burgeoning field of biosemiotics defines a "sign'" as something that an organic system associates with self-sustaining, self-referential, or homeostatic processes. At every biological level, morphologies that are coincidentally like familiar signs—that accidentally mimic them—can become new signs of different useful processes, resulting in the emergence of more complex and organized behavior.[38] Mimicry may be more appropriately recruited as the poster image for such areas of inquiry than for Darwin's gradual selection theory.

To me it appears that Nabokov's views tended in a biosemiotic direction. William Bateson's son, Gregory Bateson, is a founding figure of biosemiotics, as is René Thom, author of "catastrophe theory" (recalling Goldschmidt's work), which later inspired "chaos theory" and the orthogenesis-like evolutionary theories of the complexity sciences: these dovetail nicely with some of Nabokov's references to self-organizing wing pattern development.[39] In one of Nabokov's most poetic passages on mimicry, in "Father's Butterflies," his fictional scientist proposes the existence of an "even force" that "animates" the universe, a "thought-engendering rotation" that "gave rise in nature to the lawlike regularity of repetition, of recognition, and of logical responsibility to which the apparatus of human ratiocination, the fruit of the same agitated woodlands, is subordinate."[40]

Biosemioticians likewise argue that protosemiotic relations in nature, relations of similarity (icons) and contiguity (indices), create lawlike regularities and eventually relations of arbitrariness (symbols), from which human language emerged. Some may have supposed that Nabokov was guilty of anthropomorphism, seeing evidence of his own kind of artistry and mischievous humor in nature. But it may be that Nabokov instinctively sought to naturalize human semiosis. If Nabokov saw a primitive kind of semiotic intelligence in nature, then we cannot say he was guilty of believing in supernatural or divinelike human exceptionalism. He seems to say instead that intelligence emerges from significant coincidences. In this volume, which is partly dedicated to elucidating the relation between the arts and sciences, I argue that the artistic act is an act of nature and that science needs to understand how nature, as artist, creates. As artist and scientist, Nabokov had an insight into creativity we are only now beginning to appreciate.

Notes

1. Alexander and Salthe, "Monstrous."
2. *NB*, 219, 222.
3. Nabokov, "Vane Sisters," 622.
4. *NB*, 530.
5. *NB*, 223.
6. Nijhout, "Pattern"; Eimer, *Orthogenesis*.
7. Nijhout, "Elements," 215.
8. *NB*, 223–24.
9. The formula for assessing a similar type of probability is called the "birthday problem" formula with a "near match" variable. What is the probability of finding at least one pair in a group of n butterflies with near match patterns within k pattern elements of each other's, if there are m equally likely patterns? Wing patterns may require a more complex formula because all patterns are probably not equally likely.

$$p(n, k, m) = 1 - \frac{(m - nk - 1)!}{m^{n-1}(m - n(k + 1))!}$$

10. Alexander and Salthe, "Neutral Evolution"; H. Frederik Nijhout's work is based on a variation of Hans Meinhardt's reaction-diffusion model, according to which an activator and an inhibitor are distributed equally throughout a spokelike area called a wing cell. The activator's product tends to increase its own production. The inhibitor neutralizes the activator's product. Along the wing cell edges, which are protected from the inhibitor, a sudden increase in activation might occur spontaneously. The activator diffuses toward the center, forming a line, then retreats toward the far end. In the concentrated areas at the end of the receding line, activation increases even further. Then these areas are finally pinched off by surrounding areas of inhibitor, leaving traces of activator behind. These traces, together with any traces left on the edges, form areas of greater or lesser density. This difference in density is metaphorically referred to as the topography of the wing cell, which in turn affects the diffusion of pigment. Individual ground plan elements are determined according to whether the topography

attracts or repels pigment, condensing or stretching it into various shapes. These processes can be affected by timing, temperature, humidity, and physical shape and size of the wing cell—that is, they are not strictly genetically controlled.

11. *NB,* 282.

12. *NB,* 214.

13. Meinhardt, *Models;* Nijhout, *Development;* Goldschmidt, "Mimetic Polymorphism (Concluded)," 215, makes a similar suggestion.

14. Vila et al., "Phylogeny."

15. Punnett, *Mimicry;* Goldschmidt, *Untersuchungen, Einige Materialien,* and *Physiologische Theorie.*

16. Goldschmidt, "Mimetic Polymorphism (Concluded)," 207, 225.

17. *NB,* 178.

18. *NB,* 225.

19. *NB,* 224.

20. *NB,* 535.

21. Ritland and Brower, "Viceroy."

22. Dasmahapatra et al., "Butterfly Genome."

23. See Poulton, "Mimicry."

24. *NB,* 244.

25. *NB,* 222; Dieter Zimmer asserts that this insect is imaginary. D. Zimmer, *Guide,* 248–49, and http://www.dezimmer.net/eGuide/Lep2.1-Po-Py.htm#Ps.tschumarae. Regarding the alleged anachronistic crypsis, Zimmer notes: "The yellow color might be useful to the caterpillar even out of season, in case the local birds find a tidbit that looks like a flower not too appetizing at any time of the year."

26. Dean, "Case of Mimicry," 833.

27. *NB,* 20.

28. See Margulis and Sagan, *Acquiring Genomes;* Reid, *Biological Emergences;* and J. Shapiro, *Evolution.*

29. Turner, "Mimicry."

30. Joron et al., "Chromosomal Rearrangements."

31. Chamberlain et al., "Polymorphic Butterfly."

32. Dobzhansky, "Catastrophism," 358; Mayr, *Systematics.*

33. *NB,* 245, 248, 484.

34. *NB,* 265, 269, 270, 278.

35. Gould, *Structure,* 452.

36. *NB,* 669.

37. Favareau, *Essential Readings,* 1–77.

38. Alexander, "Poetics," and *Biologist's Mistress.*

39. Alexander, "Neutral Evolution," and "Nabokov, Teleology."

40. *NB,* 226.

Nabokov's Evolution

JAMES MALLET

*Nabokov: The highest enjoyment is when I stand among rare butterflies
and their food plants. This is ecstasy, a sense of oneness with sun and
stone.*

. . .

Interviewer: Is there any connection [of lepidoptery] with your writing?
*Nabokov: There is in a general way because I think that in a work of art
there is a kind of merging between the two things, between the precision
of poetry and the excitement of pure science.*

—Emma Boswell, *How Do You Solve a Problem Like Lolita?*

Much has been written on Nabokov's views on evolution, the origin of species, and, especially, mimicry in nature. Nabokov scholars hold a diversity of opinions, veering from defending Nabokov's heterodox beliefs about evolution to criticizing him for not fully accepting the modern evolutionary synthesis developed in the 1930s and 1940s in Europe and North America.

Nabokov the scientist was maybe a little old-fashioned in his methodology, even for his time, concerned mainly with morphological systematics. He evinced a profound dislike for applying statistics to his science. Nabokov furiously rejected a statistical criticism by F. Martin Brown of his work on "scale rows," a method he invented for the study of plebejine butterflies.[1] His lack of appreciation of applied mathematical theories of evolution probably ensured that he could not form a strong opinion about the works of Ronald Fisher, J. B. S. Haldane, and Sewall Wright, the founders of the modern synthesis of Mendelian genetics and Darwinian natural selection. However, Nabokov would certainly not have been alone among biologists of his time in his contempt for the modernization of biological science and, in particular, of mathematical or genetic approaches to systematics.[2]

Nabokov certainly understood much better the later part of the modern synthesis on species concepts and speciation, beginning in 1937 and continuing in the early 1940s largely by another triumvirate. Their leader was a fellow Russian, the geneticist and former beetle systematist Theodosius Dobzhansky. Julian Huxley from England—the grandson of "Darwin's Bulldog," Thomas Henry Huxley—and Ernst Mayr, Nabokov's own scientific colleague at the Museum of Comparative Zoology from 1953, completed the trio. Nabokov would certainly have perused the systematics-oriented book by Mayr.[3] This second wave of the modern synthesis was important in connecting systematics and natural history with the new mainstream of genetics and evolutionary thought. Nabokov criticized these ideas, mainly in unpublished documents.

Instead of attacking or defending Nabokov's views on mimicry and species concepts, I merely comment here on his views from a scientist's viewpoint. The modern synthesis was a broad church, and different players in the formation of the synthesis had varying opinions; my perception is that his views were not particularly extreme at the time. Others have already excellently described Nabokov's observations on these matters.[4] I hope to contribute a clearer understanding of Nabokov against the scenery of scientific thought in his day. Ideas not dissimilar from Nabokov's apparently heretical stances were already present in the 1940s and, indeed, have led to the progression toward today's understanding of biology.

ALTHOUGH A POTENTIALLY GREAT SCIENTIST—he might well have developed into a major systematist of butterflies had he not proved a highly successful novelist —Nabokov did not fully put his science together with his art. To many Nabokov scholars, this may seem a hard argument to swallow. Nabokov expended many words in telling his public how he embodied a fusion of art and science (see the epigraphs to this essay). And late in life he snobbishly argued that C. P. Snow's *Two Cultures* represented only a slight divide between lowbrow literary culture (presumably, he disapproved of Snow's novels) and trivial utilitarian science. Nabokov certainly was a good systematist, and yet he somehow believed that mere mechanical understanding is less sublime than the appreciation of the "non-utilitarian delights" of nature. To a scientist as well as to Nabokov, the beauty and intricacy of the underlying structure hold the greatest wonders in life.[5] To explain these mysteries is the goal of science, but this is most convincingly achieved in the mathematical language of mechanics and dynamics, and data are best analyzed by means of reference to some statistical model, rather than solely through verbal discussion.

Many biological mysteries remain, of course. Only by rational and theoretical exploration of these mysteries, rather than by rejection of the possibility of rational explanation, does science progress. Nevertheless, in science as in art, imagination and chance are keys to exploring mysteries. Without imagination, advances in scientific understanding outside the prevailing paradigm would be impossible, and science would grind to a halt.

As Karl Popper argued, the source of a scientific hypothesis is often completely nonscientific and may even stem from religious conviction. Johannes Kepler's discovery of approximately elliptical planetary motions was motivated initially by a Neoplatonic, almost religious notion: The Copernican idea that the Sun in the universe plays the same role as the Good plays in the realm of ideas. The Sun could not be expected to revolve around the Earth under this system. Orbits of mere planets like Earth (and its inhabitants) should instead move at constant angular velocity in obedient and perfect circles around the Sun. Tycho Brahe's detailed observations of the orbit of Mars, elliptical and showing variations in speed, ruled this model out, to Kepler's evident astonishment. Clearly, the initial hypothesis that led finally to Kepler's ellipses was imaginative rather than scientific or empirically based.[6] Imagination and aesthetics can thus be enormously helpful in science. But they can also hold science back. Galileo himself never accepted Kepler's eccentric planetary orbits because of the loss of perfection, just as Einstein was to reject the notion of God throwing dice with respect to quantum mechanics four centuries later. It would be unsurprising for Nabokov to have a similar weakness.

Before discussing the topics of mimicry and speciation in detail, I should like to propose that, in his dissatisfaction with the modern synthesis, Nabokov displays a blend of imagination and critical thinking likely to have been useful in making great scientific discoveries. He became a great novelist, but there is no reason, had he not given up active science and had *Lolita* not sparked a huge new readership, why he might not have become much more revered as a butterfly taxonomist instead. Among his abandoned projects were field guides. He certainly was capable of writing definitive field guides, such as to the butterflies of North America before Alexander B. Klots and the butterflies of Britain and Europe before L. G. Higgins and N. D. Riley. As it turned out, Nabokov was reduced to writing well-informed reviews of both books.[7]

IN MY VIEW, NABOKOV DID GET MIMICRY WRONG, but he certainly did not deny the importance of natural selection in general, and most of his attacks on mimicry were in any case literary flourishes not to be taken too seriously. For example, some of the literary examples of mimicry stories that seem too good to be true for the narrator in "Father's Butterflies" are simply that—Nabokov invented them for this purpose, as Dieter Zimmer has shown. In one case, a fictitious moth, complete with Latin binomen, is invented. In another, the extraordinary caterpillar of a real species, the Lobster Moth (*Stauropus fagi*) actually does go through an antlike as well as a lobsterlike stage, but Nabokov has added an extra stage that mimics bird lime, which other caterpillars (particularly of *Papilio* butterflies) do have.[8]

What formed Nabokov's views on mimicry? Like Nabokov, I became fascinated at an early age by mimicry, triggered especially by butterfly and moth adaptations. My first encounter was through books: Walter Linsenmaier's panoramic illustrations in a child's *Time-Life* book about evolution. As well as camouflage, Linsenmaier's images depicted mimetic *Heliconius* and ithomiine butterflies flying in the *Araucaria* forests of coastal Brazil. When, at boarding school, I started rearing caterpillars of moths and butterflies, camouflage became obvious. The "twig" caterpillars of the Peppered Moth (*Biston betularia*) and other geometrids were extraordinarily convincing to this young vertebrate. Even the most mundane and tiresome environmental features are copied: an astonishing array of creatures mimic bird shit. For example, moths like the Chinese Character (*Cilix glaucata*) or the Lime-Speck Pug (named after "bird-lime," of course; *Eupithecia centaureata*), or caterpillars, such as a number of young *Papilio* caterpillars. I yearned for the diverse rainforests of South America, where mimicry and camouflage seem most highly developed. It's probably no accident that I now study *Heliconius* and ithomiines in my professional career. To those of us with a similar background, it's easy to understand how Nabokov became fixated on a phenomenon first encountered in the crystalline clarity of childhood memories. He had seen the same caterpillars and imagos closely resembling leaves, bird droppings, and twigs that I had, and more.

Nabokov knew that explanations of mimicry were to be found in natural selection. As a child needing to identify his catches, he read natural history books that would have invoked natural selection to explain camouflage and mimicry. He certainly read Darwin (see *The Gift*), and he seems to have accepted Batesian and Müllerian mimicry of unpalatable species as having some utility and involving natural selection.

Nabokov's real complaint about "utilitarian explanations," as he liked to label natural selection, was with camouflage, rather than true mimicry. How could a blind mechanical process lead to such minutely detailed resemblances to such piffling features of the environment? Camouflage, sometimes dubbed "masquerade," is regarded by biologists to be distinct from actual mimicry—the copying of simple but garish warning patterns by mimic species that predators actively avoid. Detailed camouflage is instead the copying of things in the environment that predators normally ignore or overlook. The difference is that a mimic copies a signal, whereas in masquerade, the copied item (for example, a twig) is normally not in itself designed by natural selection as a visual signal.

Camouflage, the mimicry of features of the environment, such as the transparent spots in a butterfly wing imitating fungus-eaten holes in dead leaves (see Color Plate E5), to Nabokov seemed too perfect and too complex to have been effected by predators discriminating against deviants over long stretches of geological time. In his opinion, this cried out for something less chancy than a mere mechanical process of natural selection based on initially random mutations, for which "a trillion light years would hardly be sufficient."[9] Nabokov was here (at least in "Father's Butterflies") using something like the classic "argument by design" adopted by William Paley and the natural theology movement of the early nineteenth century. If you find something like a watch in a field, you know it was designed because of its function. Unfortunately for us, who would like to access Nabokov's actual beliefs, he explores the theme in much greater detail in his preserved imaginative rather than in factual writing.

Nabokov was no creationist. He was an iconoclast—not unheard of among evolutionary biologists, let alone novelists. Many other mainstream biologists have argued and still do argue against strictly selective explanations of mimicry.[10] Some scientists still doubt the camouflage explanation for the evolution of the black, or melanic, form of the Peppered Moth during the Industrial Revolution.[11]

I speculate that Nabokov was a victim of literary enthusiasm rather than a real disbeliever in the power of natural selection to effect the many minute changes that result in extraordinarily accurate mimicry. Nabokov's opinions are often strongly expressed. Perhaps Nabokov just didn't like to prevaricate or hedge enough for the scientists. In his literary treatment of mimicry, we should be aware that he was perhaps enunciating controversial views that might have been true in an alternative world, rather than always stating what he actually believed about nature.

IN 1944, NABOKOV PILLORIED THE MODERN synthesis species concept in unpublished notes by arguing that morphology should trump mere mating preference: "This is the kind of nonsense that you finally get with these subtle experiments with geneticists."[12] Here is a clear reference to Dobzhansky's view of species, the "biological species concept." Dobzhansky's concept, based on "reproductive isolating mechanisms," was triggered by his discovery of "race A" and "race B" of *Drosophila pseudoobscura*.[13] Apparently identical in appearance, they differed in chromosomal arrangements, as well as in strong mating preferences and partial hybrid sterility. To Dobzhansky, they were real species, not mere races, and race B was later named *Drosophila persimilis*.[14]

That the two are real, separate species is now widely accepted. However, the Dobzhansky-Mayr view was controversial in 1944. Even Alfred Sturtevant, who had been a senior colleague of Dobzhansky's in Thomas Hunt Morgan's laboratory, disagreed with the need to consider the two as separate species. He argued that the "race A" and "race B" designations were sufficient "in identifying wild specimens without breeding from them or examining their chromosomes."[15] Sturtevant was not a lonely Luddite; he was one of the world's best-known geneticists at the time. As an undergraduate in Morgan's lab, he had developed the key method, still used today, for mapping genes along chromosomes.

In his critique, Nabokov seizes on logical weaknesses in the biological species concept of the modern synthesis. I do not agree wholly with his attack on geneticists and their species concept, but he was not alone. In his practical application of ranks to butterflies, Nabokov's species concept seems to me close to the mainstream of the other modern systematists of his time. Today, the Mayr-Dobzhansky view of species is under more critical attack than ever. In terms of basic, practical taxonomy, many of us today hold views similar to those of Nabokov sixty years earlier.

An important way genetics impinged on Nabokov's butterfly systematics was via chromosome counts. Nabokov was always against succumbing "to the blandishments of the chromosome count" in butterfly systematics, on the grounds that their use would make species identification prohibitively difficult.[16] This is not dissimilar to Sturtevant. On the other hand, Nabokov's genitalic dissections require microscopes, albeit of lower power than those used in chromosomal studies. Both suffer from the same invisibility problem for the fieldworker. And surely it is more important to know what a taxon itself really is rather than to constrain one's viewpoint to a single mode of human perception?

Taxa with different chromosome numbers or genetic traits which appear to be reproductively isolated tend also to have subtle differences in morphology. Nabokov excelled at the study of color pattern and genitalic morphology and knew a great deal about butterflies, particularly the Polyommatini. In my somewhat inexpert opinion, the Polyommatini Blues, including *Plebejus, Aricia, Lysandra*, and *Agrodiaetus*, among others, are a mess of closely related small butterflies. Many species hybridize with one or more other species or even across genera, and they are all similar in overall morphology, though they can differ in color pattern. Nabokov did a great deal to sort this group out, especially in the Americas.

So how did he classify species in practice? Many philosophical papers have been written about species concepts, especially since the modern synthesis. Dobzhansky and Mayr in their work felt that they had rescued the pre-Darwinian idea that species were real, by which they meant objectively definable. Darwin and the early Darwinists had instead argued that although species can emerge and become more distinct over evolutionary time, there is no objective reality: we humans must decide where to cut the continuum, and what groups to demarcate as species. Mayr and Dobzhansky did not believe that species were defined by a Creator; their new objective reality was instead reproductive isolation. Both were able to maintain that species were discretely different from varieties (or races) within species, even though both agreed that species almost certainly evolve apart somewhat gradually. Mayr in particular reviewed the extensive evidence for hybridization between "good" species, and of subspecific or ecological races, incipient almost-species within these real species.[17] However, lacking molecular genetic markers (invented in the late 1960s), Mayr was able to brush difficult intermediates under the carpet. Hybridization was interpreted to be largely a pathological by-product of human-altered environments. In contrast, most or all subspecific ecological races that co-occurred in geographic overlap (or "sympatry," to use Mayr's term) were seen as plastic rather than genetic variants, or sometimes as by-products of hybridization after secondary contact between formerly geographically isolated (or "allopatric," in Mayr's terminology) good species. Mayr was adept at obfuscating the logical difficulties of his views. Although he cited reproductive isolation as the definition and concept, his species taxa were inferred largely from natural populations based on morphology. Species were deemed separate if there were few or no morphological intermediates (that is, putative hybrids) between them in areas of sympatry. If there were plenty of intermediates in areas of overlap, then the different-looking taxa were deemed to be varieties or geographic races within a single species.

One of the features of Mayr's definition that Nabokov seized on for criticism was a statement about geographically separated populations. Here the test of morphological intermediates was not possible, and so Mayr fell back on an inference of "potential interbreeding." As Nabokov realized, this meant in practice that morphological studies alone were the actual systematic tools. As for the occasional cases of hybrids in nature among good species, Mayr was somehow able to argue the "reproductive isolation" definition always held, even in taxa that didn't show it very well; to Mayr, these troublesome cases were mere exceptions that proved the rule.

The Mayr-Dobzhansky view of species became greatly celebrated, almost revered—and it still is—because it seemed to provide new and solid meaning in an area, the understanding of species, that had supposedly been laid waste by the Darwinian revolution. Nonetheless, systematists actually working at the coalface of taxonomy, like Nabokov, were not helped by these theoretical and genetic arguments for species reality. They still had to parse the actual morphological continuum into species, especially in difficult groups. Practically, the only important question about someone's systematics is not about beliefs but about what is done with those beliefs. All that really matters about Nabokov's productive work on butterflies is the question: Was he a lumper or a splitter? The American lepidopterist W. J. Holland characterized the differences between splitters and lumpers in a book that Nabokov almost certainly read:

> A "splitter" magnifies the importance of trivial details; he regards minute differences with interest; he searches with more than microscopic zeal after the little things and leaves out of sight the lines of general resemblance. . . . The labors of such naturalists may be highly entertaining to themselves, but they are, to say the least, provocative of unpleasant feelings in the minds of others who come after them and are compelled to deal with and review their labors.
>
> The "lumper," on the other hand, is a man who detects no differences. . . . Any two moths which are of approximately the same size and the same color, are, by him, declared to belong to the same species. . . . His genera are "magazines," into which he stuffs species promiscuously. The "lumper" is the horror of the "splitter," the "splitter" is anathema to the "lumper"; both are the source of genuine grief and much hardship to conscientious men, who are the possessors of normally constituted minds and truly scientific habits.[18]

In his published work, Nabokov expounds on his species concept and politely argues that it must differ from Mayr's and Dobzhansky's focus on biology:

> The strictly biological meaning forcibly attached by some modern zoologists to the specific concept has crippled the latter by removing the morphological moment to a secondary or still more negligible position, while employing terms, *e.g.,* "potentially interbreeding" that might make sense only if an initial morphological approach were presupposed. What I term species, in my department,

can be defined as a phase of evolutional structure, male and female, traversed more or less simultaneously by a number of, consequently, more or less similar organisms morphologically shading into one another in various individual and racial ways, interbreeding in a given area and separated there from sympatric representatives by any other such phase by a structural hiatus with absence of interbreeding between the two sets. . . . It may happen that two structurally distinguishable local forms belong to one species allopatrically because they racially intergrade, but at the same time belong to different species sympatrically because in some other region their structural counterparts occur side by side without interbreeding (this incidentally is the position in *Lycaeides*). In such cases one should give precedence to the all important sympatric moment and find somewhere in the spirals of racial intergradation a point at which the whole system can be elegantly . . . divided into two parts, *i.e.*, two species, using some combination of trinomials.[19]

In this, Nabokov displays a useful and informed systematic approach. He clearly recognizes logical flaws in Mayr's writings, but an underlying morphologically based practical approach is accepted (Mayr himself inherited his similar methods from David Starr Jordan, Walter Rothschild, and Rothschild's curators of birds and butterflies—Ernst Hartert and Karl Jordan, respectively—who instituted in systematics a practical version of the biological species concept in the years 1890–1915). By the time Nabokov was actively classifying butterflies, the accepted Linnaean binomial nomenclature (genus, species) had become trinomial (genus, species, subspecies). This practical revolution allowed the inclusion of geographic races or subspecies within a broader species and aided the understanding of relationships and classification. It was the nomenclatural revolution, rather than the purely theoretical reproductive isolation definition, that was the real advance of the twentieth-century species concept. Nabokov embraced this practical advance along with Mayr while maintaining his opposition to the logic used in its defense.

In summary, Nabokov, judged by his practical systematics, would have been deemed by Holland a possessor of a "normally constituted" mind and "truly scientific" habit.[20] He was a product of his times, and he was neither a lumper nor a splitter. His disagreement with genetics and chromosomal information might be criticized (although this was, in his defense, before reliable genetic markers became available), but his disagreements over species with Mayr have, in my view, considerable relevance today. Overall, he resolved his views in a way that made his revisions extremely useful, even for much later generations of butterfly researchers.

Notes

1. Nabokov, "Remarks."
2. Huxley, *Evolution: The Modern Synthesis.*
3. Mayr, *Systematics and the Origin of Species.*
4. *NB; NBl;* Zimmer, *Guide.*
5. *NB,* 399.
6. Popper, *Conjectures and Refutations.*
7. Nabokov, "World of Butterflies," and "Rebel's Blue."
8. See http://www.d-e-zimmer.de/.
9. *NB,* 223.
10. Goldschmidt, "Mimetic Polymorphism," and "Mimetic Polymorphism (Concluded)."

11. For recent discussions of this long-standing controversy, see reviews by Cook et al., "Selective Bird Predation," and Grant, "Industrial Melanism."

12. *NB,* 308.

13. Dobzhansky, *Genetics and the Origin of Species.*

14. Dobzhansky and Epling, *Contributions.*

15. Sturtevant, "*Drosophila pseudoobscura.*"

16. Nabokov, "Rebel's Blue."

17. Mayr, *Systematics.*

18. Holland, *Moth Book,* 112–13.

19. *NB,* 353–54.

20. Holland, *Moth Book,* 112–13.

Fictional Realism

Scaling the Twin Peaks of Art and Science

DORION SAGAN

The vast, bizarre cosmos displays what might be called fictional realism: it contains not only real things but all manner of (perhaps all conceivable) thoughts, lies, theories, fictions, perceptions, and imaginations. Philosopher Alfred North Whitehead faults philosophy for its fetishizing of the infinite and eternal in thinkers like Parmenides, Pythagoras, and Plato. Infinities, though they exist, do so not in another world but in this world, as the mathematical imagination suggests. Whitehead's specular pragmatism also identifies Leonardo da Vinci as intrinsically more scientific than the great early expositor of the scientific method, Francis Bacon, for Bacon was a lawyer, working more with words, while the genius Leonardo not only theorized but continuously observed—and recorded with meticulous fidelity—nature, as attested by his scientific and technical drawings. Whitehead's emphasis on the artistic, observational side of often theoretically metastasized science is manifested in the work of Nabokov, who spurns system building and novels of ideas in favor of naturalistic detail. He described himself as a landscape painter manqué. "Does there not exist," he asked, "a high ridge where the mountainside of 'scientific' knowledge joins the opposite slope of 'artistic' imagination?"[1]

And yet Nabokov, though he joined Whitehead in a dismissal of totalizing discourses (paradigmatically Marxist and Freudian, but he also expressed doubt about the power of Darwin's natural selection to explain mimicry, and even Einstein's relativity, remains skeptical, too, of observation; the "shadow of the instrument," as he poetically parses (or expands) quantum mechanics, falls over the "specimen." Compared to the scientists attempting to formulate objective "nonfiction," the fiction writer may have certain advantages. The novelist presents situated knowledges, measures and observes via the partial perspectives of individual observers. Fiction abandons the epistemological goal of telling the truth, the whole truth, and nothing but the truth (to borrow a legalistic variant whose absurdity will be familiar to lawyers and defendants alike)—arguably itself a subjective fantasy of absolute objectivity. Free to frame its observations as parochial, it reflects the phenomenological fact of individualized rather than totalizing observations. Here the novel has a genealogical tie to philosophy: Plato's dialogues took the felicitous form of real-time conversations, and reality was represented plurivocally through the perspectives of multiple personae. The Platonic dialogues' convention, owing something to drama, are especially noteworthy in light of Friedrich Nietzsche's comment in *Die Geburt der Tragödie aus dem Geiste der Musik* that Plato, originally planning to be a tragedian, burned all his plays after coming into contact with the rationalist Socrates. For Nietzsche, Euripides ruined the high tragic form, which also fulfilled a religious function far beyond recounting daily events: to showcase the ontic separation of the individual from the cosmic whole. The Nabokovian novel recovers to some extent this lost legacy; here it is perhaps noteworthy that Nabokov said that science (civilizationally launched, no less than Euripides' debased aesthetics, in Nietzsche's view by Socrates' false humility and dreams of total knowledge) "answers to philosophy, not to statistics."[2]

In his 1939 essay "John Dewey and His Influence," Whitehead writes, "At this moment scientists and skeptics are the leading dogmatists. Advance in details is admitted: fundamental novelty is barred. This dogmatic common sense is the death of philosophic adventure."[3]

Whitehead's takedown of smug would-be knowers is well taken. Supremely individualistic Nabokov, no stranger to arrogance, inured himself against the smug but passing air of authority by his less closely observing, less curious colleagues. His work harked back to an older *Naturphilosophie*, a gentleman's endeavor associated with museum collecting (see Fet, this volume), and the luxury and free time afforded to the noble class. Here science was an extension of observation for its own sake, of appreciating the beauty and naming the organisms spotted during walks in the woods. It was a cataloging of the new beasts and structures revealed by microscopes and magnifying glasses, by specimen collecting, by preservation, but also by close observation of and interaction with live organisms.[4] This allegiance to classical naturalism and study of whole organisms gave him more data points from which to speculate on evolution, and it is perhaps ironic that some of his speculations, based on a classically more rich immersion in live specimens of the field, were recently vindicated by intrinsically more reductionist molecular biology.

As Nabokov suggests, the distinct but equally necessary paths of art and science seem to scale opposite sides of the same majestic mountainscape. On the one hand, Nabokov favors science: "My passion for lepidopterological research, in the field, in the laboratory, in the library, is even more pleasurable than the study and practice of literature, which is saying a good deal. . . . The tactile delights of precise delineation, the silent paradise of the camera lucida, and the precision of poetry in taxonomic description represent the artistic side of the thrill which accumulation of new knowledge, absolutely useless to the layman, gives its first begetter."[5] On the other, science pales next to imagination-fired art: "Those miniature hooks of the male genitalia are nothing in comparison to the eagle claws of literature which tear at me day and night."[6]

The precision of art and the beauty of science; these cross-breedings and contradictions reflect Nabokov's view of art-science as a continuum. Nature is no more inert than perception is universal, as shown by Nabokov's childhood synesthesia, which he shared with his mother, and which flooded him with a productive excess of possible meanings. Nature and art were a continuum for Nabokov; the artist can discover, and nature, a subtle deceiver, gave the poet a template. If Jupiter moon-observing Galileo ignored ecclesiastic authorities to peer directly into the "Book of Nature," announcing a beautifully changing universe out of step with human notions of celestial eternity, so Nabokov took cues from biological cunning to speculate on the *potustoronnost'* (Russian for otherworld) that his wife, Véra, suggested was of serious importance to him: "To myself I appear as an idol, a wizard, bird-headed, emerald gloved, dressed in tights made of bright blue scales."[7] We humans "are the caterpillars of angels."[8] It is possible that Véra missed some of Nabokov's literary subtlety here; the heroes of Ada, he says, "die into the book," a quasi-cabalistic trope repeated in the poem "Cubes": "Let's fold / our wings, my lofty angel."[9] Death may be, as Tolstoy wrote, "a kind of border," but one surprisingly close, not outside of space-time but only inches from our nose, in the otherworld of characters living again in a book.[10] When one reads, "the leopards of words, / the leaflike insects, the eye-spotted birds," one imagines that for Nabokov nature and art form a single, self-mirroring, self-mimicking substance.[11]

In some cases, Nabokov's reversals, superficially contradictions, are better viewed as perspectival shifts, specifically a shift from the linear temporal view to a(n infinitizing) view above or beyond time. Nabokov confesses that he doesn't believe in time.[12] But even from within time he invokes the figure of the spiral to disturb our conventional view. The spiral, he says, is Hegelian, a "spiritualized circle": the arc is a thesis, it curls around like an antithesis, and then reappears outside the original curve as the synthesis, moving in the same direction but with the experience of having been here before at another level.

Corresponding with the above figure, we might say that science trumps art, which trumps science. This is not so much illogic or contradiction as an implicit recognition of partial perspectives within time, which brings first one then the other to the fore. Similarly, on a sentence level, Nabokov indulges in "patterned antinomy"; saying something and disavowing it is not the same thing as not saying it.[13] Rather, it leaves a trace, an image or a semantic shadow. By averring and retracting something, uncertainty and indeterminacy are established, and the self-contradictory nature both of people, who change their minds, and nature, which changes depending on our perspective on it, is better modeled. Like Whitehead accepting infinity and the eternal as real parts of a single, human-containing world, Nabokov embodies the eternal, and nontraditional phenomenological modelings of "subjective" time, within fiction. It may also be that the occasional doubts Nabokov voiced concerning evolution was not because of cryptic creationism or scientific naïveté but reflected his skepticism of linear time, a skepticism voiced also by certain mystics (for example, Meister Eckhart, Cusanus) and scientists (for example, Kurt Gödel, Einstein).

One might say, in a Nabokovian register, that our world is neither natural nor supernatural but rather both: this quality I call Todorovian, after literary critic Tzvetan Todorov's category of the *fantastique,* which simultaneously admits of natural and supernatural explanations but is confined to readings of literature, not reality.[14] I was recently reading a defense of parapsychology that criticized a too-high standard of improbability being set for awarding a million-dollar prize for ESP (for, say, clairvoyance or telepathy, and not just luck), when I noted that the name of the critic of parapsychology debunking was an anagram (RADIN) of the skeptic (RANDI) offering the prize; I was relating this in my mind to the old SANTA/SATAN anagrammaticity when, scrolling down, I came across an Internet respondent making the exact same point of the adversaries' interanagrammaticity, and referencing, as if having read my mind, or foretelling just this future to which I was coming. I thus could attribute the improbability to mere chance or something more significant. This productive, unforeclosed confusion would be an example of the Todorovian. As life is not a repeatable experiment on the personal level, we do not seem to have the inductive prowess to distinguish between, say, rare occurrences and psychic powers, and perhaps the difference is even a reflection of our outlook.

Nabokov, despite his strong opinions and apodictic poses, seems to embrace—and produce—a similar undecidability and productive duplicity. For example in lines 647–61 of *Pale Fire,* Nabokov's John Shade lays to rest supernatural explanations of creaking and thuds coming wintrily from the shutters, dismissing the possibility of a dead daughter returning via a "keyboard of dry wood" (Ouija board). He accepts rather the secular modern reality of "death's abyss" (l. 647). In a live game of chess, his "knight is pinned," but a page later he casually reports that "one night I died" (ll. 661, 682). As Brian Boyd intimates, this initial disavowal of the transcendental works all the better to embrace potustoronnost', later on, upon another reading, at another level.[15] This ability to "hang" with multiple, even diametrically opposed ideas is, as argued above, a gift of philosophical discourse and the legacy of the novel. Plato's dialogues privilege the rational over the irrational; they develop a chorus-less conversational realism; they are like drama in their form but not mythical or so dramatic, taking place on the page rather than on the stage. But novels still allow for multiple and opposed points of view to be performed simultaneously. Their form, in other words, preserves undecidability, productive contradictions, curiosity, and a foregoing of conclusions that, no matter how strong the characters' or narrators' (or author's) opinions, is essentially scientific.

This undecidability is also a hallmark of quantum physics, where superpositionality contains, in a liminal state as it were, multiple possibilities that can be differently but validly construed upon measurement. From the novelist's point of view, reality depends on the character's perceptions. In my own case, with my mother having died less than a year before, I stood next to the recreational vehicle in the driveway of my son's house, where I was housesitting, when I noticed the improbable arrival of a red balloon, reminding me of the famous French movie *Le Ballon rouge*, with which I was smitten after having been brought to it by my mother as a child: Could I be certain, in the absence of absolute knowledge, that the buoyantly full balloon had arrived via natural causes or was not pushed by those subtle Nabokovian ghosts whom the author portrays as nearly but not quite impotent in affecting happenstances in our world? This is not science, of course, but an example of keeping open multiple possibilities for reality of the kind that science espouses. As stated, Todorov erected the category of the fantastique for effects the causation of which, either natural or supernatural, remains winkingly open.

THE MODELING OF PARTICULAR INDIVIDUALS in fiction is mirrored by the devotion and attention of particular scientists. Nabokov reasoned that the ancestors of the *Polyommatus* Blues navigated the Bering Strait across the north polar region to populate South America. Nabokov's reasoning was deeply informed by facts, but those facts were collected via the uncommon attention he devoted to finding, viewing, and chronicling lepidopteran minutiae (and reputedly ruining his eyesight along the way)—as the drawings collected in this volume attest. His cabinet at Harvard's Museum of Comparative Zoology lovingly housed the genitalia, complex and distinct under the microscope, from the male Blues he dissected.

With Nabokov's contribution to butterfly systematics one can never overemphasize the affective register, especially as it relates to his observational acumen. Recent research by neuropsychology shows the importance of emotional valence in attention, consciousness, and memory; how we have enhanced perceptual vividness for emotionally salient things.[16] This certainly applies to Nabokov's butterfly fetish–scientific prowess. Nabokov's biographical details, from the viewpoint of *then* contingent and stochastic, but from the viewpoint of *now* fateful and inevitable, honed to a laser focus the lens of his lepidopteroman[t?]ic fetish. The two things he loved most were not fame and sex, let alone shopping and sports, but "the most intense [pleasures] known to man: writing and butterfly hunting."[17]

And both for him were a reflection, dealing with "childhood memories of strange/

nacreous gleams beyond the adults' range," to quote John Shade, destroyed by a bullet meant for another in the-poem-within-the-commentary-within-the-book that is Nabokov's ghostly *Pale Fire* (ll. 633–34). The circumstances of Nabokov's upbringing made him uniquely attentive to butterflies fleeting by nature and thus a double challenge—classical, of the collector, and, for Nabokov, as Lolita-like objects of desire, whose elusiveness (a cipher perhaps for mysteriously passing time) was inversely proportional to the aesthetic reward of their capture in the net of his literature.

Intellectually, Nabokov's antitotalizing stance, privileging the individual (threatened by communism), reluctantly embraced its would-be opposite, the kitsch, the vulgar, the democratic common denominator of America, made unique, irreplaceable, and irresistible as the seduction of the "nymphet" (nymph also refers to the immature form of damselflies and certain other insects) Lolita by the culturally and the chronologically more mature European intellectual Humbert Humbert. This vengeance of the old and Old World and stamping of its impress on the new and New World might be interpreted as a simultaneous attraction-repulsion for the mob individualism and crass commercialism of America, which in real life provided Nabokov with sanctuary and a meritocratic ladder with which to rescale the social heights of which he and his family were deprived by the Russian Revolution when he was eighteen.

The experientially reinforced distaste for the general, even the democratization of science, is relevant to understanding Nabokov the scientist, since the essence of science is to not multiply causes, to simplify as much but no more than possible, to seek simple, impersonal principles (in the parsings, respectively, of William of Ockham, Albert Einstein, and Steven Weinberg). These run counter to the artist's taste for uniqueness, detail, perplexity. Despite his family's privilege, Nabokov's father was a liberal, and Nabokov himself an emulator and carrier on of the high literary tradition of Pushkin, whose maternal great-grandfather Abram Petrovich Gannibal (1696–1792) was a fully African former slave.[18] Like that bullfighter whose goring left him paralyzed for life with a permanent physical self-image in the splayed position of his bovine revenge, Nabokov's nostalgia was visceral. Butterflies for him became a kind of placeholder, beings real and alive but whose evanescence gave them the status of transitional objects after the loss of his family, including his seldom-mentioned, gay stutterer of a brother, Sergei (who died in a concentration camp—*mariposa*, butterfly, is vulgar Spanish slang for homosexual). Beautifully elusive butterflies were living objects whose markings, fragility, and unfoldings lent them the status of unique signs, primarily of themselves. While imprisoned, his father relayed a message for him to his mother: "Tell him there are no butterflies here in the prison yard except *rhamni* and *P. brassicae*. Have you found any *egerias*?"[19]

His concern with future memory, his father's imprisonment, his mother's admonition to "Now, remember" moments of evanescent natural beauty, his art tutor Mstislav Dobuzhinsky's emphasis to depict detail from memory, his filially encouraged love of Lepidoptera, exemplified by his bringing a butterfly to prison to show his father and last but not least, the botched assassination in Berlin which took that beloved father from him forever, depriving the family of money and making it impossible for him even to afford to go to his own mother's funeral later in Prague—all these things must have reinforced the preciousness of the moment, the particular, the unique, the personal, the irreducible, and the true over any generalized, one-size-fits-all, presumptuous notion of truth or taste, be it political, philosophical, or intellectual.[20]

IN HIS ESSAYS ON THE ORIGINS of modern science, Whitehead identifies as a crucial move what might seem surprising to some: not an embracing of theory and thought or a will to explain but what Whitehead identifies as science's distinctly "anti-intellectualist" strain.[21] Theoretical strength correlates, not with sophistication, but with simplicity, a kind of tragic determinacy or determination: for Whitehead the essence of modern science was latent in Greek tragedy, which unveiled in narrative "the relentless working of things." In order to apply this determinacy to nature as a productive explanatory schema, scientists had to observe what Whitehead called (taking his cue from a letter from William James to Henry James) the "irreducible and stubborn facts."[22]

Among such facts in biology are the fabulous and increasing evidence for the evolution of new species through unexpectedly rapid bacterial and interspecies unions, as well as genome rearrangements involving transposons and other elements in rapid response to environmental change.[23] Nabokov's novels, stories, and poems are rife with secret liaisons, incestuous unions, and intergenerational lust. However frowned on, such "inappropriate" desires leading to trysts seem to reflect a wider tendency in nature to join up with itself, relatively quickly, in sometimes surprisingly productive evolutionary ways. Indeed one could even adduce this closely observed tendency for real people to flout approved categories as a point in favor of Nabokov's skepticism of Neo-Darwinism, narrowly conceived as a gradual process of speciation by randomly accumulated mutations, as *the* means of evolution understood by his contemporaries. We now know that Darwinian evolution involves symbiogenesis, the evolution of new species, relatively suddenly, by the metabolic and then genetic merger of radically different organisms, whose initial attractions to one another may be for food or as hosts to infect. From the ennucleation experiments of Kwang Jeon, in which a new species of amoeba was seen to arise through infection, to the speciation of fruit flies due to possession of *Wolbachia* symbionts, dangerous liaisons take place not only in Nabokov's novels but in the book of life, where the progeny is strikingly nonfictional.[24] (Groundbreaking early symbiogenetic work, done in Nabokov's Russia, suggests an intellectual climate of broad evolutionary speculation that may have informed Nabokov's scientific attitudes.)[25]

George Gaylord Simpson, an authoritative proponent of Neo-Darwinism whom Nabokov read, emphasized (in part, as today, to distinguish it from the superficial fantasies of anthropomorphic creationism), the purposelessness of the Neo-Darwinian evolutionary mode: In the *Meaning of Evolution,* he writes, "Man is the result of a purposeless and natural process that did not have him in mind."[26] That may be largely true. However, insofar as organisms are attracted to one another, and are not above crossing species and higher-taxa lines to look for food and love, we may admit some purposiveness beyond blind chance and mechanical action at work in evolution's quirky panoramic history. All of us (and we may include lepidopterists with Lepidoptera here) come in part from "illicit" cross-kingdom, cross-"species" unions. And these were not gradual or casual relationships, but permanent genetic marriages, organisms taking up residence inside one another forever.

WHITEHEADIAN REALISM EXCLUDES neither organisms' inner sensibilities nor mathematicians' infinitizing imaginations from our account of the real world as revealed by science. Beautiful theories are not enough. Insular mathematics is no substitute for the shrewd eye, pattern-finding abilities, and sheer fact checking of the artistic builder and scientist observer.

As Blackwell points out, after a virtuoso passage in *Lolita* where Nabokov's Humbert imagines himself as a predatory spider lowering invisible strands of proprioceptive silk around the Haze house to provide subtle vibratory feedback as to his still-virginal charge's whereabouts—concluding from this exercise that she is gone—the "beautiful warm-colored prey" appears with a knock on his door to inform him, "Don't tell mother but I've eaten *all* your bacon."[27] In Blackwell's reading, and despite Nabokov's proscription against symbol searching in his work, the salty pink meat refers not just to itself but also to the undeconstructed proper name of the lawyer and philosopher of science: the sly pedophile's deductions, as Bacon underscored, are no match for the inductive evidence of fact, in this case Lolita's fleshly appearance.

Nabokov's success in retrodicting the evolution of the *Polyommatus* Blues arguably comes not so much from immersion in theory but from a wise reluctance to favor the evidence of his own eyes over then-current and fashionable explanations. His amateur status made butterfly watching, cataloging, capturing, and measuring their own rewards. An experienced litterateur well attuned to the writer-magician's tendency to be duped and surprised by the slow unveiling, or evolution if you will, of his own words, Nabokov had an inductive database that stretched from science into the aesthetic experiences of alpine butterfly hunting for its own sake and into the found teleologies of art. (For more on the relationship of probability to purpose, see Alexander, this volume.)

The discovery of evolutionarily fruitful interspecies and interkingdom encounters testifies to the specificity, variety, and even (from a Judeo-Christian viewpoint) perversity of nature. Nabokov was not aware of the new symbiogenetic, natural genetic engineering, and hybridiogenic research in evolution, but his ability to reserve judgment, to resist being buffeted about by the winds of academic orthodoxy and intellectual fashion, kept him receptive to multiple possibilities. One can interpret a comment of his such as that organic traits look "invented by some waggish artist precisely for the intelligent eyes of man"[28] to be not a serious hypothesis (say of a whimsical mega-Nabokovian creator) so much as a refusal to rush to conclusion in the face of nature's mysterious structures and aesthetic largesse.

John Shade, as he goes unconscious after a poetry reading in Nabokov's *Pale Fire,* describes "a system of cells interlinked within / Cells interlinked within cells interlinked / Within one stem" (ll. 704–6). This emphasis on cellular linking, and despite Nabokov's obsession with demarcating species, was an outstanding premonition of what biology is rapidly becoming. Today, symbiogenesis, hybridism, bacterial gene transfer, and other rapid evolutionary processes are increasingly becoming recognized.

In "The Century of Genius," Whitehead valorizes the artist who records and measures, dwelling in the things themselves, rather than theories that can too often become a form of latter-day scholasticism. It is this close attention of the artist—repeating the lesson of the art student, who discovers the added clarity and detail of drawing from life as opposed to depicting, say, a face as an abstract oval with stylized eyes from memory—that combines with the bold, lucid speculations to make modern science. For science, theory is necessary but not sufficient; for art, theory may not even be necessary. Whitehead credits Francis Bacon chiefly with the "explicit realisation of the antithesis between the deductive rationalism of the scholastics and the inductive observational methods of the moderns. . . . It was implicit in . . . Galileo [and other scientists of the seventeenth century, but] Bacon was one of the earliest . . . and also had the most direct apprehension of the full extent of the intellectual revolution which was in progress." Whitehead adds that the greatest anticipator of this revolutionary scientific view was "the artist Leonardo da Vinci, [who demonstrated] that the rise of naturalistic art was an important ingredient in the formation of our scientific mentality. Indeed, Leonardo was more completely a man of science than was Bacon. The practice of naturalistic art is more akin to the practice of physics, chemistry, and biology than is the practice of law."[29]

In Whitehead's comments we can identify the brilliant, aesthetically attuned, but ever skeptical and curious—and theoretically inconclusive—Nabokov as a genius not only of the artistic but of the scientific kind. But then, as he said, at the heights to which both the artist and the scientist can sometimes scale, the landscape is one and the same.

Notes

1. Nabokov, "World of Butterflies."
2. Nabokov, "Remarks," 75–76.
3. Whitehead, *Essays*, 91–92.
4. For, e.g., Darwin's interactions with orchids, see Hustak and Myers, "Involutionary Momentum."
5. *SO*, 78–79.
6. *SO*, 190.
7. Nabokov, *Selected Poems*, xxiii.
8. Nabokov, *Selected Poems*, xvii.
9. Nabokov, *Selected Poems*, 13.
10. Nabokov, *Selected Poems*, 61.
11. Nabokov, *Selected Poems*, 133.
12. *SM*, 125.
13. Rowe, *Nabokov*, 61–68.
14. Berger and Sagan, "Brief History," 40–43.
15. "Since 1979, when Véra Nabokov declared the beyond (*potustoronnost'*) Nabokov's 'main theme.'" Boyd, *Nabokov's "Pale Fire,"* 253.
16. Todd, "Affect-Biased Attention."
17. Cited in K. Butler, "Nabokov's Legacy."
18. Barnes, *Stolen Prince*; and see Nepomnyashchy et al., *Under the Sky*.
19. Cited in Grayson, *Vladimir Nabokov*, 11.
20. Vladimir Dmitrievich Nabokov, assassinated at a public meeting in Berlin, March 28, 1922, with bullets intended for Pavel Milyukov, wrote to his son less than twelve months before, in May 1921, a letter to Cambridge University, in which the elder VN signs off, "And lots of big kisses on the lips right in the middle of Cambridge." Grayson, *Vladimir Nabokov*, 11. This is a vivid testimony of his father's undying humor and love, playing with the different standards of public affection in stiff-upper-lipped Great Britain, and underscoring his allegiance to both his son and their vanishing culture.
21. Whitehead, *Science*, 15.
22. Whitehead, *Science*, 10.
23. See, e.g., Margulis and Sagan, *Acquiring Genomes*. Donald Williamson even theorizes the possibility of hybridogenesis (the origin of new species by cross-species unions) in the origin of butterflies (from onychophorans [velvet worms] and flying insects); see, e.g., Ryan, *Mystery of Metamorphosis*. J. Shapiro, *Evolution*.
24. Jeon, "Amoeba and x-Bacteria."
25. Khakhina, *Concepts of Symbiogenesis*; Sapp, *New Foundations*; Kozo-Polyansky, *Symbiogenesis*.
26. Simpson, *Meaning of Evolution*, 345.
27. Blackwell, "New or Little-Known Subtext," 51–55.
28. Nabokov, *The Gift*, 108.
29. Whitehead, *Anthology*, 42–43.

Mountains of Detail

On the Trail with Nabokov's Blues

LAUREN K. LUCAS, MATTHEW L. FORISTER,
JAMES A. FORDYCE, AND CHRIS C. NICE

Walking in the footsteps of a master is always a humbling experience in science. The humility is doubled when the master is best known for work in a field far removed from one's own. We have walked in Vladimir Nabokov's footsteps, in ways both literal and conceptual. Here we describe the travels we have taken, sometimes following trails that Nabokov walked in the mountains of the West and sometimes following his intuition regarding hypotheses of hybridization or species boundaries. Our findings have often confirmed what Nabokov inferred or suspected, though we have access to an ocean of information that no one in his time could have imagined, such as detailed genomic data from wild-caught animals. It is interesting to ponder what Nabokov, the master of detail in fiction and in butterflies, would have made of the variation that we can now quantify. Before attempting that speculation, we (humbly) note some convergence in our travels and general approaches to the study of butterflies.

In many ways, our habits in the field have converged to those of Nabokov. While in the United States, he spent several summers at high elevation, particularly in the Rocky Mountains, because of the diversity of alpine butterflies and because the vegetation reminded him of his childhood expeditions in Russia.[1] He was often chasing the hint of an idea that some locality might contain intergrades between species, but he also felt that the pursuit of montane butterflies was both a restful and a productive time for reflection on literature and science. He had broad interests in butterfly diversity, but his focus seems most often to have been little blue butterflies in the genus *Lycaeides*. We have spent over a decade of summers in the high elevations of the West studying *Lycaeides* butterflies, primarily investigating the causes and consequences of hybridization. Nabokov wore shorts in the field, as we do, but at a time when grown men rarely did so.[2] He became accustomed to drawing "morbid interest" from passing cars, "broad-minded vacationists," and horses while in the field, as we have.[3] He killed his butterflies, as we do, in the European, or Continental fashion, by pinching the thorax and then storing them, as we do, in glassine envelopes in a Band-Aid box in his shirt pocket.[4] While chasing butterflies in the Swiss Alps in 1975, Nabokov suffered a severe fall down a steep slope, which began a two-year decline in health from which he never recovered. We hope not to suffer the same fate but would not be terribly surprised if it worked out that way. The love of the alpine comes with some peril.

Fortunately for us, Nabokov left detailed records of his expeditions. For example, Nabokov's descriptions led us to a hybrid population of *Lycaeides* on Blacktail Butte in Wyoming (Color Plate E2). Nabokov wrote that he "had not solved the problem of the *L. melissa* strain so prominent in some colonies of *L. argyrognomon longinus* (i.e. Black Tail Butte near Jackson). [He] had conjectured that hybridization occurs or had occurred with wandering low elevation *L. melissa* that follows alfalfa along roads as *Plebeius saepiolus* does clover." Over the past few summers we have searched for other hybrid populations in the vicinity of Blacktail Butte, as far south as "the altogether enchanting little town of Afton" Nabokov described.[5] Similarly, we followed his description of the path to a population of *sublivens*, a disjunct population of *L. idas* whose genitalia have morphology intermediate between *L. melissa* and *L. idas*, fifty-eight years after he wrote about the location. Indeed, we found the putative hybrid *Lycaeides* population above Telluride, Colorado.[6] We were able to add yet another putative hybrid *Lycaeides* population to our ongoing study of hybridization during the summer of 2012 by looking for "clumps of Douglas fir, ant-heaps, and an abundant growth of *Lupinus parviflorus*" near Alta, Utah (Nabokov's description of the habitat of *L. m. annetta*).[7] We have seen Nabokov's collections at Cornell and Lausanne and were impressed by the great range of his travel. It is humbling (to say the least) that all of these expeditions were made, along with detailed field notes, while some of the masterpieces of twentieth-century fiction were being conceived and written in a language that was not Nabokov's native tongue.

UNDER THE MICROSCOPE, we have similarly followed Nabokov's lead. During his time spent in the United States, Nabokov was a taxonomist who categorized specimens based on precisely observed and measured morphology. In 1940, William P. Comstock, a fellow lycaenid butterfly taxonomist at the American Museum of Natural History, taught him the fundamentals of genitalic dissection, which involved boiling the abdomen in a strong base to loosen and dissolve the soft tissues and removing the sclerotized structure of the male genitalia.[8] Nabokov spent much time observing variation among butterfly genitalia and considered the taxonomic observations made by others before deciding that the internal clasping structure of the *Lycaeides* genitalia, the "uncus," was ideal for differentiating species in the genus. His descriptions of the morphology of the uncus could be understood by both the scientist and the layperson. "The male armature consists of a dorsal (in regard to the body) portion (the uncus) and of a ventral one (the valves). The two are hinged to each other somewhat in the way of the lids of a shell and appear 'closed' when viewed in situ. . . . [The] most conspicuous thing about the upper portion is the presence of a pair of formidable semi-translucent hooks (the subunci or falces—of a peculiar shape not found in allied genera), . . . [that emerge] from the opposite side of the distally twinned uncus and . . . [face] each other in the manner of the stolidly raised fists of two pugilists (of the old school) with the uncus hoods lending a Ku-Klux Klan touch to the picture" (refer to Color Plate E3 for a labeled photograph).[9] Once dissected, he would make morphological sketches (for example, Figure 39) and measure four particular features of the uncus, features he named F, H, E, and U: "F for the length of the upright portion, or *forearm*, of the falx measured from its distal point to the apex of its elbow; H for the length of the *humerulus* of the falx, from the apex of its elbow to the apex of its shoulder; [E for the width of the elbow;] and U for the length of the *uncus lobe* from its distal point to the apex of the shoulder of the falx" (see Color Plate E3).[10] He recorded his measurements for each collected series on index cards (Figure 92). We admire Nabokov's accurate drawings and the time and skill required to produce them. Today we use the same dissection technique as Nabokov but need relatively little artistic skill to visualize *Lycaeides* morphological variation. We use a camera embedded in a microscope to image *Lycaeides* genitalia and computer software to make the same four measurements of the "fists" and the "hoods." Nabokov dissected and measured about 1,500 specimens over seven years.[11] By comparison, we have dissected approximately 1,900 genitalia during our fourteen years of *Lycaeides* research. We have made sure to give our eyes a break among bouts of microscope work to try to avoid the eyestrain from which Nabokov suffered during his time at the microscope.[12] We have certainly learned from Nabokov in more ways than one.

In following Nabokov's lead, both in the field and under the microscope, we have come to many of the same conclusions while using techniques that would have seemed like science fiction in the middle of the twentieth century. Our conclusions are discussed further below, but first it is important to understand the intellectual context in which Nabokov's scientific interests were pursued. What preparation allowed a prominent literary figure to write papers on butterfly morphology that would still be read almost four decades after his death? As Robert Dirig and Victor Fet describe in their essays in this volume, he had a lifelong interest in butterflies and aspired from an early age to discover and name a new species. He was tutored at home, where his father passed on to him the basics of lepidoptery and natural history of the local fauna. From a young age he read field guides and technical literature, and he started drafting manuscripts for publication at age seven. At Cambridge University, he had only brief exposure to zoology and none to entomology.[13] Whenever intellectuals launch themselves from one field to another, there is a temptation to speculate on the effect of the "outsider" status on accomplishments made in the new field, and we find it impossible to resist that temptation here. With little formal training in his pocket, he took on an exceedingly ambitious project, a taxonomic revision of a large group of butterflies, a group, moreover, that was known to be taxonomically complex. It is interesting to ask why he felt that he could be so ambitious. Perhaps it was simply because he had confidence in whatever the field of endeavor. Or it might be that he was catching up for lost time, having dreamed of taxonomic work for so many years. It might also be the case that his experience as a novelist (and not just any novelist but the author of notoriously complex works) gave him a certain appetite for complexity that was undaunted by a taxonomically "messy" group. Perhaps he craved the challenge.

With respect to theoretical issues in evolution, it seems he had a distrust for the apparent simplicity of natural selection as a mechanism generating butterfly mimicry and perhaps the biological diversity around us (for example, as expressed fictitiously in "Father's Butterflies"). The degree to which Nabokov was familiar with the development of population genetics theory in the early 1900s and the modern synthesis between 1936 and 1947 is unknown. His views on species concepts or species boundaries also are not well documented. We do know that he was skeptical of the biological species concept (see Introduction), a species definition in which geographically separated forms are considered as part of the same biological species if they are potentially able to interbreed. Perhaps the apparent simplicity of that perspective did not comport with his view of the world; how could the complexity of closely related butterflies, with differences barely discernible to the naked eye, be captured by the same "rules" that define larger and cruder organisms, such as birds or mammals? Or perhaps he actively favored Darwin's definition of species, which allowed some intergradation and relied on quantifying morphological complexities.[14] If Nabokov had had more formal education in population genetics, it is possible that he would have seen things differently. As it was, his appreciation of the complexity of the world must have, at least in part, led him to take on such an ambitious project, and his outsider status may have left him more latitude to be open-minded about hybridization between species.

Nabokov made surprisingly insightful conclusions by paying attention to the messy details embedded in the New World polyommatine butterflies. From 1942 to 1948 he curated the butterfly collections of the Harvard Museum of Comparative Zoology, during which time he wrote four key papers in which he rearranged the classification within the genus *Lycaeides* based on male genitalic morphology and male and female wing patterns. Nabokov met his childhood goal of naming newly discovered taxa. Most notably, he was the namer of the Karner Blue (*L. melissa samuelis*). He pulled together genitalic morphology (Figure 40), wing morphology (Figure 41, Color Plates 31 and 48), as well as biogeographical, historical, and habitat information to determine that the name *scudderi* (by which the Karner Blue had previously been known) properly applied to a subspecies of *idas* in Manitoba, not the *L. melissa* in northeastern United States, leaving the northeastern *melissa* without a valid scientific name. Nabokov filled the nominal opening with *samuelis*, after Samuel Scudder, a famous entomologist for whom the Karner Blue populations had originally been named (a historical and taxonomic reversal of fate that must have appealed to Nabokov's love of word-play). He noted that they were "sharply cut off from the western bulk of melissa" and found only "in isolated colonies, and only in association with lupine (thus on sandy soil)."[15] He further alluded to their looming endangered status when he wrote, "owing to various causes (building, farming, fires, etc.) old colonies die out, while new ones founded by wandering females in quest of lupine, may not always thrive beyond one season."[16] He subsequently and informally revised his opinion to consider it a distinct species.[17] The Karner Blue was declared federally endangered in 1992.[18]

Nabokov named other *Lycaeides* subspecies: *L. argyrognomon longinus*, *L. argyrognomon sublivens*, *L. melissa pseudosamuelis*, and *L. melissa inyoensis*. Three of these groups were a challenge to classify because they contained transitional forms. He said himself, "While studying the nearctic organs, I have come across an extraordinary case not easily paralleled in the annals of speciation."[19] *Lycaeides melissa* and *L. idas* are sympatric in British Colombia and the western United States (California, Oregon, Washington, Idaho, Montana, and Wyoming). He made his rules of classification clear; in sympatry, where there is no apparent interbreeding or morphological merging, there are "absolute specific distinctions" between species, but in cases "when there is spatial contact between two different forms at the limits of their distribution, with some morphological merging there, we have either relative subspecific distinction between the two or some minor racial distinction not requiring a quadrinomial designation."[20] In addition, a subspecies had to be separable from any other intraspecific form by at least two morphological characters, one of which had to be a male genitalic character.[21] Thus he described *L. argyrognomon longinus* from northwestern Wyoming, an intermediate form of *Lycaeides* between *L. argyrognomon* and *L. melissa* (notice the intermediate shape of Figures 35–37 and 39) and noted those specimens forming transitions to *L. m. melissa,* such as those we have seen from Blacktail Butte, Wyoming.[22] Similarly, he gave new subspecific names to two intermediate forms in Colorado, *L. a. sublivens* and *L. m. pseudosamuelis* but was honest about the morphological complexity here, "a resemblance which suggests further investigation (it is not unlikely that a state of affairs similar to the Jackson Hole tangle may be discovered in S. Colo.)"[23] Again we are reminded that Nabokov disliked the impulse to reduce the complexity of the natural world. Between 1948 and 1959 Nabokov continued to collect specimens in the West with his wife, Véra, and son, Dmitri, during most summers. We wish to know how Nabokov thought these collected specimens fit into the *Lycaeides* puzzle, but apart from short pieces, he published nothing on *Lycaeides* butterflies after 1949.

Despite Nabokov's extensive fieldwork and opportunities for observation of *Lycaeides*, few of his descriptions of taxa and revisions of the taxonomy are based on characters other than morphology. Except for the Karner Blue delineation, he made no use of ecological characters—habitat characteristics, host-plant associations, life-history variation—in his classification of *Lycaeides*. Nabokov also has been criticized for ignoring developments in statistics from the late nineteenth and early twentieth centuries. F. Martin Brown pointed this out in his commentary on Nabokov's 1949 monograph on *Lycaeides* taxonomy.[24] Brown noted that even though Nabokov often, though irregularly, reported his measurement data, he made no attempt to compare the various measurements in a statistical way. Thus, although he realized the potential for detailed and careful measurements to distinguish taxonomic units, Nabokov never took advantage of the tools that would have put his morphological taxonomy on a solid statistical footing. What is particularly interesting to the modern scientific reader is Nabokov's response in which he suggests that natural science need not answer to statistics: "After all, natural science is responsible to philosophy—not to statistics."[25] The absence of statistical analysis in Nabokov's work and the neglect of ecological data in defining taxonomical units prompted us, in part, to use statistical analyses to evaluate his taxonomic designations as well as his claim that *Lycaeides* species boundaries were not discrete.[26] We measured F, H, and U of the male uncus of approximately nine hundred North American *Lycaeides* and assigned them to species and subspecies based on wing pattern and geography and used a multivariate statistical analysis (Canonical Discriminant Analysis) to assess the reliability of Nabokov's and more recent taxonomic delineations. Our statistical analyses validated Nabokov's morphologically based species designations; approximately 97 percent of the specimens were correctly classified to Nabokov's recognized species, *L. melissa* and *L. idas*. The Canonical Discriminant Analysis correctly classified only half of the individuals to the correct nominal subspecies, but a number of the misclassifications were hybrid individuals or subspecies he did not describe or recognize (for example, *L. i. azureus*). However some of his recognized subspecies were much less distinguishable, particularly *L. i. alaskensis*, with only 16.7 percent correctly classified. In the end, we agree with Robert Robbins, curator of butterflies at the Smithsonian Institution and a lycaenid specialist, when he stated: "Nabokov did a very nice job—as good or better than most of the professionals of his time," especially given that his decisions rested solely on his eye for subtle morphological differences.[28]

WHEREAS NABOKOV SPREAD AND PINNED his butterflies upon return from the field, we store our collections in −80° C freezers to preserve the specimens' DNA for extraction and genetic sequencing. We have collected genetic data from hundreds of *Lycaeides* across North America to ask a number of questions, including: (1) Is genetic variation congruent with morphology?[28] (2) Are patterns of DNA sequence variation consistent with the hypothesis that fragmentation during the Pleistocene had significant impacts on processes of differentiation within the genus?[29] (3) Are alpine populations of *Lycaeides* in the Sierra Nevada homoploid hybrid species?[30] (4) Has hybridization between Rocky Mountain *L. idas* and *L. melissa* led to the formation of a hybrid zone or the establishment of an isolated hybrid lineage?[31] (5) Has there been a history of gene flow between the Karner Blue and *L. melissa*?[32] and (6) Do differences in male genitalic morphology constitute a barrier to gene flow between *L. idas* and *L. melissa*?[33] The answers to these questions about the evolutionary history of North American *Lycaeides* have shed further light on Nabokov's taxonomic designations and hypotheses about hybridization in the genus. We have helped to confirm Nabokov's convictions made fifty-seven years earlier about hybridization at Blacktail Butte. In fact, *L. melissa* and *L. idas* have hybridized extensively in the Jackson Hole valley, leading to the formation of a fairly wide zone of hybridization.[34] We have estimated the genotypes of thousands of nucleotides (loci) in *Lycaeides* genomes from hybrid and putative parental populations (for a visualization of example data, see Color Plate E4). Nabokov presumably would have been interested to know that a greater proportion of the Jackson Hole hybrids' genome was inherited from *L. idas* than from *L. melissa*, indicating that *L. idas* alleles were favored because *L. idas* and Jackson Hole *Lycaeides* populations occupy similar habitat and feed on the same larval host plant.[35] As further evidence of his powers of observation despite the statistical vacuum in which he worked, Nabokov correctly assigned Jackson Hole *Lycaeides* (subspecies *longinus*) to *L. idas*, despite their male genitalia being more *melissa*-like.

Nabokov was a lumper when it came to the range of morphological variation he included in the subspecies *L. m. melissa*. Whereas he recognized specimens from Gold Lake and Mammoth Lake, California, as "striking local races embossed as it were on its rather monotonous morphological texture" and created other subspecies based on similar criteria, he continued to recognize them as *L. m. melissa*.[36] We used male genitalic morphology and genomic data to classify these populations and other high alpine *Lycaeides* of the Sierra Nevada as a homoploid hybrid species.[37] A homoploid hybrid species is one in which hybridization between "parental" species produces a new lineage (the hybrid species) without a change in the number of chromosomes, which is distinct from the similar process frequently observed in plants that leads to changes in the number of chromosomes. We left this homoploid hybrid species unnamed in our demonstration of its hybrid origin (some, however, refer to it as *L. m. fridayi*). This group of butterflies is known for its adaptive radiation, and a number of new lineages have arisen relatively recently. We currently are navigating the delicate balance between oversubdivision and naming taxa with obviously distinct variation.

IN ADDITION TO ADDRESSING SPECIES DELINEATIONS and hybridization, we have painted a more complete picture of the evolutionary history of the Karner Blue than Nabokov was able to accomplish with the tools he had available in the mid-1900s. Recently, we used both genomic information (hundreds of loci, or genetic regions) and sophisticated statistical and computational methods (necessitating weeks of computer time) in addition to information about the Karner Blue's host plant association, morphology and life-history variation. We found levels of gene flow between the Karner Blue and *L. melissa* to be comparable to levels of gene flow between *L. melissa* and *L. idas*, which are recognized as distinct species. Rates of gene flow between the Karner Blue and *L. melissa* are also lower than rates detected between populations of *L. melissa*.[38] Nabokov apparently regretted his original description of the Karner Blue as a subspecies rather than a full species. We are confident that he would have appreciated our conclusion that the current subspecific designation for the Karner Blue is inappropriate and that it should be elevated to *Lycaeides samuelis* Nabokov.

Other researchers like Naomi Pierce and her collaborators have spent part of their careers using modern methods such as DNA sequencing to revisit Nabokov's phylogenetic hypothesis of New World Blues. Nabokov's phylogenetic hypothesis is one of unidirectional and parallel change toward larger or more robust forms in both the Old and New Worlds from a smaller ancestral state. In pursuit of this hypothesis, Nabokov inferred the ancestral state and even claimed to have observed the ancestral state in *Paralycaeides* from South America.[39] This exercise is difficult to interpret, since there is a notable absence of formalized methods in Nabokov's descriptions of the process. Nor can this be translated into modern methods. However, even with this fuzziness, Nabokov's ideas of parallel morphological evolution bear an eerie resemblance to recent molecular phylogenetic work that shows seemingly parallel evolution of lineages of *Lycaeides* in the Old and New Worlds.[40]

We have found cases in which Nabokov's taxonomic classifications have not stood the test of time, but how was he right so frequently? One part of the answer involves his choice of focusing his taxonomic decisions on the male genitalic morphology. The use of male genitalic morphology in insect systematics has a long history. Nearly a century before Nabokov published his taxonomic papers on *Lycaeides*, naturalists had recognized that variation in male insect morphology tended to be quite species specific and therefore highly useful in delineating species boundaries. A functional interpretation was quickly added: the lock-and-key hypothesis, which suggests that male and female genitalia fit together in species-specific ways and evolve to reinforce species boundaries.[41] This notion of "mechanical isolation" is observable in the modern synthesis, for example in Theodosius Dobzhansky's description of isolating mechanisms.[42] However the lock-and-key hypothesis of mechanical isolation does not always predict hybridization between differentiated taxa. For example, Adam H. Porter and Arthur M. Shapiro rejected the lock-and-key model for pierid butterflies involved in a hybrid zone in Chile based on lab pairing and prevalent natural hybridization.[43] Nabokov was skeptical of the lock-and-key model to describe variation in the Blues but admitted that he may not have found a correlation between male and female genitalic structures because female genitalia are difficult to observe (they are not sclerotized like male genitalia). Based on inferences made from relations between genetic distance and phenotypic divergence and patterns of variation in male genitalia in admixed *Lycaeides*, male genitalic structure contributes to isolation between *L. idas* and *L. melissa*.[44] In this sense, Nabokov, in part, made so many accurate conclusions about *Lycaeides* because he chose the right morphological character.

We have discussed thus far that Nabokov made a number of sound conclusions about *Lycaeides* taxonomy because he started his lepidopterist endeavors at a young age, he collaborated with entomologists at the American museums in which he volunteered, and he read work by other lepidopterists. In addition, his lack of formal advanced education in the biological sciences may have led to open-mindedness about hybridization among species, and he may have been somewhat lucky in choosing to focus his taxonomic decisions on male genitalic morphology, a reproductive isolating trait. Yet it is also possible that all of these facets of Nabokov as a scientist pale in comparison to the fact that he had great insights owing to his passionate attention to detail. He stated, "The thrill of gaining information about certain structural mysteries in these butterflies is perhaps more pleasurable than any literary achievement."[45] Nabokov was undaunted by the task of dissecting hundreds of tiny genitalic structures or counting scale rows on wings; he welcomed the challenge. We can imagine Nabokov smiling over "the thrill" of gaining access to the genomes of each *Lycaeides* species and specimen, with all of the truly formidable complexity that entails.

We can picture him nodding in approval as we comb through millions of DNA sequences from each wild-caught individual and use computational methods to test hypotheses about levels of differentiation among populations. Furthermore, his talents in both science and art reinforced each other. He was able to reproduce accurate images of the morphology he saw under the microscope (for example, Figures 32, 35). And those butterflies helped him in his writing. "Very often when I go and there are no butterflies, I am thinking. I wrote most of *Lolita* this way. I wrote it in motels or parked cars."[46] As in many other ways, we have also been inspired by Nabokov's artistic sketches of whole butterflies to include sketches of butterflies in our own publications.[47] Last, his "extreme achievement of memory" may have helped him piece together all the places in which he collected and specimens he observed.

We continue to follow humbly in Nabokov's footsteps. We are currently working on comparing genomic data from multiple hybrid populations to which Nabokov led us in recent summers to shed light on the repeatability of evolution and gain a better understanding of the evolutionary processes that have been associated with repeated hybrid speciation in *Lycaeides*. Also, we are delving into the details Nabokov described about *Lycaeides* wing patterns ultimately to understand more about the genetics and evolution of butterfly wing patterns. In the process, we have poured over his 1944 monograph, "Notes on the Morphology of the Genus *Lycaeides* (Lycaenidae, Lepidoptera)," in which he "enjoyed playing with his verbal descriptions as much as doing the science."[48] We will test his thoughts (Figure 41) and the thoughts of his contemporaries, such as Boris Schwanwitsch, regarding wing pattern development and will pay close attention to what he calls the "critical cell" of the hind wing, in which black spot "Cu1 has proved to be most valuable in giving as it were a summary of the main variational characters in a race" (Color Plate 48).[49] It is clear that we remember Nabokov often—when we use his morphological observations to test new hypotheses about the evolution of *Lycaeides* and during the "highest enjoyment of timelessness . . . when [we] stand among rare butterflies and their food plants" along one of his trails in the mountains.[50]

Notes

1. Boyd, "Nabokov, Literature," 28.
2. Pyle, "Between Climb and Cloud," 52.
3. *SM*, 131.
4. Boyle, "Absence of Wood Nymphs."
5. Nabokov, "Butterfly Collecting," *NB*, 492, 490.
6. *NB,* 501–2.
7. Nabokov, "Nearctic Members," 535.
8. Boyd, "Nabokov, Literature," 40–41.
9. Nabokov, "Notes on the Morphology," 108.
10. *NB*, 322.
11. Blackwell, *Quill*, 22.
12. Boyle, "Absence of Wood Nymphs," *NB*, 536.
13. Pyle, "Between Climb and Cloud," 21.
14. Mallet, "Species Definition."
15. Nabokov, "Nearctic Members," 540.
16. Nabokov, "Nearctic Members," 540.
17. Nabokov, "Novelist as Lepidopterist"; Nabokov to Robert Dirig in *NB*, 713–14.
18. *Federal Register* 57 (240): 59236–44.
19. Nabokov, "Nearctic Members," 482.
20. Nabokov, "Nearctic Members," 480.
21. Nabokov, "Nearctic Members," 480.
22. Nabokov referred to *L. idas* as *L. argyrognomon*, but see Higgins, "Correct Name"; Nabokov, "Nearctic Members."
23. Nabokov, "Nearctic Members."
24. Brown, "Measurements and Lepidoptera."
25. Nabokov, "Remarks."
26. Lucas, Fordyce, and Nice, "Patterns of Genitalic Morphology."
27. Pyle, "Between Climb and Cloud."
28. Nice et al., "History and Geography."
29. Nice et al., "History and Geography."

30. Gompert et al., "Homoploid Hybrid Speciation."

31. Gompert et al., "Secondary Contact."

32. Forister et al., "After Sixty Years."

33. Gompert et al., "Geographically Multifarious Phenotypic Divergence."

34. Gompert et al., "Secondary Contact."

35. Gompert et al., "Genomic Regions."

36. Nabokov, "Nearctic Forms."

37. Gompert et al., "Homoploid Hybrid Speciation"; Lucas, Fordyce, and Nice, "Patterns of Genitalic Morphology."

38. Forister et al., "After Sixty Years."

39. Nabokov, "Notes on Neotropical *Plebejinae*."

40. Talavera et al., "Establishing Criteria."

41. Dufour, "Anatomie générale."

42. Dobzhansky, *Genetics.*

43. Porter and Shapiro, "Lock-and-Key."

44. Gompert et al., "Geographically Multifarious Phenotypic Divergence."

45. Boyle, "Absence of Wood Nypmhs," *NB,* 530.

46. Boyle, "Absence of Wood Nypmhs," *NB,* 536.

47. For example, Forister and Scholl, "Use of an Exotic."

48. Robert Robbins, pers. comm.

49. Nabokov, "Nearctic Members," 483.

50. *SM,* 139.

Nabokov's Morphology

An Experiment in Appropriated Terminology

STEPHEN H. BLACKWELL

The pleasures and rewards of literary inspiration are nothing beside the rapture of discovering a new organ under the microscope or an undescribed species on a mountainside in Iran or Peru.

—Vladimir Nabokov, interview for the *Paris Review*, 1967

Asked by an interviewer about C. P. Snow's *Two Cultures* debate, Nabokov said, "I welcome the free interchange of terminology from any branch of science and any raceme of art." With this bold sweep, he authorized any appropriation of scientific vocabulary by the student of literature. Two sentences later, however, he quipped: "Aphoristicism is a sign of arteriosclerosis." In short, such terminological exchanges do not come without risks—not least the risk of becoming ridiculous. Nevertheless, morphology has a heritage that makes it a perfect fit for such an experiment.[1]

As a taxonomist, Nabokov was, more than anything else, a morphologist: a scientist interested in the exact shapes that make up an organism. Wing patterns, body form, and the shapes of various external and internal structures—especially the genitalia: these were the features he used to discern which butterflies were more closely or more distantly related. Evolution means the evolution of species, but it also implies, and is based on, the evolution of the forms, large and small, that make up an organism. The study of morphological variety allows scientists to imagine the course of evolution and to understand what makes a species distinct from its relatives. Nabokov once said that literary inspiration is "nothing beside the rapture" of morphological discovery, but he also—at the peak of his scientific work—considered the quest for scientific knowledge to be a "very apt" example of what transpires in the study of literature. These statements suggest that rapture—he also called it "inestimable happiness"—is to be found in the discovery of new shapes, forms, features—in short, new "organs" of fiction, "mysterious fields" of their interrelationships. The seeking of such forms is what drives science forward, and it is also, by extension, what drives the progress of human aesthetic activity. This essay sketches, in brief, what a morphological approach might bring to the study of Nabokov's art.

In his research articles, Nabokov wrote of the "all-important morphological moment."[2] Writing to Edmund Wilson in 1945, he criticized his friend for emphasizing sociology over morphology in evaluating and understanding literary art (echoing the Russian critic Iulii Aikhenvald, with a scientific twist): "In other words two butterfly populations may breed in vastly different environments—one, say, in the Mexican desert and the other in a Canadian bog—and still belong to the same species. Similarly, I do not give a hoot whether a writer is writing about China or Egypt, or either of the two Georgias,—what interests me is his book. The Chinese or Georgian features are intraspecific ones. What you want me to do is to give superiority to ecology over morphology."[3] This direct analogy between scientific and literary activity suggests that in his artworks, as well, Nabokov was guided by a minute attention to form—only here it was from the creative, rather than analytical, point of view.

The term "morphology" was coined by another literary scientist—Goethe—and it was apparently first recruited for the study of literature by Andrei Bely, a Russian poet, novelist, and theorist whom Nabokov revered as one of the four greatest literary artists of the twentieth century.[4] Nabokov's empirical focus on form, shape, and metamorphosis over time, transferred into the world of aesthetic activity, challenges his readers to consider whether, and how, he might have thought about formal structures as meaningful components of literary art. The critical movement known as Russian Formalism, precursor to much of the West's literary scholarship in the twentieth century, was a mode of study that strove to apply empirical tools to art. Although Nabokov was not part of that early proto-structuralist movement, he was aware of it. Nevertheless, we should consider his approach to morphological properties to be essentially independent of any existing scholarly trends, even when he is in conversation with them.

There are a variety of ways to apply morphological vocabulary to art, and to literature in particular. When Vladimir Propp published his *Morphology of the Folk Tale* in 1928, he broke a huge corpus of traditional narrative into a set of plot units from which folktales consistently selected within a strict sequential framework. Thus, there was a morphological template for *all* folktales, and then a given folktale would have a morphology, defined by which of the genre traits it deployed. This method is comparable to the way Nabokov thought about taxonomic groups when he wrote of "omissions, gaps, fusions, and syncopatic jerks" creating the signature "rhythm" of a species or genus.[5] This is macro-level morphology, where traits are simply enumerated but not studied deeply in themselves or as specifically arranged subsets of variations. Similar approaches to more traditional narrative began around the same time in the practice of other scholars, such as Viktor Shklovsky in "Art as Device" and "Sterne's *Tristram Shandy* and the Theory of the Novel," or Boris Eikhenbaum's "How Gogol's 'The Overcoat' Is Made" and *Young Tolstoy,* and so on.

Like these writers, Nabokov was interested in the building blocks of stories—interested both as a craftsman and as a passionate student of the world around him. We can tell from his novels which areas of structure were receiving his greatest attention at a given time: from early on, and certainly from his second novel, *King, Queen, Knave,* it is clear that pattern was a structural concern of dominant interest for Nabokov. Weaving various significant patterns through his works, in order to encode vital meanings that were not part of the surface narrative, was one of his most characteristic—perhaps even signature—practices. These patterns occur in natural repetitions (butterflies, circus decorations, dogs, trees), in hidden literary allusions (notoriously ubiquitous), in sound features (emphasis on certain verbal textures, rhymes, assonances, and the like), and in typographical features (anagrams, acrostics, cryptograms, typos, misspellings, italics). By deploying and continuing to discover new methods of encoding pattern, Nabokov infused his works with a morphological complexity that came closer and closer to matching that of the natural world.

SIMILARLY, THE STRUCTURE AND MORPHOLOGY of the narrator of a fictional text was clearly a constant source of fascination and creative variation for Nabokov. Beginning roughly with *The Eye* in 1931, Nabokov's novels include ever more intricate elaborations on the implicit identity of the narrator. Corollary to this line of morphological work is his exploration of the reader's perception of the narrator and overlap of this structural concept with the concept of "author"—specifically, the implied author but also the public figure of the author in the world beyond a text. (The private figure of the author remains a hidden theme, too.) In the late Russian novel *The Gift* and in its successors *The Real Life of Sebastian Knight, Bend Sinister, Lolita,* and *Pnin,* a series culminating with *Pale Fire,* one finds increasing experimentation with and elaboration of troubled boundaries for these textual storytelling voices. In these novels, narrators are not simply unreliable: they tempt readers with a false assurance of identity—*who they are*—and then, subtly, erode that assurance to a point of radical doubt and uncertainty.

Authenticity and sincerity are fraught concepts in modern human life. For Nabokov, fiction's narrators were a place to examine and play out the ways that individuals experience the voices of others who are telling stories, describing events, or attempting to persuade an audience of a perspective. Nabokov's narrators offer a study of the contingent and often specious authority of anyone telling a story or speaking about the world in any way at all—including scientists. John Ray, Jr., PhD is the "author" and narrator of the fictitious Preface in *Lolita* (a fact that is obvious but is sometimes missed even by acute readers); he attempts to convey his own professional authority as a psychologist who simply passes on a case study, but readers certainly doubt the legitimacy of his comments on Humbert Humbert's text. Humbert himself is explicit about his deceptive habits: he enjoys entrapping psychotherapists with fake oedipal dreams. Yet he constantly parades his sincerity, his desire to be perceived as sincere, even as he simultaneously undermines it with descriptions that are clearly biased and even cruel, though glossed with a veneer of sophistication and aestheticism that tricks most readers, for a while, into sympathizing with his controlling, misogynistic perspective.

In a still more extreme experiment, Nabokov's *Pale Fire* has an inner and an outer narrator, and the outer one, Charles Kinbote, is either mad, delusional, someone else's imaginative or delusive projection, or the deposed King of Zembla. And choosing among these options is not a simple task, either for scholars or for creative and passionate readers. By developing these variations on narrative ambiguity, Nabokov compels his reader to consider the relationship of each human being to a world that is saturated from all directions by the words and stories of all sorts of speakers: some near, dear, and authentic, others more distant and authoritative—but all deeply fallible. How are we to orient ourselves toward these words and the alleged truths that they urge upon us?

NABOKOV'S NOVELS FREQUENTLY EMBODY inner fragmentation, or division into distinct component parts. Evolving from conceptions of narrative that rely on unity (for example, the Aristotelian unities of space, time, and action; chapters, or plot elements, like complication, dénouement, and climax), component parts of a literary work grew to encompass the play within a play, story within a story, inserted texts such as letters, poems, and diaries, and stream of consciousness. In Nabokov, these inner segmentations push into new territory: Nabokov's works include biographies (*The Gift, The Real Life of Sebastian Knight*), summarized novels (*Invitation to a Beheading, The Real Life of Sebastian Knight*), summarized translations of plays (*Bend Sinister*), class lists, forewords, hotel registers, a philosophical treatise, and a narrative poem with commentary and index. Such innovations in structural form invite attention to the communicative promise and failings of various genres. The rising structural complexity of the works (at least through *Ada,* and with *Pnin* partially excepted) derives specifically from Nabokov's interest in the evolution of form itself, and his awareness of the aesthetic properties or potential of formal elements in nature as in fiction. In the early 1940s, even as he was beginning serious study of the morphological variety of butterfly wings and genitalia, Nabokov also began experimentation with the role of chronologies and indexes in his quirky book on Nikolai Gogol. Though not a work of fiction, it was a highly aestheticized portrait of the writer and his works, and Nabokov added to his characterization's communicative power by reversing the life-death teleology, by including at least one (apparent) non sequitur in the formal chronology that follows the main text (concerning Elizabeth Barrett Browning's door preserved in the Wellesley College Library), and by including absurd and nonsensical entries in the index (for example: "Nose, Gogol's").[6] *The Original of Laura,* his last, unfinished novel, appears to have been working toward a complication of the interbraiding of fragments of inner and outer text (on the model of a novel within a novel), perhaps itself to be communicated as a series of fragments and erasures. As Nabokov knew well, the fragment had a genealogy stretching back to the Romantic era as a literary feature meaningful by nature of its morphology: truncated, torn out from some imagined whole, gappy, hinting at something "more" beyond itself. Perhaps it is no surprise that he returned to this morphological theme as he imagined coming closer to death.

THE GIFT, BEND SINISTER, AND *PNIN* all involve increasingly elaborate play with the concept of a work's fictitious author, mentioned as such within a novel or story. This personality is often distinct from the narrator, but usually the two overlap in mischievous ways. The idea of an author figured explicitly within a work draws a special kind of attention to the problems of authorial identity, creative control, and the nature of textual sincerity. In between two once-famous "deaths of the author" in aesthetic philosophy, Nabokov began exploring the paradoxes raised by the fact that an author desires to have control over a creation's interpretations and meanings but could only do so by becoming, as it were, biographically connected to the fictitious world.[7] Such a connection is impossible in life—at some point we humans demand that life is life, and fiction is fiction—but it can be portrayed within fiction and given the illusion of plausibility, an exercise that serves simply to underscore the hopelessness of the author's desire to retain control. This struggle for mastery is an evolving process in Nabokov, for what in *The Gift* is a fairly strong illusion of authorial control becomes in *Bend Sinister* a world no author could want to imagine or validate, whose author-deity ends up granting his main protagonist insanity. Finally, in *Pnin,* the "author's" attempts to control his narrative and his main character are thwarted when the final chapter devolves into a series of attacks on the narrator-author's trustworthiness and the hero's escape out from under the "author's" pen. There are further constructions based on the author-as-character trope in *Despair,* and perhaps *Lolita* and *Ada,* though not, I suspect, in *Pale Fire.* However, Nabokov largely concealed the morphology of his tales—like the hidden and microscopic genitalia of butterflies—and so discovering where the fictitious author is hidden is one of the implied games in some of these later works. Regardless of how subtle or blatant the authorial theme is in a given case, all the novels that feature it offer the chance to reflect on questions of individual autonomy in the face of a world filled with preexisting statements by others, on the one hand, and with audiences who will twist and distort a book's intended meanings, on the other, effectively removing those meanings from the author's control.

There are of course many other features of a literary work that can be studied morphologically—beginnings and ends; digressions; games; romance; moments that I have elsewhere called textual ruptures; the representation of time (as Nabokov did in his lecture on *Anna Karenina*) or of consciousness (which he studied in Tolstoy and Joyce); or of all sorts of inserted texts (lists, letters, diaries, torn notes, and so on). Shortly after Nabokov began exploring the morphology of the fictitious preface (in *Lolita*'s John Ray, Jr.), he created a no-man's-land between fiction and reality in his many forewords and interviews and in the Afterwords to *Lolita* and *Ada*. The outward creep of fiction into reality, presaged in "Vivian Darkbloom's Notes to *Ada*," came to fruition in Alfred Appel Jr.'s *Annotated Lolita*, in which the annotator serves nearly as a ventriloquist's doll for Nabokov's mischievous play. Deliberate typos or editing comments and errors became a morphological element in *Lolita*, much more fully evolved in *Pale Fire* (which includes even "double italics") and *Ada*. Although these errors have received excellent attention in specific studies by Brian Boyd, James Ramey, Alexander Dolinin, and others, "error" or mistakes as a major morphological component in Nabokov's work has yet to be fully explored. Nabokov incorporated and developed variations on every potentially meaningful quirk of text he could imagine, in a clear engagement with the knowledge that artworks are built up out of such units of form and that these forms change and diversify as human thought progresses over the course of cultural history.

IF THE NATURE OF ART EVOLVES, and can be seen to do so even in the work of one individual, then the term "art" always ought to represent something with a genealogy, an inheritance that links it continuously to a human cultural past. Nabokov famously rejected the suggestion that he was "influenced" by prior writers, and he was often challenged and occasionally mocked by critics on these grounds. However, the term "genetics" can be used to depersonalize the connection between generations, and it offers a neater way of encapsulating how a living artist works as part of a continuum. Nabokov disliked the word "influence" because it suggested limits to his creative autonomy or, worse, that the essence of his creativity was no more than a mixture derived from great writers of the past or present. Describing individuality in a lecture for his students, Nabokov identified "agent X" as the most significant factor.[8] What this "agent X" may be is an open question—it might be similar to what modern theorists call emergence (as discussed by Victoria Alexander and Stanley Salthe); it might indicate (as many suspect) Nabokov's idealism and his belief in the irreducible existence of individual, autonomous selves—a controversial topic in philosophy and especially in literary criticism for the past few decades. However, Nabokov also refers to "heredity" and "environment" as real, if lesser, factors that determine an individual's form in the world, and the same mix will be present in works of art. Nabokov acknowledges his "fathers" in Russian literature, a term that focuses attention to *genetic* inheritance more than environmental or social influence.[9]

Confident of his "agent X," Nabokov allowed his works to teem with signs of genetic connections to a hereditary past (and with clear signs of the environmental present, although there is not space here to elaborate, other than to provide a shorthand example: Freud). In part, it is the extensiveness and density of his works' allusions that promote the scholarly quest for "influence" and its attendant "anxiety." There are also a great many clandestine allusions, and when readers dig deeply into a work like *Pale Fire* and find detailed connections to Robert Browning, Alexander Pope, Sir Walter Scott, T. S. Eliot, Edsel Ford, Robert Frost, Mary Shelley, Proust, Francis Bacon, Shakespeare, Conan Doyle, Lewis Carroll, Hans Holbein the Younger, and many more, it raises the suspicion—in some readers—that this display of erudition might be just fancy strutting, meant to flaunt the artist's dominance and the reader's subservience. All of Nabokov's novels have this quality, sometimes more, sometimes less openly. In fact, it not infrequently happens that a single moment in Nabokov's works will be shown to have probable and relevant connections to two or more literary precursors.[10] There is a deep genetic structure within Nabokov's creations: he writes in a way that deliberately acknowledges, incorporates, and develops hereditary traits from past artworks. This approach to creation reverberates with his scientific approach to taxonomy: as he looked at specimens of the *Polyommatus* Blues, he was imagining how they embodied and emerged from past forms, and how each of several modern forms manifested a unique set of relations to what he imagined to be the early realization. In part this practice is due to the way he saw the present: as always infused with the past, a feature he emphasized while describing his impressions of Cambridge (in *Speak, Memory*) and elaborated more fully and playfully in *Transparent Things*, his second-to-last completed novel.

In some ways, art's evolution is far more complex than natural evolution, since each new work can be genetically connected to many more than just two immediate progenitors—although newer models of evolution, elaborated by Lynn Margolis and discussed in this volume by Victoria Alexander and Dorion Sagan—restore nature to parity in this regard. The totality of the genetic pool from one generation (and all previous generations, too) is available for recombination and mutation during the next. But what Nabokov's fiction reminds us is that a story is never just a story, no matter how guileless it may seem. Simplicity is an illusion, and a dangerous one, especially in storytelling. Every artifact, every utterance, every story is a complex, multilayered communicative event. Nabokov takes care to incorporate that deep complexity into the structure and fabric of all of his works—even when, as sometimes happens, they appear simple enough on the surface.

NABOKOV WAS A NATURALIST AND TAXONOMIST, and he revered the naturalists (and some of the taxonomists) who came before him. Another important feature of his creations, as of most artworks, relates specifically to the work of the naturalist, the explorer of nature who goes off in search of new species, who "name[s] the nameless at every step," as he wrote of his fictitious naturalist in *The Gift*, Konstantin Godunov-Cherdyntsev.[11] Naming the nameless, for a scientist, means learning more about the contents and complexity of the world. For an artist, it means the same thing, but the sphere of observation is shifted to the world of human consciousness and its experiences. Nabokov once quipped that "all novelists of any worth are psychological novelists," and by this he meant that every work of fiction deals, inescapably, with the data of human perception and the vagaries of conscious cognition first and foremost.[12] This trait of art is probably one of the main reasons it moves nearly all humans so powerfully: the artist sees and describes the world, and human experience, in ways that have not yet been expressed. In this sense the artist gives name to the nameless (or, in the nonverbal arts, gives explicit form to something previously implicit, perhaps merely intuited or sensed indirectly). Nabokov is revered by readers for his ability to craft sentences that feel new and somehow uncannily perfect in their communication of something that may have been felt but had never been uttered before. Tolstoy presents this phenomenon (which he also practiced as a writer) in his description of the artist Mikhailov's portrait of Anna Karenina: "It was strange how Mikhailov could have discovered just her characteristic beauty. 'One needs to know and love her as I have loved her to discover the very sweetest expression of her soul,' Vronsky thought, though it was *only from this portrait that he had himself learned* this sweetest expression of her soul. But the expression was so true that he, and others too, fancied they had long known it."[13]

In Tolstoy, it is the artist who sees things and translates them for others to see: Vronsky recognizes the expression when shown it, but he had never perceived it on his own. In Nabokov's case, this perceptive gift is often called style and Nabokov himself a cold stylist. But style and craftsmanship are tools for expanding the domain of what can be perceived and expressed by individuals: naming the nameless facets of human perception, thought, and emotion, and also naming connections in the natural world that can appear only to the human mind (like the butterfly organ in which Nabokov sees a "striking resemblance to the head (frontal view) of a caterpillar," in Figure 14: the self-referential butterfly!; and like the new structure, the sagum, that he identified in South American Blues [see Figure 44]). To the extent that all art offers a broadening and deepening awareness of the natural world and the human world within it, artistic activity is one of the primary modes of discovery and the advance of knowledge. Nabokov's claim for the kinship between art and science was rooted in this sense of their twinned epistemological importance. His readiness to follow unlikely analogical trails (which, as in the case of the caterpillar mentioned above, can seem almost Gogolian) represents a kind of intellectual exploration of the interconnectedness of things: sometimes such trails lead to traps or dead ends; sometimes they lead to revelatory new ways to view the world—paradigm shifts, in Thomas Kuhn's later phrase.

AFTER DISCOVERY COMES TAXONOMY: the establishment of clear lines of relationship and clusters of species. Although Nabokov was, it turns out, a world-class taxonomist, this is a discipline that crosses the boundary into the sphere of human culture with difficulty and peril. If the professional field is sometimes divided into lumpers and splitters, and Nabokov attempted to find his proper place among those tendencies, in human life and art he was decisively a splitter. A radical splitter, in fact: as his mentor and friend Iulii Aikhenvald wrote, "There are no movements: there are only writers."[14] Like Aikhenvald, Nabokov believed that every work of art was its own unique species, thanks to "agent X" and the uniqueness of the creative moment. In his study of butterflies, Nabokov was interested in the variation within species, or within subspecies—that is, he was drawn to individuality as much as to the patterns the groups of individuals revealed. In his treatment of human beings within his fiction, this tendency receives its fullest expression: just as Hermann is wrong to think Felix is his exact double in *Despair*, Humbert is mistaken in thinking that twelve-year-old Dolores Haze is the perfect reincarnation of his childhood love, Annabel Leigh. He is equally wrong in thinking that "nymphet" is a taxonomic category that is both valid and significant to human life. It is not hard to detect the classifier's bent for precision in Humbert's "original description" of this genus:

> Between the age limits of nine and fourteen there occur maidens who, to certain bewitched travelers, twice or many times older than they, reveal their true nature which is not human, but nymphic (that is, demoniac); and these chosen creatures I propose to designate as "nymphets." It will be marked that I substitute time terms for spatial ones. . . . Neither are good looks any criterion; and vulgarity, or at least what a given community terms so, does not necessarily impair certain mysterious characteristics, the fey grace, the elusive, shifty, soul-shattering, insidious charm that separates the nymphet from such coevals of hers as are incomparably more dependent on the spatial world of synchronous phenomena than on that intangible island of entranced time where Lolita plays with her likes.[15]

In Humbert's case, his creation of a taxonomic description of nymphets is part of his self-delusion, a way to suppress his awareness of his own monstrosity. But this categorization, or the blindness it fosters in Humbert Humbert, causes Dolores-Lolita immense, irreparable harm and years of suffering. The danger of taxonomy in the study of human beings—so evident in the Holocaust immediately before *Lolita*'s composition, and in all similar category-based discussions of humans or groups of people—lies for Nabokov in its overwhelming tendency to gloss over individual traits, the variety of existence, the beauty of an infinite source of "agent X" among an infinitely varied populace. There is a boundary in his approach to the world that begins with the variety of human life: Nabokov groups insects into "peaks of speciation" along a flow of diversity, apparently with some reluctance; but human beings all deserve full attention to their unique features. "What the artist perceives is, primarily, the difference between things. It is the vulgar who note their resemblance," says the artist Ardalion in *Despair*.[16] It is the credo of the committed (but responsible) splitter—the assumption that there is always something new and different to be found, right around the next corner.

Nabokov's works invite the reader to approach them like scientists. As he told his Wellesley students, scientific study makes a "very apt" comparison to the study of literature, because "whichever subject you have chosen, you must realize that knowledge in it is limitless. Every subject brims with mysteries and thrills."[17] If Nabokov's creations are beautiful yet, beneath the surface, complex to the point of being baroque, it is because he felt that the creative act, like the perceptive, analytical one, offers limitless opportunities to explore the intricacies of life in the inexhaustible world.

Notes

1. *SO*, 78.

2. *NB*, 335, 353.

3. *NWL*, 241.

4. Belyi, "Sravnitel'naia morfologiia."

5. Nabokov, "Notes on the Morphology," 137.

6. Nabokov, *Nikolai Gogol*, 170.

7. One implied by French philosopher Hippolyte Taine in the 1860s (in the introduction to his *History of English Literature*); the other pronounced by Roland Barthes in 1967 ("The Death of the Author") and Michel Foucault in 1969 ("What Is an Author"). The same idea was already implicit in the work of W. K. Wimsatt and Monroe Beardsley in their essay "The Intentional Fallacy."

8. *LL*, 126: "Three forces make and mold a human being: heredity, environment, and the unknown agent X. Of these the second, environment, is by far the least important, while the last, agent X, is by far the most influential. In the case of characters living in books, it is of course the author who controls, directs, and applies the three forces." This formulation itself is strongly reminiscent of Iulii Aikhenvald, a famous critic who was Nabokov's close friend in the 1920s. See also Alexander and Salthe, "Monstrous Fate."

9. *LRL*, 11.

10. See Blackwell, "Notes"; Pekka Tammi's related study of "polygeneticity" in *Russian Subtexts*; and Senderovich and Shvarts, "Juice."

11. Nabokov, *Gift*, 119.

12. *SO*, 174.

13. Tolstoy, *Anna Karenina*, part 5, chapter 13, Constance Garnett trans., emphasis added.

14. Aikhenval'd, *Siluety russkikh pisatelei*, cited in Blackwell, *Zina's Paradox*, 27.

15. *Lolita*, 16–17. Here, the word "synchronous" even echoes one of Nabokov's scientific papers, "Nearctic Forms," 87.

16. Nabokov, *Despair*, 41.

17. *NB*, 398–99.

Swift and Underwing, Boulderfield and Bog

How Nabokov Drew the World from Its Details

ROBERT MICHAEL PYLE

Everywhere the details leap like fish.

—Jane Hirshfield, "The Stone of Heaven"

In the year 2005, I taught place-based writing to instructors and students from several universities in Tajikistan, Kyrgyzstan, and Kazakhstan. This took place in Dushanbe, the Tajik capital, sponsored by the Aga Khan Project for the Humanities. English skills varied widely, the only lingua franca among the participating writers being Russian. Much of the workshop was presented in English, translated into Russian, and the responses and writings then translated back again—sometimes with Kyrgyz, Tajik, or Kazakh detours along the way. There were many moments of levity and mystery, and sometimes deep connection. But as for true transcendence—that moment when everyone present completely melded—nothing matched the recording of Vladimir Nabokov reading his poem "The Swift," first in English and then in his native Russian.[1] The listeners liked the English version well enough, but when I played back the Russian reading, in his own sonorous, accented voice, my students were transfixed.

"The Swift" describes a moment of high romantic tension, when the appearance of a certain swift over a misty bridge cements the moment "forever" in the minds of the young lovers. What captured that instant, the poet and his love, and the hearts and minds of my students, was the extraordinary attention given to one small *detail* of the physical world, at that precise time and place. I believe it is this same, almost preternatural attention to detail that informed Nabokov's remarkable hand art, his literary work throughout, and ultimately, his science as well.

In *Nabokov's Butterflies,* Brian Boyd and I did not include "The Swift," since it is an ornithological reference (though a thin case can be made that it is also a sneaky lepidopteran allusion: see below). But another short bit that we missed definitely refers to a moth. Merely a snippet in the *News* of the Lepidopterists' Society about *Catocala fraxini* feeding on dead fish in Russia, it was attributed to Nabokov by the editor, C. L. Remington, when they were both at Harvard's Museum of Comparative Zoology in 1947.[2] This "beautiful Palearctic species," in Remington's words, is commonly known as the Clifden Nonpareil. It belongs to a group of Owlet Moths whose forewings match the bark of trees, but whose alarming hind wings ("underwings") are usually banded with scarlet, pink, or orange, like bright petticoats peeking out beneath a dull cloak. On this species alone, the bands are palest blue. I've described the similar White Underwing (*Catocala relicta*), a visitor to autumn porch lights in North America, as "a big gray delta nearly two inches to a side [with] fox-gray fur, the pile deepest on the thorax, lightened into a silvery pelt knitted with black-and-white bars, spots, and zigzags across the wings."[3] Nabokov described the Clifden Nonpareil in a rhyme of Fyodor's in *The Gift:* "Your blue stripe, Catocalid, shows from under its gray lid." While he never drew it, to my knowledge, you can see a similar chalky pallor, gray striations, and cool bluish shades in Color Plate 56, drawn with a fine nib and wash to catch the subtle hues of a blue's scales.

That Nabokov would take notice of such a thing, while it would fly right over most people's attention, is not unusual. He spoke often of the importance he placed in what he called "the individuating detail." As much as any other element of his output, scientific or literary or combined (for he *did* combine them), it is this extreme attention to exactly which elements distinguish one thing from another that so enlivens his fiction, ennobles both his poetry and his science (even unto its latter-day vindication), and elevates his drawings from doodles to art. Maybe the very fact that the world he saw and described is largely invisible to his readers helps to render their thrall so complete.

Nabokov did, indeed, discover the individuating details everywhere he went, and by no means always pretty ones. On weekend collecting trips away from Cambridge or Ithaca he was as likely to be repulsed by the stench of a rancid clam-fry shack as he was to be entranced by Early Hairstreaks on the wing.[4] When "Collecting in Wyoming," in between moments in motels or on the road that would find their way into *Lolita*, he could not help but notice the low notes: "I visited a remarkably repulsive-looking willow-bog, full of cow-merds and barbed wire."[5]

"No one's poems, essays, stories, and novels," I wrote elsewhere, "are more richly dressed in nuance than those of Nabokov, who exalted 'the individuating detail' in service to his art: as Humbert and Lolita passed through all those motor courts, the American landscape rolled by in the first-person particular, trained under the same microscope that the author brought to bear upon butterfly scales and genitalia."[6] It was this extreme sensitivity to detail that allowed Nabokov to pick out one Blue from hundreds he encountered in the Alps one summer; and though it turned out to be a hybrid between two species rather than the new species he initially suspected, its subtle distinction was enough to catch his eye, later to be celebrated in both a scientific paper and the often-quoted poem, "On Discovering a Butterfly." It was this same acute perceptivity that allowed him to spot subtle likenesses and distinctions among specimens in his drawers and on his microscope slides, and to deduce real relationships between North and South American Blues and those in the Holarctic, in spite of precious little material at hand.[7]

By that same "good eye," as the old lepidopterists admiringly called such a subtle talent, he plucked particularity from every object or setting that engaged his interest and frequently hung his writing raiment on those pegs. For example, in his poem "Lines Written in Oregon," he notices not only the pale, saprophytic phantom orchids in the forest but also "peacock moth on picnic table."[8] The Phantom Orchid is a real plant (*Cephalanthera austinae*). The "peacock moth" may have been a silver-marked plusiine that reminded him of a related moth he thought he had discovered anew as a boy but had been gazumped by another lepidopterist long before. Nabokov got his own back in *Laughter in the Dark* by naming a blind man Kretschmar, after "his" moth's original namer.[9] He may have evoked that moth in the poem, also, in the final words "Esmeralda, *immer, immer*"—because the closest thing to "his" moth in the main handbook and checklist of the time was then called *Chrysoptera moneta esmeralda*.[10] If I am right about that (and it is mere conjecture), then from this moment "in the bewitched and blest / mountain forests of the West," Nabokov penned a haunting lyric that both evoked a sharp disappointment of childhood and inspired extensive debate as to the (maybe multiple) ID of "Esmeralda"—all because he *noticed* a random moth on a picnic table. Taking extreme notice is what he did.

When I visited the Berg Collection at the New York Public Library while working on *Nabokov's Butterflies* and confronted the extensive body of drawings archived in the mostly unworked area labeled, at the time, "Leppy Stuff," I was struck by Nabokov's ability to transfer his attention to detail into graphic as well as verbal form. I was already well acquainted, of course, with the *Lycaeides* scale-row drawings from his scientific papers and with the commemorative fancies he contrived for certain inscribed copies of his books, several of which I had examined at Wellesley College and elsewhere. But I had no idea of the breadth, range, or number of the drawings. I was especially taken with a coppery green rendition of remarkably anthropoid-looking male genitalia. Not only was the drawing beautiful, but it showed certain distinctive features of this minute structure that prompted Nabokov to conclude the close connection between the Blue genera *Scolitantoides* and *Glaucopsyche* (Color Plate 22).

Another drawing that strikes home for me is that of a related, European Blue, *Lycaeides ismenias* (Color Plate 18). It brings to mind, in shape and in its shimmery dark coloration, the unrelated *Erebia magdalena*. The Magdalena (or Rockslide) Alpine is a denizen of the high alpine talus slopes in the Rocky Mountains. This ebony-black butterfly pops up in a botched story related by first biographer Andrew Field; and in the corrected and fleshed-out version, told humorously by John Downey (Nabokov's legatee for his work on the blues) in an oral history transcribed in *Nabokov's Butterflies*.[11] It reappears in a poem written by a character in a novel.[12] In between, Nabokov actually encountered the Magdalena Alpine during a summer in which he often felt, as he wrote Edmund Wilson, "some part of me must have been born in Colorado, for I am constantly recognizing things with a delicious pang."[13]

Nabokov's first meeting with Magdalena took place in the summer of 1947, when he and Véra spent a month collecting out of Columbine Lodge, situated along the Peak-to-Peak Highway between Estes Park and Nederland in the Colorado Front Range. Columbine Lodge is just south Long's Peak Lodge, built and formerly run by the famous nature writer and early national park advocate Enos Mills. (Today, Columbine Lodge is a retreat owned by the Salvation Army, known as High Peaks.) There, on the morning of July 13, the Nabokovs met Charles Remington, a PhD student and later Nabokov's "fellow sufferer" at Harvard, where Nabokov curated the Lepidoptera.[14] As Remington described the day to Harry Clench in a letter, he left the University of Colorado's Science Lodge near Ward, then "drove over to Estes Park and picked up Nabokov and his son and wife and took them on a jaunt to Tolland, the famous locality for *Brenthis*."[15]

The party drove south to Rollinsville, then west beside the railroad tracks toward the East Portal of the famous Moffat Tunnel. There they entered Tolland Bog, the type locality for *Boloria selene tollandensis* (then known as *Brenthis myrina*). "We had a very pleasant day and excellent collecting," wrote Remington, "Nabokov and his son are both skillful net wielders." This butterfly is the Rocky Mountain subspecies of the Silver-bordered Fritillary, a species Nabokov knew in the bogs of Vyra. It was likely the one referred to when he "stooped with a grunt of delight to snuff out the life of some silver-studded lepidopteron throbbing in the folds of my net," in the stilling, time-traveling, penultimate paragraph of chapter 6 of *Speak, Memory*.[16] Similar silvery studs adorn the Northern Metalmark in his rendering (Color Plate 59).

That Tolland outing with Remington took Nabokov back to his boyhood bogs and their smells and sights. But his actual encounter with that most mysterious of alpines, Magdalena, almost certainly took place right near Columbine Lodge itself. The cabins back right up to Mount Meeker, a southeastern arm off Long's Peak in Rocky Mountain National Park and an impressive 13,911-foot peak in its own right. The entire front of Mount Meeker consists of perfect Magdalena habitat, so much so that I rechristened it "Magdalena Mountain" in a work of fiction.[17] It is likely that Nabokov climbed the scree behind the lodge to collect this butterfly. No account of this event seems to survive. However, we know that he saw it just a little to the north, in the lap of Long's Peak itself. Concrete evidence of this encounter appears in another work of fiction: Nabokov's novel *Look at the Harlequins!* where the character Vadim relates a visit to Long's Peak with his daughter Bel:

> From Lupine Lodge, Estes Park, where we spent a whole month, a path margined with blue flowers led through aspen groves to what Bel drolly called The Foot of the Face. There was also the Thumb of the Face, at its southern corner. I have a large photograph taken by William Garrell, who was the first, I think, to reach The Thumb, in 1940 or thereabouts, showing the East Face of Longs Peak with the checkered lines of ascent superimposed in a loopy design upon it. On the back of this picture—and as immortal in its own little right as the picture's subject—a poem by Bel, neatly copied in violet ink, is dedicated to Addie Alexander, "First woman on Peak, eighty years ago." It commemorates our modest hikes:

> Long's Peacock Lake:
> the hut and its Old Marmot;
> Boulderfield and its Black Butterfly;
> And the intelligent trail."[18]

As a long-time student of the all-black *Erebia magdalena*—the quintessential resident of the Long's Peak boulderfields, along with the pikas and the marmots—I realized that Bel's "Black Butterfly" referred to this species. I published this identification in *Nabokov's Butterflies*, adding that "Nabokov's initial encounter with this butterfly did not occur in the manner suggested by biographer Andrew Field."[19] In fact, it occurred that summer of 1947 in the Rockies. And, through Bel's poem and this black butterfly, *Look at the Harlequins!* refracts a key moment behind the epiphanic scene of *Lolita* itself.

In the context of Nabokov as a visual artist of words, what Bel's poem shows me is his capacity for *deep refraction:* to skate from a moment's attentive encounter with pure physical detail, across the broad rink of reality and time, to arrive at the critical creative moment several removes away. A profound expression of what I mean arose from another high place with butterflies. Such rarified redoubts in thin, clear air held a powerful claim on Nabokov's affections. Again and again he sought the alpine haunts, in Russia, the Crimea, the Rockies, the Alps. It was in one such montane locale where he achieved two of his greatest summits—one scientific, the other artistic, both wrought of the same stuff, and both sprung from the same details that he had not failed to notice. And happily, *via* Bel's poem and a tiny mote of hand-script, it is the Magdalena Alpine once again whose broad black wings carry us to that very locale.

Circling back, in his postscript essay "On a Book Entitled *Lolita*," Nabokov wrote that "the locality labels pinned under these butterflies will be a boon to some twenty-first century scholar with a taste for recondite biography."[20] Well, here goes, because a pin label for a specimen of *Erebia magdalena* I saw at Cornell University reads: "Telluride, COLO. / Alt. 10,000 ft. / July 15, 1951 / V. Nabokov." Though collected four summers after his first encounter with the black butterfly on the other side of the Continental Divide at Long's Peak, this locality label connects us directly to *Lolita*; and it connects Bel's poignant poem to Humbert Humbert's agonized epiphany—both built from Nabokov's impressions while striding high ridges among these dusky butterflies, the first female of a certain Blue and his first Magdalenas.

Since well before he'd met John Downey in a canyon above Alta, Utah, Nabokov had been engaged in parsing the systematics and geographic plasticity of the scintillated, orange-bordered blues in the genus *Lycaeides*. Many of the drawings of scale rows and genitalia concern this group, as do his longest, most detailed scientific papers (Color Plate 17). He managed to deduce major relationships and biogeographically iconoclastic (but correct) conclusions for these and other groups of Nearctic and Neotropical Blues, based on relatively little pinned material. He also correctly separated out the taxon *Lycaeides samuelis* from the bunch, now known as the Karner Blue, which has become a conservation cause célèbre that drives a great deal of Midwest land management (and which has finally been elevated to full species status, as he suspected it someday would).[21] Yet his large series and extensive field studies of western *Lycaeides* left him flummoxed as to some of the evolutionary directions involved (or as flummoxed as Nabokov ever allowed himself to be). As he wrote about a particularly challenging tangency of these Blues: "In the Jackson Lake region such an intergradation actually does occur, apparently within the same colony or array of connected colonies. At this point of its development *argyrognomon* does turn into *melissa* (from which, however, only 300 miles to the west, it is sharply separated in all characters). That it wavers here at the crossroads of evolution and may select another course, is proved by the *ismenias*-like genitalia of the paratypes."[22]

One of his most persistent personal challenges was to find the previously undiscovered female of *L. argyrognomon* (now *idas*, as he also suspected) subspecies *sublivens*, which he had described as a new taxon from the male.[23] In July 1951, he wrote, "I bungled my family's vacation, but I got what I wanted."[24] When he did, it was just where he expected it to be: in southwest Colorado's San Juan Mountains, above the (then) small mining town of Telluride—where he also found the Magdalena that now resides in a specimen cabinet at Cornell University.

"The colony I found was restricted to one very steep slope reaching from about 10,500 to a ridge at 11,000 feet and towering over Tomboy Road between 'Social Tunnel' and 'Bullion Mine,'" he wrote in a paper describing the female of *L. a. sublivens*.[25] "The livid tones of the butterflies' undersides nicely matched the tint" of their lupine host plants, he wrote, on which they rested during the frequent dull and wet weather with which both butterfly and collector had to contend. And in a letter to his sister Elena that fall, he wrote, "It will not be hard for you to understand what a joy it was for me to find at last my exceedingly rare god-daughter, on a sheer mountainside covered with violet lupine, in the sky-high, snow-scented silence."[26]

But it wasn't all silent up there above Tomboy Road, as we know from another letter Nabokov sent to Edmund Wilson about the same time. What he revealed in that letter proved in advance something Nabokov would write one year later: "Does there not exist a high ridge where the mountainside of 'scientific knowledge' joins the opposite slope of 'artistic imagination'?"[27]

If Nabokov's discovery of his "exceedingly rare god-daughter" were not enough in itself to demonstrate his balancing act on that high ridge (literal, in this instance), there is one particular detail of that day that ties it deeply to his best-known work of literature. For when Nabokov discovered the missing female of what came to be known as Nabokov's Blue—a lynchpin in his science—he also discovered perhaps the most important scene in what many consider to be his most important book.[28]

It happened like this: as Nabokov scaled the stony, flower-flecked habitat of the Marshall Basin, which feeds the San Miguel River, and that the Dolores River, and finally the Colorado River, he noticed many details. Among them, a plenitude of blue: Colorado Blue Columbine, the western Jacob's Ladder known as Sky Pilot, Alpine Forget-me-nots, a sea of lupine, and bursts of blue butterflies: Boisduval's (for which he had erected the genus *Icaricia*), Arctic, Silvery, and Greenish Blues among them. It was on the lupines where Véra first spotted the brilliant blue male, and then, at last, mirabile dictu, Nabokov netted the heretofore unknown female of *Lycaeides idas sublivens:* "rather peculiar, smooth, weak brown, with an olivaceous cast . . . more or less dusted with cinder-blue scales."[29] Of course, he knew her on sight, because he had taken notice, exquisitely so, of her exquisite details. It had been fifty years since the male's discovery by a collector in Telluride and three years since Nabokov had described it, based on eight specimens that he had found in the Harvard collection. Now, at last, he knew it as a living creature.

But the Blues, the lupines, and the troublesome weather were not the only specifics he noticed up there above Telluride, under a sky of columbine and cloud, that rainy and windy summer month of 1951. For here is what he wrote to Edmund Wilson about the experience: "I went to Telluride (*awful* roads, but then—endless charm, an old-fashioned, absolutely touristless mining town full of most helpful, charming people—and when you hike from there, which is 9000', to 10000', with the town and its tin roofs and self-conscious poplars lying toylike at the flat bottom of a *cul-de-sac* valley running into giant granite mountains, all you hear are the voices of children playing in the streets—delightful!" And that is the exact detail of the scene that enables the climacteric of *Lolita,* in the sixth paragraph from the end of the novel, when Humbert Humbert, bereft, wretched, and sick, high on a mountain road, hears "a melodious unity of sounds rising like a vapor from a small mining town that lay at my feet, in the fold of the valley." "Reader!" he says. "What I heard was but the melody of children at play, nothing but that . . . and then I knew that the hopelessly poignant thing was not Lolita's absence from my side, but the absence of her voice from that concord."[30]

This is the very moment where Humbert Humbert acknowledges the enormity of his crime, the nature of his monstrosity. Martin Amis, in a reconsideration of the novel *Lolita,* wrote that "it has often been suggested that the 'morality' of *Lolita* is not inherent but something tacked on at the end." His essay shows why such an assumption is exactly wrong. It takes a careful reading to see why *Lolita* may be the greatest moral statement we have on pedophilia: a subject rendered almost banal today through its ubiquity in the news but undiscussed in 1956, even unmentionable. "Even sophisticated readers," wrote Amis, "still think Nabokov had something to feel guilty about."[31]

For a writer of such a story at that time, or at any time, to leave the moral payoff until the very end of a long book, was an astonishing artistic risk to take. Yet it pans out: as Amis wrote, "In Nabokov art itself provides the reproach and the punishment." In that one alpine wail of a sentence about the voices of the children at play, any careful reader who has suspended his or her judgment thus far is bound to hear the judgment meted out to Humbert as intended and to grasp the inherent morality of *Lolita*. This paragraph is the most heartbreaking thing I've read, and ties only with Darwin's final paragraph in *On the Origin of Species* as the most beautiful and affecting passage I know in the language.[32] It is "Humbert on the hillside begging forgiveness of Lolita and the American landscape," in Amis's words, and it "can still make the present reader shed tears as hot as Humbert's." This reader, too.

All this, because the author pursued a little butterfly up this particular high ridge, "somewhere between climb and cloud," and brought back something else: something that he had not expected but that he certainly figured out how to use.[33] Gary Snyder, in a poem called "What You Should Know to Be a Poet," enjoined writers to follow "your own six senses, with a watchful and elegant mind."[34] This exactly describes Nabokov's artistic sensibility.

When he wrote about "The Swift" that captured the young lovers' momentary attention on the bridge at sunset, Nabokov might have actually *meant* a swift, or a swallow, or even a moth. The swift in question is commonly assumed to be the bird of that name (*Apus apus*, family Apodidae). But in the original Russian version, it is called *Lastochka*, which is better translated as "swallow" (family Hirundinidae), "swift" in Russian being *strizh*. In his translation of *The Gift* (where the poem first appeared) into English, the bird became the unrelated but similar swift, and so it is called in the collection and recording referred to above. Yet in an interview, in which Nabokov refers to this verse as "probably my favorite Russian poem," he calls the creatures *swallows*.[35] More than likely, the author's taxonomic flip-flop was a tactical, artistic decision involving scansion—"Lastochka" working better in the Russian, "swift" in English. I like to think that in switching to swift, Nabokov was also punning on the moths of the family Hepialidae, known in Europe as "the swifts." The Swift (or Ghost) Moths appear at dusk, with a phantasmal shimmer that is difficult to forget. No matter. Whether Nabokov had in mind one bird or the other, a moth, or all three, the point is the same: even when his subject's biological identity is fluid in this archtaxonomist's mind, its role as a particular atom of experience is unequivocal. As he explained it to the interviewer, "The boy turns to the girl and says to her, 'Tell me, will you always remember *that* swallow?—not any kind of swallow, not those swallows, there, but that particular swallow that skimmed by?' And she says, 'Of course I will,' and they both burst into tears."[36]

Individuating details: they are what Vladimir Nabokov mined, from which to build both his science and his art—often exploiting the same ore deposit for both endeavors. His fine lines, the pen strokes that made his drawings, were just one more medium he employed. Maybe even more directly than the poems and the papers, the drawings cut to the heart of the beholder. Both bodies of Nabokov's work, his words and his images, display the breadth of his superacute vision, in every sense. Taken together, they tell us how he rebuilt the world—Terra, Antiterra, Telluride, all of it—out of its own delicious yet all too often overlooked elements.

Acknowledgments

Lars Crabo kindly helped me to pin down the likely species of moth that Nabokov found on an Oregon picnic table. Warmest thanks go to Drs. Deborah Piot and Larry Gall for finding and providing the letter from Remington to Clench that fixed the date for Nabokov's Tolland expedition, in the superbly curated Charles L. Remington archive at the Yale Peabody Museum of Natural History, also thanked. Dr. Andrew D. Warren, McGuire Center for Lepidoptera and Biodiversity, deserves keen thanks for assisting me in the finding of Nabokov's Telluride specimen of *Erebia magdalena* at Cornell.

Kind thanks to Stephen Blackwell for guiding this essay and helping me to notice and navigate Nabokov's interesting "swift/swallow" ambivalence in his poem "Lastochka"; and to Brian Boyd for furnishing the Martin Amis essay, and always for his richly instructive collaboration on *Nabokov's Butterflies*. All of my other acknowledgments in that book apply here, too.

Notes

1. "The Swift" (1963), in Nabokov, *Gift*, 94; original Russian in Nabokov, *Sobranie*, 4:277.

2. Nabokov, "Note," 34.

3. Pyle, *Sky Time*, 191.

4. *NB*, 397.

5. Nabokov, "Lines Written in Oregon," 50.

6. Pyle, "Lookee Here!," 56.

7. Nabokov, "*Lysandra cormion*," in *NB*, 265–67; Nabokov, "On Discovering a Butterfly," in *NB*, 26.

8. Nabokov, "Lines Written in Oregon," 28.

9. *SM*, 134.

10. McDunnough, "Check List," 114.

11. Field, *Nabokov*, 202; *NB*, 51–53.

12. Nabokov, *Look*, 144.

13. Nabokov, "Note," in *NB*, 403.

14. *SM*, 127.

15. Remington to Harry Clench, July 15, 1947, Remington Archive, Yale Peabody Museum of Natural History, New Haven.

16. *SO*, 138.

17. Pyle, "Walking," 3–12.

18. Nabokov, *Look*, 144.

19. *NB*, 51.

20. *Lolita*, 312.

21. Pelham, *Catalogue*, online version, species #558.

22. Nabokov, "Nearctic Members," in *NB*, 429.

23. Nabokov, "Nearctic Members," in *NB*, 429.

24. Nabokov, "Female," 35.

25. Nabokov, "Female," 35.

26. Nabokov to Elena Sikorski, September 6, 1951, in *NB* 479.

27. Nabokov, "World of Butterflies," cf. *SO*, 329–30, and *NB*, 487–88.

28. Brown, Eff, and Rotger, *Colorado Butterflies*, 161.

29. Nabokov, "Female," 32.

30. Nabokov to Wilson, early September 1951, *NB*, 478. The connection between mountainside and novel is discussed also by Boyd in *VNAY*, 203.

31. Amis, "*Lolita* Reconsidered," 115.

32. Darwin, *Origin of Species*, 490.

33. Nabokov, "Notes on the Morphology," 137.

34. Snyder, "What You Should Know."

35. *SO*, 14.

36. *SO*, 14.

Enchanted Hunting

Lolita and Lolita, Diana and *diana*

BRIAN BOYD

An eager supporter of this volume from the first, I am delighted to add a coda to Robert Dirig's chapter. Bob, one of the lepidopterists I've most enjoyed working with intermittently for twenty years, asked me for feedback on his essay. I could do little more than tell him how enchanting I—and Nabokov would have—found it, especially to be led by a knowledgeable naturalist on guided tours through the McLean Woodlands near Ithaca, the Cove Woodland Trail in the Great Smoky Mountains, and 880 Highland Road, Ithaca, Nabokov's home at the time he conceived of *Pale Fire*.

We are as enchanted walking with Bob Dirig as *Pale Fire*'s Charles Kinbote is irritated, rambling in the hills with John Shade and wanting to talk only of Zembla, while Shade instead marvels at "the extraordinary blend of Canadian Zone and Austral Zone that 'obtained,' as he put it, in that particular spot of Appalachia." Kinbote grumbles that "a humble admirer is considerably more interested in discussing with him literature and life than in being told that the 'diana' (presumably a flower) occurs in New Wye together with the 'atlantis' (presumably another flower), and things of that sort."[1] Readers attuned to Nabokov's ways and Kinbote's woes will suspect a joke here and quickly confirm the suspicion that (*Speyeria*) *diana* and *atlantis* are not flowers but butterflies.

But Nabokov loves the covert surprise he can hide behind an almost overt joke. Dirig convincingly shows that Nabokov here has in mind the great lepidopterist W. H. Edwards's home in the southern Appalachians, Coalburg, where in the 1870s he worked on the southern *diana* and in whose neighborhood he was astonished to find that an apprentice had caught its northern congener, *atlantis*. Nabokov "defocalizes" *Pale Fire*'s New Wye[2]—it is in Appalachia, "in New England," yet it would also seem to be in "New Y" State *and* "at the latitude of Palermo"[3]—but knows of, and all but documents, at least *one* Appalachian environment that could support all the flora and fauna of his novel's central locus.

Coalburg, as Dirig shows, is also the type locality, established by Edwards, the butterfly's discoverer, of *Pieris virginiensis*, on which Nabokov and Shade, in a fusion of poetry and precision, of art and science, bestow the superbly apt common name Toothwort White.[4] Nabokov could hardly be further from impatient Kinbote: he marks and commends Shade as a fine naturalist, the direct heir of his ornithologist parents, and a spiritual descendant of his Appalachian predecessor Edwards, whose *Butterflies of North America* Nabokov thought "one of the finest works on butterflies ever published."[5] Dirig also develops my 1999 suggestion that Shade and Nabokov have made the Toothwort White an emblem of the retiring Hazel.[6] He proposes that by defining the fauna and flora of New Wye—which of course has much in common with Ithaca, New York, where toothworts grew and *virginiensis* flew—in terms of the type locality of this butterfly, Nabokov underscores in another way the centrality of Hazel not just to Shade's poem but also to his maker's novel.

But Dirig's discovery of Nabokov's use of Coalburg, the fabled locality of one of America's great lepidopterists, suddenly struck me when he explained that in Coalburg Edwards also worked out that the small orange fritillary and the large blue fritillary that flew with it were the male and female, respectively, of *diana*: for "Diana" and Coalmont play important roles in *Lolita*.

AFTER CHARLOTTE'S DEATH Humbert Humbert, set on taking Lolita to the Enchanted Hunters Hotel and plying her there with sleeping pills so he can avail himself of her overnight, tells her he is driving toward distant Lepingville, where her mother is in the hospital; only because they cannot reach Lepingville that day, supposedly, do they stop at the Enchanted Hunters. Humming with anticipation, Humbert's scheming fancy savors the hotel's name when he first thinks of this as a destination for possessing Lolita: "Was he not a very Enchanted Hunter as he deliberated with himself over his boxful of magic ammunition?"[7] Lying beside her on the bed at the hotel, frustrated that the "magic" of the sleeping pills has not, after all, put her at his disposal, he nevertheless has moments of hope: "Now and then it seemed to me that the enchanted prey was about to meet halfway the enchanted hunter."[8]

But playwright Clare Quilty is also at the hotel. He has met Charlotte, and spots Dolores Haze, who has already appealed to his pedophiliac eye, and knows that Humbert is not her father. Over the coming months he writes a play, *The Enchanted Hunters,* whose heroine, Diana, named after the virgin goddess of hunting, is the enchantress. When Beardsley School stages the play, Lolita wins the role of Diana, and when the school invites the author to monitor the play's progress, Quilty, relishing an opportunity of intimacy with young girls, responds to the invitation, only to find, as if by enchantment, the girl who inspired *The Enchanted Hunters* already playing the role of Diana. Dolly once had had the famous playwright's picture on her wall; now she swiftly becomes his lover and excitedly agrees to his plan to have Humbert drive her across state lines to Quilty's ranch. They settle on Elphinstone as the town where he will whisk Lolita away from her brooding captor.

When Lolita absconds from Elphinstone, Humbert vainly searches the taunting trail that Quilty has laid for him. But it leads nowhere. Time passes. Humbert picks up Rita in meager consolation and eventually reports: "A curious urge to relive my stay [at the Enchanted Hunters] with Lolita had got hold of me. I was entering a phase of existence where I had given up all hope of tracing her kidnaper and her. I now attempted to fall back on old settings in order to save what still could be saved in the way of *souvenir, souvenir que me veux-tu?*"[9] Unable to get a room in the Enchanted Hunters, he visits the town's public library to scour the local newspaper at the time of his earlier visit. He misses there clues to Quilty; but more important for our quest, he also writes a verselet to amuse Rita:

> The place was called *Enchanted Hunters.* Query:
> What Indian dyes, Diana, did thy dell
> endorse to make of Picture Lake a very
> blood bath of trees before the blue hotel?[10]

As if his incorporating "Diana" into his own "Enchanted Hunters" poem acts as a magic releaser, Humbert finds on his return to New York that the girl he has hunted in vain has hunted him down: a letter awaits him from "Dolly (Mrs. Richard F. Schiller)," return address "'General Delivery, Coalmont' (not 'Va.,' not 'Pa.,' not 'Tenn.'—and not Coalmont, anyway—I have camouflaged everything, my love)."[11] He heads straight for Coalmont, hoping to find and kill the man who abducted his Lolita, but instead learns from her lips that the man who took her from him was the author of *The Enchanted Hunters* and therefore the creator of the character of Diana she was to play. Now Humbert can understand the clues in the cryptogrammic paper chase that turned him into such a cursedly enchanted hunter. He heads off to hunt down Quilty.

Nabokov links two key destinations in Humbert's and Lolita's (and Quilty's) lives, through the unusual names of Lepingville and Elphinstone. "Lepingville" and "Elphinstone" match oddly but insistently: a three-syllable place name, with the letters *e, l, p,* in the first syllable, a second syllable in *-in,* and a third syllable, ending in *e,* meaning the name of a small community, *town* (slurred to *-ton,* as in Kingston, rather than retained as in Queenstown) or *ville.* Butterflies confirm the link. "Lepingville" plays on "lepping," butterfly hunters' slang (and a favorite Nabokovian term) for their chase. "Elphinstone" commemorates a former butterfly genus, *Elphinstonia,* now known, following "the inexorable law of taxonomic priority," as *Euchloe.*[12]

On the way to Lepingville, at the Enchanted Hunters, Lolita turns the tables on the man who has hunted her: *she* becomes the huntress, *she* proposes that they do what she and Charlie did together at Camp Q. Quilty, also at the hotel, turns his glimpse of Dolores into the inspiration for his play, "*The Enchanted Hunters,* . . . in which [in the Beardsley production] Dolores Haze was assigned the part of a farmer's daughter who imagines herself to be a woodland witch, or Diana, or something, and who, having got hold of a book on hypnotism, plunges a number of lost hunters into various entertaining trances before falling in her turn under the spell of a vagabond poet."[13] That forms a central part of the "hunter hunted" motif that pervades the novel and that I have discussed elsewhere.[14] The play itself becomes the basis of Quilty's and Dolores's being thrown together by fate and drawn together by desire, and their plotting to escape from Humbert at Elphinstone.

And at Coalmont, just after he has written his little poem about Diana and the Enchanted Hunters, Humbert discovers to his complete surprise but immediate recognition that the abductor was the man who wrote *The Enchanted Hunters* and created the everyday Diana who enchants her suitors—or becomes the object of enchantment: "A seventh Hunter . . . was a Young Poet, and he insisted, much to Diana's annoyance, that she and the entertainment provided (dancing nymphs, and elves, and monsters) were his, the Poet's, invention."[15]

Coalmont's final syllable, like *-ton* or *-ville,* is a regular toponymic suffix, like the *-burg* of W. H. Edwards's Coalburg (in German, *-burg* has the same role, matching *-ton* or *-ville,* while *-berg* matches *-mont,* as in the recurrent German place-name Kohlberg).[16] "'Coalmont' (not 'Va.,' not 'Pa.,' not 'Tenn.' . . .)" would seem by the mapping of excluded states to be in West Virginia, like Edwards's Coalburg, which Dirig describes in his essay as "internationally known as the locus of Edwards's pioneer butterfly work," a "'legendary' locus in the history of American butterfly scholarship"—a Lepingville if ever there was one, since there was nothing much there but Edwards's imposing Italianate villa, where he worked intensively on the butterfly *diana.*

"Coalmont," in paying oblique tribute to an enchanted butterfly hunter, extends the pattern of the *Enchanted Hunters* and Diana, goddess of the hunt. But why? Why is this pattern so insistent?

Nabokov put his butterfly hunts around the United States, and the roadside cabins he stayed in en route, to astonishing literary use when he reworked his old, unsuccessful Russian novella *The Enchanter* (written 1939, published 1986) into *Lolita*, turning it from static and Paris-based into the first and greatest of American road novels.[17] He had already written about his passion for butterflies in his autobiography, at the end of the 1940s, but word of his lepidopteral quests reached a huge audience when the essay he wrote on *Lolita* in 1956 became an appendage to the multi-million-selling novel.[18] There, after noting that "every summer my wife and I go butterfly hunting," he discloses that one of "the secret points, the subliminal co-ordinates by means of which the book is plotted" is the scene of "the tinkling sounds of the valley town"—Elphinstone, in the novel, Telluride, Colorado, in real life—"coming up the mountain trail (on which I caught the first known female of *Lycaeides sublivens* Nabokov)."[19]

In a 1960 article, "*Lolita* Lepidoptera," Diana Butler rightly took this as a hint. (And perhaps influenced by her, or simply by Nabokov, John Fowles, novelist and amateur lepidopterist, wrote his 1963 novel *The Collector,* about a butterfly collector who captures a young girl.) Nabokov later dismissed "the unfortunate Diana" and in 1966, to Page Stegner, writing the first book on him, he had Véra report: "My husband wants to repeat that there is no connection whatsoever, either in his work or his mind, between entomology and humbertology."[20] The next year he mentioned to Herbert Gold, interviewing him for the *Paris Review,* "a young lady who attempted to find entomological symbols in my fiction. The essay might have been amusing had she known something about Lepidoptera. Alas, she revealed complete ignorance and the muddle of terms she employed proved to be only jarring and absurd."[21]

Nabokov may not have liked Butler's explanations, or an obvious link between hunting post-nymphic butterflies and hunting nymphets, but he does not deny a relation between *Lolita* and Lepidoptera. And just as he suggests the link in his afterword to the novel, so he does in the novel's Foreword, written as it is by one "John Ray, Jr." As if to match the way Humbert Humbert's very European name echoes itself, John Ray's "Jr."—a very American addendum—echoes the initials of his first and last names. But John Ray, Sr., as it were—the original John Ray (1627–1705), the English naturalist—laid the pre-Linnaean foundations for modern taxonomy—another hint that Diana Butler recognized.[22]

Nabokov had been a taxonomist for a decade before he began writing *Lolita,* and found the first known female of *sublivens* in 1951, while half way through composing the novel.[23] It would be remarkable if his most intense years as a lepidopterist and an explorer of American landscapes had no serious repercussions in his next novel, his first set in America, crisscrossing the country and saturated by the rhythms of the chase.

After he developed the first germ for an expanded American version of *The Enchanter,* in 1946, Nabokov had written to Edmund Wilson, "The longer I live the more I become convinced that the *only* thing that matters in literature, is . . . that the good writer is first of all an enchanter."[24] In 1948, he wrote in a serialized chapter of his autobiography that writing and butterfly hunting were both "a form of magic, both were a game of intricate enchantment and deception."[25] In 1943, he had described Edwards's *Butterflies of North America,* written at Coalburg, as "one of the finest works on butterflies ever published"; in 1948, he wrote to two fellow lepidopterists he admired that the three of them should write together "a new 'Butterflies of North America.'"[26] Instead in 1950, he began to write *Lolita,* where he has Humbert find, at Coalmont, the enchantress he has been hunting for years.

WHEN QUILTY ARRANGES WITH LOLITA to have Humbert drive her across state lines to his ranch out West, they plan to coincide in Wace, Continental Divide, before heading off together from Elphinstone. Quilty has specified Wace because there Lolita can see him at a performance of the play, *The Lady Who Loved Lightning*, that he has co-authored with Vivian Darkbloom.

"Vivian Darkbloom" is a by now famous anagram for Vladimir Nabokov, and Elphinstone in real life is Telluride, the mountain town above which this very Nabokov, as his Afterword notes, caught the first known female of *Lycaeides sublivens*. In his short paper on catching these females, written as he composed *Lolita*, Nabokov reports the "daily electric storm, in several installments, accompanied by the most irritatingly close lightning I have ever encountered anywhere in the Rockies."[27] He was after the *female* of *sublivens*: after, if you like, a female who loved lightning, since this was her type locality (and since Nabokov was able, even with this lightning to contend with, to catch ten females here). With his penchant for verbal play Nabokov would also have recognized the near-pun embedded in the species name he bestowed before he ever found the species in situ: *sub-livens* (dark bluish) almost spells out "under lightning," since *levin* is an archaic word for "lightning," still used by nineteenth-century poets he knew well, like Sir Walter Scott, Henry Wadsworth Longfellow, and A. C. Swinburne.[28] Vladimir Nabokov had written in 1951 and published in 1952, as he was writing *Lolita*, his report on this female who lived among lightning, in curious conjunction with the Vivian Darkbloom who had written with Quilty *The Lady Who Loved Lightning*, staged at another location high in the Rockies. Quilty is on the way to capturing at Elphinstone the female he has dreamed of and has even written into *The Enchanted Hunters* as Diana; Nabokov has captured the female *he* had wanted to find for years (since he first named the species, from male specimens, in 1949) on those lightning-struck slopes from where, in the topography of fiction, Humbert will bemoan Lolita's absence.

"Vivian Darkbloom" within the novel is "a hawklike, black-haired, strikingly tall woman," but Lolita feigns that it is the coauthor, Clare Quilty, who is "the gal author."[29] As we have seen, one of the surprises of W. H. Edwards's lepidopterological work at Coalburg is that he ascertained there that two strikingly differently colored fritillaries were the male and female of the same species, *diana*. One of the teases of the Quilty story is that both his first name, Clare, and that of his coauthor, Vivian, can be male or female, and that Lolita, who through her rehearsals for the part of Diana knows exactly who—and from intimate evidence, what sex—the author of *The Enchanted Hunters* is, misleads Humbert with a false scent, in his hunt for his possible rival, when she tells him at Wace that Clare Quilty is the "gal author" of *The Lady Who Loved Lightning*. Only at Coalmont will Humbert at last find from the girl who played Diana that the person he has been trying for three years to hunt down is Clare Quilty, not the gal but the guy author of *The Lady Who Loved Lightning* and the sole and priapically male author of *The Enchanted Hunters*.

Two parallel scenes remove any doubt that the Diana of *The Enchanted Hunters* is named after not only the goddess of hunting but also after the butterfly named in her honor, *Speyeria diana*, and that Nabokov also has in mind the Lepingville at Coalburg where Edmunds determined that two seemingly unrelated butterflies were sexual morphs of the same *diana*. In the first scene, at Hourglass Lake, Jean Farlow emerges from the greenery where she has been trying to finish a lakescape, and surprises Charlotte and Humbert, who had thought themselves unwatched:

> Charlotte, who was a little jealous of Jean, wanted to know if John was coming.
>
> He was. He was coming home for lunch today. He had dropped her on the way to Parkington and should be picking her up any time now. It was a grand morning. She always felt a traitor to Cavall and Melampus for leaving them roped on such gorgeous days. She sat down on the white sand between Charlotte and me. . . .
>
> "I almost put both of you into my lake," she said. "I even noticed something you overlooked. You [addressing Humbert] had your wrist watch on in, yes, sir, you had."
>
> "Waterproof," said Charlotte softly, making a fish mouth.
>
> Jean took my wrist upon her knee and examined Charlotte's gift, then put back Humbert's hand on the sand, palm up.
>
> "You could see anything that way," remarked Charlotte coquettishly.
>
> Jean sighed. "I once saw," she said, "two children, male and female, at sunset, right here, making love. Their shadows were giants. And I told you about Mr. Tomson at daybreak. Next time I expect to see fat old Ivor in the ivory. He is really a freak, that man. Last time he told me a completely indecent story about his nephew. It appears—"
>
> "Hullo there," said John's voice.[30]

Jean's effusiveness had been just about to divulge the "completely indecent" gossip about fat old Ivor Quilty's nephew, Clare: presumably, a story about the playwright's penchant for young girls that would have instantly put Humbert on the alert. That implication is confirmed when Lolita at Coalmont at last discloses Quilty's name to Humbert. Humbert conceals the name from us for the moment but gives us this clue: "Waterproof. Why did a flash from Hourglass Lake cross my consciousness? I too, had known it, without knowing it, all along."[31]

But notice Jean's dogs, Cavall and Melampus. As I have noted elsewhere:

> Cavall was not only King Arthur's favorite hound but the first of his hounds to turn the stag in a hunting episode in *The Mabinogion*. Melampus is the name of the first hound of Actaeon, in Ovid's telling of the story of Diana and Actaeon in his *Metamorphoses*. The precision of these allusions startles: two hounds from different literary traditions that are the first to chase or turn a stag. Actaeon, remember, is the hunter who spies Diana, the virgin goddess of hunting, naked. Diana, enraged, transforms him into a stag, and his hounds pursue him, Melampus leading, and tear him to pieces.[32]

Nabokov obliquely introduces the myth of Diana as huntress just as he almost divulges to Humbert the name of the man who will hound him across America and whom he will then hunt for desperately.

A second, matching, near-disclosure of Quilty's name brings together Diana the huntress and butterfly names. Just as she and Humbert start their cross-country drive to Elphinstone and away from Beardsley and her role as Diana as enchanting huntress, Lolita has to cut a conversation short and to throw Humbert off the scent of Quilty:

As we pulled up, another car came to a gliding stop alongside, and a very striking looking, athletically lean young woman (where had I seen her?) with a high complexion and shoulder-length brilliant bronze hair, greeted Lo with a ringing "Hi!"—and then, addressing me, effusively, edusively (placed!), stressing certain words, said: "What a *shame* it was to *tear* Dolly away from the play—you should have *heard* the author *raving* about her after that rehearsal—" "Green light, you dope," said Lo under her breath, and simultaneously, waving in bright adieu a bangled arm, Joan of Arc (in a performance we saw at the local theatre) violently outdistanced us to swerve into Campus Avenue.

"Who was it exactly? Vermont or Rumpelmeyer?"

"No—Edusa Gold—the gal who coaches us."

"I was not referring to her. Who exactly concocted that play?"

"Oh! Yes, of course. Some old woman, Clare Something, I guess. There was quite a crowd of them there."

"So she complimented you?"

"Complimented my eye—she kissed me on my pure brow"—and my darling emitted that new yelp of merriment which—perhaps in connection with her theatrical mannerisms—she had lately begun to affect.[33]

The person who has directed Dolly as Diana in the Beardsley production of Quilty's *The Enchanted Hunters* has such a bizarre cognomen because she too is named after a butterfly, *Colias edusa*, an old name for the Clouded Yellow.[34] Humbert comments here on Lolita's "theatrical mannerisms" and will soon lament: "By permitting Lolita to study acting I had, fond fool, suffered her to cultivate deceit."[35] Nabokov introduces the cryptic butterfly name just after Edusa's effusiveness threatens to identify the name or at least the sex of the person who will soon be play-hunting Humbert across America and just before quick-acting Dolly throws Humbert off the scent by identifying the playwright as "Some old woman, Clare Something," anticipating her identification at Wace of Clare Quilty as "the gal author" of *The Lady Who Loved Lightning*.

Nabokov shows here how central a part *diana* as butterfly name, as well as Diana as goddess of the hunt, plays in his design, and how central to that role of *diana* is the confusion of sexes disentangled in real life by Edwards at Coalburg and, within the novel, at last dispelled by Lolita at Coalmont. To follow the implications of this pattern and its intersection with others might take a whole book. Meanwhile, thank you, Bob, for the Coalburg lead.

Notes

1. *PF*, 169 (commentary to l. 238).

2. In 1957, Nabokov described to Jason Epstein his early plans for a novel called *Pale Fire*, situated "somewhere on the border of Upstate New York and Montario . . . on Sundays the Hudson flows to Colorado. Despite these . . . little defocalizations. . . ." *SelL*, 213.

3. *PF*, 139, 19.

4. For the elegance of the name, see Dirig, in this volume, and Boyd, *Nabokov's "Pale Fire,"* 276n5.

5. *NB*, 261 (from Nabokov, "Some New or Little Known Nearctic *Neonympha*").

6. See also Boyd, *Nabokov's "Pale Fire,"* 135–37.

7. *Lolita*, 109.

8. *Lolita*, 131.

9. *Lolita*, 261.

10. *Lolita*, 263.

11. *Lolita*, 266, 267.

12. Nabokov, *Ada*, 57.

13. *Lolita*, 200.

14. See Boyd, "*Lolita*: What We Know," 331–34.

15. *Lolita*, 201.

16. Nabokov himself, after his most ambitious butterfly expeditions to date, in the Pyrenees in spring 1929, spent the summer at a plot of lakeside land he briefly owned at a Kolberg in Germany: see *VNRY*, 290–91. Dieter E. Zimmer does a brilliant job of identifying the Kolberg in question in "Nabokov zwischen den Ziestseen," 2014, http://www.d-e-zimmer.de/NabokovZiestseen/NabokovZiestseen.htm.

17. *Lolita* was published in 1955 (Paris: Olympia Press); Jack Kerouac's *On the Road* appeared in 1957 (New York: Viking).

18. Nabokov, "On a Book Entitled *Lolita*," originally in *Anchor Review,* but thereafter published as an afterword to the novel.

19. *Lolita*, 312, 316.

20. *NB*, 646, 644.

21. Interview with Herbert Gold, *SO*, 96. For a good example of Diana Butler's serious confusions, see the discussion of "nymph" in section 2 of D. Zimmer 2012, http://www.d-e-zimmer.de/eGuide/Lep2.1-M-O.htm.

22. D. Butler, "*Lolita* Lepidoptera," 63.

23. For the biographical details in this paragraph, see *VNAY*.

24. Letter of November 27, 1946, *NWL*, 203.

25. *SM*, 125.

26. *NBl*, 261, 406–7.

27. Nabokov, "Female"; *NB*, 481.

28. Or did he intend the pun from the first? He had already caught butterflies in the high Rockies, already been exposed to lightning there, when he named the species, identifying as the type locality Telluride, in the high Rockies: see Nabokov, "Nearctic Members," 513–16; see also *NB*, 425.

29. *Lolita*, 221.

30. *Lolita*, 88–89.

31. *Lolita*, 272.

32. Boyd, *Stalking*, 333.

33. *Lolita*, 208–9.

34. Now *Colias crocea*. See D. Zimmer, *Guide*, s.v. "Edusa Gold," http://www.d-e-zimmer.de/eGuide/Lep2.1-D-E.htm#EdusaGold. Zimmer points out that "the lady would seem to be named after the Eastern Bath White (»*Pontia edusa* Fabricius, 1777) if it were not for her surname, Gold. *Pontia edusa* is not at all golden but white, with dark markings. So her name might rather be a reference to an obsolete name of the Clouded Yellow (»*Colias crocea* Fourcroy, 1785)." Nabokov confirmed he had the Clouded Yellow in mind; *Lolita*, 408n233/1.

35. *Lolita*, 229.

Nabokov's Notes and Labels from the Museum of Comparative Zoology

Boon for a Recondite Biographer or Data for a Serious Systematist?

NAOMI E. PIERCE, RODNEY EASTWOOD, ROGER VILA,
ANDREW BERRY, AND THOMAS DAI

*My work enraptures but utterly exhausts me; I have ruined my eyesight,
and wear horn-rimmed glasses. To know that no one before you has seen
an organ you are examining, to trace relationships that have occurred to
no one before, to immerse yourself in the wondrous crystalline world of
the microscope, where silence reigns, circumscribed by its own horizon, a
blindingly white arena—all this is so enticing that I cannot describe it.*

—Vladimir Nabokov to Elena Sikorski, November 26, 1945

Nabokov first came to Harvard's Museum of Comparative Zoology (MCZ) in 1941. He was
teaching at Wellesley College, a few miles away from Cambridge, and needed to compare
butterfly specimens he had collected on a trip to the Grand Canyon with the relevant holdings of the museum. That initial visit would evolve into a research position at the MCZ until
his departure for Cornell University in 1948.

Founded in 1859 by the great (if anti-Darwinian) Swiss naturalist Louis Agassiz, the
MCZ is unusual by the standards of university museums in both its size and the scope of
its ambitions. At the time when Nabokov came to the museum, the MCZ had at least the
seventh largest insect collection among North American institutions and, with more than
seven million specimens, the largest university collection.[1] The number of insect "types"
in the MCZ, the specimens from which a species was originally described and ultimate
source of authority on the identity of that taxon, is second only to that of the Smithsonian
Institution. Nabokov was thus joining not only a major scientific institution but one with
all the advantages of a leading university. He wrote to his good friend Edmund Wilson in
August 1942, "It is amusing to think that I managed to get into Harvard with a butterfly
as my sole backer."[2] This was the ideal arrangement: he could do science by day and talk
literature, philosophy, and politics with colleagues at night. For example, Harry Levin, the
literary critic and professor of comparative literature, and his wife were regular guests at
the Nabokovs' home.

Despite the museum's reputation, Nabokov was dismayed by what he found in the Lepidoptera room. The collections were in poor repair, and he discovered evidence of an in-
festation of that bane of natural history collections, dermestid beetles, which can literally
eat their way through a collection of dead insects. He spoke to the head of entomology,
Nathaniel Banks, and was pleased to see, as they talked, that a paper he had written on the
butterflies of the Crimea was open on Banks's desk. Banks referred him to Thomas Barbour,
a herpetologist and the director of the MCZ, who was happy to provide improved cases for
the butterfly collection and who offered Nabokov the opportunity to work in his spare time

on reorganizing and curating the collection. Nabokov would be working as a volunteer, but with the unlimited access to the collection the position promised, Nabokov was excited to take it on. Soon, in 1942, he was appointed research fellow in entomology, which came with a salary of $1,000, a sum that was increased to $1,200 in subsequent years.

In October 1942, Nabokov and his wife, Véra, moved from Wellesley to Cambridge, which meant that Nabokov would have to commute to Wellesley on teaching days. They moved into 8 Craigie Circle, Apartment 35, and invested all of $100 to populate it with used, dowdy furniture. They stayed there until their departure for Cornell in 1948. As biographer Brian Boyd has pointed out, this six-year period represents an unusual block of domestic stability in what was otherwise a peripatetic existence in the United States.[3]

Nabokov's particular interest among butterflies was the Lycaenidae, the family comprising Blues, Coppers, and Hairstreaks. He was especially interested in the smallest (and most taxonomically confusing) members of this group, the grass Blues (or polyommatines). Because the species in this group vary little in color or pattern (they are collectively called "Blues" for a reason), Nabokov had to rely on non-color- or pattern-based indicators more than other contemporary butterfly systematists. It is for this reason that he was forced to explore and emphasize "alternative" methodologies.

Insect genitalia have long been recognized as being especially informative organs in taxonomy. Two butterfly species that are otherwise virtually identical on a general morphological assessment may differ considerably in the structure of their genitalia. Alternatively, two close relatives, with very different wing patterns—this trait is adaptively labile, so different environments may result in very different selection pressures—will still retain similarities in their genitalic structures derived from their relatively recent common ancestor. There are several good evolutionary reasons for this focus on the morphology of genitalia, including the lower probability relative to other more apparent characters that features of the genitals would undergo convergent evolution. This process, whereby natural selection causes two unrelated species to have similar features (typically because they are undergoing adaptation to similar environmental factors), is the bugbear of systematics, whose underlying premise is that similarity reflects recent common ancestry. Under convergent evolution, wing color, for example, may be the same in two unrelated species because they have evolved that color independently for adaptive reasons. In general, the structure of genitalia is less likely to undergo convergent evolution than an "obvious" character such as wing color. Two further reasons for Nabokov's focus on genital morphology stand out.

First is sexual selection, which is the process first described by Darwin whereby selection operates in favor of traits that facilitate access to members of the opposite sex (even if the sexually favored trait is deleterious to the survival of its possessor). The peacock's tail is the canonical example. How can sexual selection apply to concealed features of an animal, such as the genitalia of an insect? William G. Eberhard has argued effectively that many female insects assess the value of a male partner on the basis of the interaction between his genitalia and her own.[4] Typically, it is the female who is choosy (rather than the male) because she is the one investing heavily, in the form of eggs, in each bout of reproduction. Aspects of the genitalia are therefore prone to evolve rapidly, because the coupling of female preference with variation in the trait in males results in an escalating evolutionary process in which difference fosters more difference. Rapid evolution is useful for the taxonomist because it permits closely related species to be distinguished.

Second, there is a "lock-and-key" element to genitalic evolution. If two closely related species hybridize, the resulting offspring will likely be at some kind of selective disadvantage: sometimes, for example, they are sterile, or maybe they are poorly adapted to their environment. Selection therefore favors the evolution of mating discrimination such that an individual is less likely to make the wrong reproductive choice. One way this can occur is through genitalic evolution, where hybridization is prevented by a simple mechanical failure in sexual engagement. Again, this is an instance in which the evolution of genitalia outstrips that of other components of the organism.

This evolution of specialized genitalia may contrast with evolution of other aspects of the animal, which are often under stabilizing selection, whereby selection acts against the extremes of the distribution of variation in a species, effectively ensuring a lack of evolution. Natural selection commonly results in this kind of selection: if a trait "works" in its environment, then stabilizing selection operates to ensure that it remains that way. In short, a great deal of natural selection operates to promote evolutionary conservatism, involving little or slow rates of change. Here, then, is the value of genitalia to the taxonomist: whereas many aspects of the organism are highly constrained in evolution (and are therefore uninformative from a systematic point of view), some traits, and genitalia in particular, often evolve relatively rapidly. A group of closely related butterflies may thus have identical wing coloration but distinct yet related genitalic morphologies.

As Brian Boyd has described, "At other manual work Nabokov felt himself all thumbs, but when he began to dismantle a butterfly he found he suddenly developed very delicate hands and fine fingers and could do anything. He would prize apart under the microscope one hooked lobe of the somewhat triangular genitalia, remove the genitalia from the butterfly, coat them in glycerine, and place them in a mixture of alcohol and water in a separate labeled vial for each specimen. This, he had found, enabled him to turn the organs around under the microscope to obtain a three-dimensional view impossible on a conventional microscope slide."[5] The second methodology that Nabokov developed, and placed special emphasis on, was analysis of fine structural detail. Take, for example, a trait that is under stabilizing selection, such as the pattern of blue pigmentation on the hind wing of a butterfly. Nabokov appreciated that the blue pattern could be formed through multiple means. The scales on the wing could be oriented in different ways; the number of rows of scales could vary; the specifics of the wing venation could differ. In other words, it is possible to find significant differences underlying apparent similarity. This led to Nabokov's detailed work on wing morphology: the patch of blue might look the same, but one group of presumably closely related species dedicated four rows of scales to its construction whereas another group used five rows for the same purpose. The argument here is that the fine detail of the trait is selectively neutral (and therefore not subject to the conservative force of stabilizing selection): as long as the patch of blue is present and the trait itself is subject to stabilizing selection, which would eliminate individuals whose patch is either too big or too small, the details of how it is constructed are inconsequential (and therefore free to evolve). Once again, freedom from selective constraint results in evolutionary change that may be useful to the taxonomist.

NABOKOV APPLIED HIS APPROACH to a range of taxonomic problems while at the MCZ.

What impressed us most in reviewing the combination of the Nabokov holdings in the MCZ and the drawings from the Berg Collection at the New York Public Library is the painstaking, meticulous rigor of Nabokov's work. There were, for him, no shortcuts. He did not make his broad-scale comparison among taxa on the basis of published accounts; rather, his preference was to examine in person each and every relevant taxon. Although his immediate interest was in New World species, he was truly global in his perspective, choosing species to study from Asia and elsewhere to ensure that his was a comprehensive knowledge of the morphology of the Blues. We see Nabokov the mental-databaser at work here. Beside many of the drawings are notes identifying similar or possibly related genital forms he had seen in other species. This is the mark of a true naturalist: essentially, he had created, in his mind, a vast 3-D database of every set of butterfly genitalia he had ever seen. He then had the capacity to scan through that database with every new set he encountered, identifying in the process forms that were similar and therefore likely to be related.

Considering the assembled notes and drawings, it seems clear that Nabokov eventually intended to revise all of the Old World polyommatines, a particularly large and intractable group. His drawings include species of *Maculinea*, *Leptotes*, *Brephidium*, and other groups well outside his beloved Plebejinae. Had he continued working as an entomologist and carried out this massive reorganization, which would have included species much more familiar to the majority of working entomologists, it also seems likely that his ideas would have been more quickly evaluated, challenged, and incorporated in one way or another into mainstream systematic usage.

While working at the MCZ, Nabokov made a series of collecting trips that provided the background for some of the road-trip vignettes of *Lolita*. In 1947, briefly flush with cash after the advance for *Bend Sinister*, Nabokov spent June through September in the Colorado Rockies. In a letter to Edmund Wilson, he wrote, "I am having a wonderful, though somewhat strenuous time collecting butterflies here. We have a most comfortable cabin all to ourselves. The flora is simply magnificent, some part of me must have been born in Colorado, for I am constantly recognizing things with a delicious pang."[6] He loved pursuing fast-flying lycaenids in particular and relished their preferred mountain habitats. Nabokov exhausted anyone who went collecting with him and noted that he lost twenty pounds during a summer's collecting trip. A total of 2,472 specimens belonging to seventeen families collected by Nabokov (1,615 by Vladimir, 71 by his son, Dmitri, 785 by both, and 1 by Véra) are deposited in the MCZ (Table 2). Of these, 802, or 32 percent, are in the family Lycaenidae.

Despite his rigor in observing and measuring morphological characters, Nabokov's presentation of pinned specimens did not reflect a similar attention to detail (and to be fair, the specimens bearing his labels may have been prepared by an assistant rather than by Nabokov himself). The wings were generally well spread; however, the antennae were left in random positions or pinned under the wing, and even Lycaenidae did not receive much special attention. Almost all of the butterflies and moths that he collected while traveling in Utah and Colorado are spread, but nearly all the moths and many of the butterflies collected in the eastern states are simply pinned and labeled.

Table 2. Nabokov's Lepidoptera in the MCZ, sorted by family, state (North America), and year (1942–58)

Family	N	State	N	Year	N
Arctiidae	28	Alberta	1	1942	715
Crambidae	1	British Columbia	2	1943	697
Drepanidae	8	Colorado	815	1944	2
Geometridae	236	Georgia	20	1045	21
Hesperiidae	151	Massachusetts	166	1946	8
Lasiocampidae	25	Montana	8	1947	792
Limacodidae	3	Nebraska	3	1948	0
Lycaenidae	802	New Hampshire	7	1949	20
Lymantriidae	3	Oregon	52	1950	0
Noctuidae	608	South Carolina	20	1951	27
Notodontidae	22	Utah	695	1952	124
Nymphalidae	450	Vermont	531	1953	52
Papilionidae	22	Wyoming	140	1954–57	0
Pieridae	102			1958	6
Pyralidae	2				
Sphingidae	8				
Tortricidae	1				
TOTAL	2,472				

Note: Total numbers in the second and third columns do not equal the first column due to missing data

Nabokov's labeling was also variable. Most of his specimens have adequate locality data, but his dates of collection are often limited to the year. Presumably he knew that the dates could be inferred from records of his vacations. Many handwritten labels on torn scraps of paper, especially specimens from the East Coast, have only the location and his name, but again the approximate dates can often be inferred from his well-documented travels. Nabokov's lack of consistent date-keeping is reflected in *Lolita*, where temporal anomalies have been the subject of contention and speculation.

One anecdote must stand in for many concerning his extensive contributions to the MCZ. From February 1953, Nabokov was again resident in Cambridge, departing on April 7 for yet another trip west in his Oldsmobile to work on the final draft of *Lolita*. He and Véra ended up in Oregon, where they stayed at the house of Arthur Tailor in Ashland. Three pinned and forty-seven papered specimens of *Plebejus anna ricei* from Keene Creek, eighteen miles east of Ashland, as well as one *Pieris marginalis* and one *Anthocharis lanceolata* (both Pieridae) are in the MCZ Collection. Three papered specimens from Oregon in the MCZ are still in a round tin of "Charlotte Charles Sherry Pralines," and on a label inside, written in red grease pencil (in Nabokov's hand), are the words "Cut them out." All the butterflies are in stiff glassine envelopes that are folded close to the specimens, so the wording "Cut them out" is likely an instruction to cut the envelopes around the specimens to avoid damage that may result from unfolding them when removing the precious butterflies. One can imagine the scene at Keene Creek in 1953: following a picnic lunch, Nabokov, with his net, exploring the grassy slopes and flowers in search of Blues. Returning with his catch, he would seek a sturdy container—what better than a tin of Charlotte Charles Sherry Pralines, perhaps brought by Véra as refreshment? The name "Charlotte" was to appear later as Lolita's mother and Humbert's wife, Charlotte Haze.

OF THE DISTRIBUTION OF BLUES in the new world, Nabokov wrote, "I find it easier to give a friendly little push to some of the forms and hang my distributional horseshoes on the nail of Nome rather than postulate transoceanic land-bridges in other parts of the world." [7] In the margins of the MCZ's copy of *Check List of the Lepidoptera of Boreal America* by Barnes and McDunnough, Nabokov revised and annotated two checklists of polyommatine butterflies.[8] These appear to represent his views on a systematic reorganization of the North American Blues before the publication of his paper on Nearctic Polyommatina.[9] These notes and the names he applied to new genera enable us to see the evolution of his thinking about the systematics of this complex group: (1) he proposed the name *Pseudolycaeides* for the monotypic genus that contains *emigdionis* but in his 1945 revision of this taxon he published it as *Plebulina*; and (2) he split *Icaricia* into three genera: *Polyommatides*, *Plebejus* and *Nearicia*, reserving *Polyommatides* for the monotypic *P. icaricioides*, placing *saepiolus* in *Plebejus*, and grouping the remaining species in *Nearicia*. His decision to split *Icaricia* into three genera is insightful. *Icaricia* is one of the genera examined in a recent paper by Talavera et al. on the systematics of polyommatines using DNA data to establish taxonomic boundaries. The group of species that Nabokov calls *Nearicia* is old enough (between four and five million years) to be considered a genus based on the criteria proposed in the Talavera et al., paper.[10] Nabokov also recognized that the sister taxa to *Nearicia*, namely, *saepiolus* and *icarioides*, should each be in different genera. Thus results from DNA analysis again corroborate Nabokov's sense that this group falls close to the limits of the genus-species boundary. If at some point this group is formally reclassified, the name *Nearicia* would be highly appropriate.

All three generic names, *Pseudolycaeides*, *Polyommatides* and *Nearicia*, proposed by Nabokov in his notations on the *Check List* are valid in the sense that they have not been formally applied and are therefore not "preoccupied" by other taxa. Thus they are new names that have never been published. Nabokov later changed his mind about two of the names when he published his revision of this group, but his generic divisions are yet another example of Nabokov's remarkable insight into butterfly systematics.

Another surprise in Nabokov's marginalia: he attributes a New World species to the well-known Old World genus *Maculinea*. He later changes his opinion, and in one of the lists crosses out his own writing. The "*Maculinea*" taxon in question is *pardalis* (Berh. 1867), which is now considered a subspecies of *Icaricia icarioides*. There is no doubt that it is not a *Maculinea*, and Nabokov himself realized this (probably when he looked carefully at the genitalia). However, the external morphology is strikingly similar to *Maculinea*, and it is easy to see how he could have been misled on first glance.

Nabokov's detailed analyses are impressively meticulous and extensive, but perhaps his big-picture thinking is most impressive. Both his classification of species and his inferred timing of evolutionary events are remarkably accurate by today's DNA-informed standards.

One wonders at his ability to do this without modern analytical tools, but the conclusion seems simple: Nabokov's encyclopedic, firsthand knowledge of the group as a whole, including both Old and New World species, coupled with his facility for synthetic insight, together provided the essential ingredients. Nabokov's achievement leaves no doubt that he was a great synthesizer.

NABOKOV'S MOST EXTRAORDINARY scientific contribution during his years at the MCZ was his reassessment of Neotropical Polyommatina (= Plebejinae) Blues. This is a remarkable paper. He divided the group into nine genera, seven of which were entirely new (only *Itylos* and *Hemiargus* had previously been described, and these he reorganized). He also proposed a detailed hypothesis for the evolutionary origins of these groups. He asserted that high-altitude taxa from the Andes are not, as is typically the case for high-altitude specialists, each derived from its respective nearby lowland forms, although they are related to the lowland Caribbean forms (*Cyclargus, Hemiargus, Echinargus,* and *Pseudochrysops*). Rather, they come from the first of a series of five invasions by Asian ancestors, which colonized the Americas via the Bering connection. The most singular features of this paper are Nabokov's explicit scenario whereby this process unfolded, as well as the language with which he described it:

> One can assume, I think, that there was a certain point in time when both Americas were entirely devoid of *Plebejinae* but were on the very eve of receiving an invasion of them from Asia where they had been already evolved. Going back still further, a modern taxonomist straddling a Wellsian time machine with the purpose of exploring the Cenozoic era in a "downward" direction would reach a point—presumably in the early Miocene—where he still might find Asiatic butterflies classifiable on modern structural grounds as Lycaenids, but would not be able to discover among them anything definitely referable to the structural group he now diagnoses as *Plebejinae*. On his return journey, however, he would notice at some point a confuse adumbration, then a tentative "fade-in" of familiar shapes (among other, gradually vanishing ones) and at last would find *Chilades*-like and *Aricia*-like and *Lycaeides*-like structures in the Palaearctic region.
>
> It is impossible to imagine the exact routes these forms took to reach Chile, and I have no wish to speculate on the details of their progress, beyond suggesting that throughout the evolution of *Lycaenidae* no two species ever became differentiated from each other at the same time in the same habitat (*sensu stricto*), and that the arrival of *Plebejinae* in South America preceded the arrival in North America (and differentiation from Old World ancestors) of the genera *Icaricia* and *Plebulina* (and of the species *Plebejus saepiolus*) while the latter event in its turn preceded the invasion of North America by holarctic species which came in the following sequence: *Lycaeides argyrognomon* (subsequently split), *Agriades glandon, Vacciniina optilete.*

By the standards of today's scientific literature, this is an extremely whimsical presentation. Yet it is worth scrutinizing the language Nabokov uses here. It is in fact extraordinarily compact and succinct, and the time travel device is without question a supremely efficient way in which to communicate such a complex scenario, one that sprawls in both time and space.

Here, in Nabokov's notes recovered from the Berg Collection, is an earlier version of the same passage:

> A modern taxonomist straddling a Wellsian time machine to explore the Cenozoic era would reach a point, presumably in the early Miocene, where he still might find butterflies classifiable on structural grounds as lycaenids, but possibly would not be able to discover among them anything distinctly referable to the structural group he now diagnoses as Plebejinae. Indeed it is highly probable that by comparison with the orderly categories he has learnt to distinguish the "blues" which he would collect (with the help of some bright *Pithecanthropus* boy) would reveal a strange confusion of structures some of them vaguely familiar (i.e. resembling perhaps his Catochrysopinae or Everinae) others having no niche in his system or showing complex mosaic intrusions on the part of other "families"; and since it is quite possible that at different periods of phylogenetic time nature stressed different aspects of structure in lycaenids, the taxonomist would have to devise a completely new classification adapted to each given period.

The earlier version of this passage in his notes shows how heavily it was rewritten before publication and how carefully each word was chosen in his quest for precision. Butterflies are never lightly cited in his writing, and a deep respect and fascination for their study is evident throughout. Nabokov applied the same exacting standards to his scientific writing as his literary work. For some reason—maybe he deemed it too frivolous an inclusion in a published scientific paper—he elected to exclude his charming reference to "some bright *Pithecanthropus* boy" in the final version.

One of the most obvious ways in which Nabokov the scientist limited emotionalism and fantasy in his research is reflected by his choice of scientific names. Virtually all the scientific names Nabokov proposed were highly technical and made reference to other morphologically similar taxa. This policy forms an interesting contrast with some of the names proposed by other lepidopterists for new taxa in the group of butterflies studied by Nabokov. A number of these have been named after characters in Nabokov's literature and are not objectively related to the taxon in question. As such, they are a lasting tribute—but perhaps more to Nabokov the artist than to Nabokov the scientist.

In his paper on Neotropical Blues, Nabokov's conclusions are based on two kinds of data. First was the systematic information he assembled on these groups: Who is related to whom? Second, in determining the history of the multiple waves of immigration, he used a simple evolutionary logic. An older group is likely, on average, to have produced more descendant species than a younger one. New species vary in the rate at which they are produced, but, regardless, we expect more to be produced over a long period than over a short one. Nabokov used this to add a chronological component to his scenario. Thus, the oldest colonization is that of the diverse Neotropical lineage, which generated nine genera and many region-specific species. After that came the *Icaricia-Plebulina* stock, with two New World genera and a handful of species. The *Lycaeides* group, which diversified into a few species but did not generate a new genus came next, and, more recently, the *Agriades* group with its mildly differentiated species. The final (that is, most recent) invasion involved *optilete*, which has apparently not even had time to differentiate from Old World populations.

Possibly because Nabokov's proposition for the evolution of the New World *Polyommatina* blues was so unconventionally written from the point of view of practicing systematists, or even because it was advanced for its time, it remained forgotten for a surprisingly long time. Had Nabokov revised a more familiar group, or one that was more commonly studied by other systematists, his work might have also been assessed and reevaluated more quickly by mainstream taxonomists. As it was, the work was largely ignored until three butterfly specialists, Kurt Johnson, Zsolt Bálint, and Dubi Benyamini, focused their attention on Nabokov's work and masterfully revitalized his views.[11]

Moreover, the advent of molecular approaches to understanding evolution has meant that we can now simply sequence a homologous stretch of DNA in multiple species and count the differences among the DNA sequences. Closely related taxa—that is, those sharing a relatively recent common ancestor—have relatively similar DNA sequences; distant taxa—those whose common ancestor is relatively far back in time—have relatively different DNA sequences. We no longer need to rely on a lifetime's experience of comparing the relative similarities and differences in aspects of genital morphology.[12]

It turned out that collecting fresh specimens of the many taxa that Nabokov included in his analysis was surprisingly difficult. These are obscure species from remote locations, often high in the mountains of Central and South America. But, specimens finally assembled, we were able to do the DNA sequencing and then the analysis of the sequences. It was an extraordinary moment when, at last, all the results were in place. The conclusion was clear: Nabokov's "Wellsian" scenario was correct in every detail. It is sobering to imagine that what took a modern laboratory using state-of-the-art tools almost ten years and six expeditions to South America to complete was even feasible for Nabokov as he squinted through his microscope at butterfly genitalia. Nabokov's paper on the evolution of the Neotropical *Plebejinae* was a scientific tour de force and also a monument to his scientific intuition.[13]

Our study has permitted us to elaborate on Nabokov's original scenario in ways that he would surely have approved. First, there is the matter of using the molecular clock to put real numbers on the relative dates that Nabokov was able to derive. We conclude that the first invasion event occurred around eleven million years ago and that the final (fifth) one occurred around one million years ago.

Second, we have been able to infer paleoclimate data for the Bering region, which Nabokov argued (and we concur) was the key step in each of the Asia-to-Americas invasions.[14] We can also use computational techniques to reconstruct the ancestral state of a character for which we have information among a group of related taxa. In this case, we reconstructed the temperature tolerances of the ancestral taxa that braved the Bering region in the course of each of the five invasion events. Over the relevant time period (the past twelve million years), the climate of the Bering region has cooled. The thermal tolerances of the ancestral invading taxon in each case correspond well with the Bering temperatures in the appropriate period.

THAT NABOKOV SHOWED SUCH INSIGHT despite working with such apparently limited tools at his disposal argues that he was more than just a competent taxonomist. His work on the Neotropical *Polyommatus* Blues stands out as a bold and brilliant scientific advance. Nor is this a lone example: extensive molecular analyses of Nabokov's hypotheses regarding relationships in the genus *Lycaeides* made by the team of Chris Nice, Art Shapiro, Jim Fordyce, Matthew Forester, Zach Gompert, and Lauren Lucas, including tests of hybrid species complexes and the subspecies status of the Karner Blue, have repeatedly confirmed his reputation as a remarkable systematist.[15]

Nabokov deemed his years in the MCZ "the most delightful and thrilling in all my adult life."[16] We, in turn, have been delighted and thrilled to follow in his scientific footsteps and to find evidence of true scientific greatness dwelling in the cabinets, on the bookshelves, and, occasionally, in the quiet corners of the MCZ's Lepidoptera room.

Acknowledgments

We thank Stephen Blackwell and Kurt Johnson for inviting us to participate in this volume. Kurt encouraged us both early on in our research and at several critical junctures, and Steve's advice and edits of our manuscript were invaluable. Bob Pyle and Brian Boyd provided insight and guidance. Scott Walker prepared the map of Nabokov's and Humbert Humbert's travels and deftly fielded questions and suggestions from five people at once. Gabriel A. Miller took the photographs of Nabokov's genitalia cabinet, marginalia, reprints, and specimens from the MCZ shown in Color Plates E6, E7, and E8.

Notes

1. http://www.mcz.harvard.edu/Departments/Entomology/holdings.html.

2. *NWL*, 76.

3. *VNAY*, 58–59.

4. Eberhard, *Sexual Selection;* Eberhard, *Female Control.*

5. *VNAY*, 58–59.

6. *NWL*, 218.

7. Nabokov, "Notes on Neotropical," 44.

8. Barnes and McDunnough, *Check List.*

9. Nabokov, "Notes on the Morphology."

10. Talavera et al., "Establishing Criteria."

11. See *NBl* and its complete list of references to Johnson, Bálint, and Benyamini's papers.

12. Another advantage of using DNA in this work is that it permits temporal analysis via a so-called molecular clock. Imagine that we know from unrelated, independent studies that a one thousand base pair stretch of DNA accumulates on average one new mutation every hundred thousand years. Suppose we have two species that differ for that stretch of DNA at six sites: because genetic divergence is usually a symmetrical process, we posit that three mutations arose along each lineage since their common ancestor, implying that that common ancestor existed three hundred thousand years ago. Not only, then, would DNA analysis allow us to reconstruct the evolutionary history of these butterflies, but it would also give us a timeframe for the relevant events.

13. Nabokov, "Notes on Neotropical."

14. This was done using the Zachos method for estimating changes in local temperatures from changes in oxygen isotopes of seawater. Zachos et al., "Trends."

15. Forister et al., "After Sixty Years"; Gompert et al., "Homoploid Hybrid Speciation."

16. *SO*, 190.

Chronology of Nabokov's Life

1899: Nabokov is born in Saint Petersburg on April 10 (old calendar)

1906: Begins butterfly collecting

1909: Writes to the *Entomologist* (London) attempting to name first species

1916: Publishes his first poem in a "thick" Russian literary journal

1918: Escapes with his family to Crimea

1919: Moves to London and then begins studies at Cambridge University; his family settles in Berlin

1920: Publishes first lepidopterological paper, in the *Entomologist*

1921: Publishes his first short story, "The Wood Sprite"

1922: In March his father is murdered during an assassination attempt on Pavel Miliukov in Berlin; settles in Berlin after graduating from Cambridge University

1923: His translation of *Alice's Adventures in Wonderland* into Russian is published

1925: Marries Véra Slonim

1926: *Mary* (*Mashenka*), his first novel, is published

1929: *The Defense,* his third novel, is published

1931: Publishes his second lepidopterological paper in the *Entomologist*

1934: *Invitation to a Beheading,* his eighth novel, is published

1934: His son, Dmitri, is born in May

1937: Moves with his wife and child from Berlin to Paris

1937–38: Publishes his ninth and final Russian novel, *The Gift,* without its fourth chapter, which was refused by the literary journal *Sovremennye zapiski* (Contemporary annals); the complete novel is published only in 1952

1938: Writes his first English novel, *The Real Life of Sebastian Knight* (published 1941)

1939 or 1941: Writes "Father's Butterflies," a fictitious but unused addendum to *The Gift*

1940: Fleeing Hitler's army, sails with wife and child to the United States in May

1940–41: Works at the American Museum of Natural History, New York

1941: Volunteers at the Museum of Comparative Zoology, Harvard University; publishes first new species name, *Lysandra cormion* (later determined to be a hybrid rather than a distinct species)

1942: Begins teaching Russian language and literature at Wellesley College; begins paid curating position at MCZ; gives mimicry manuscript (now lost) to William DeVane, dean of Yale College, for submission to the *Yale Review*

1943–44: Writes biography of Nikolai Gogol

1941–49: Publishes twelve lepidopterological papers, including the three major revisions

1947: Publishes *Bend Sinister,* his first novel written in the United States

1948: Departs MCZ and Wellesley for a full-time, tenured position at Cornell University as professor of Russian literature

1948–53: Summer travels and collecting trips while researching and writing *Lolita*

1951: *Conclusive Evidence,* the first version of his memoir, is published

1952–63: Works on translation of and commentary to Alexander Pushkin's *Eugene Onegin* (published 1964)

1955: *Lolita* is published in France

1957: *Pnin* is published

1958: *Lolita* is published in the United States

1959: Takes leave from Cornell and sails for Europe

1961: Settles "permanently" in the Palace Hotel, Montreux, Switzerland

1962: The film *Lolita,* directed by Stanley Kubrick, is released

1962: *Pale Fire* is published

1963–65: Works on an enormous project, "Butterflies of Europe"; abandons it when his publisher's commitment waivers

1966: *Speak, Memory: An Autobiography Revisited* is published

1969: *Ada, or Ardor* is published

1972: *Transparent Things* is published

1973: *Strong Opinions* is published

1974: *Look at the Harlequins!* is published

1977: Dies on July 2 in Lausanne, Switzerland, of a mysterious infection

Bibliography

Abbreviations

BA   Butterflies of America, Vladimir Nabokov Papers, Berg Collection, New York Public Library (seven boxes, specific locations approximate due to library practices: BA[box]-[card holder][slot])

LL   *Lectures on Literature.* New York: Harcourt, Brace, Jovanovich, 1980.

Lolita   *The Annotated Lolita.* Rev. ed. Edited by Alfred Appel Jr. New York: Vintage Books, 1991.

LRL   *Lectures on Russian Literature.* New York: Harcourt, Brace, Jovanovich, 1981.

NB   Brian Boyd and Robert Michael Pyle, eds., *Nabokov's Butterflies: Unpublished and Uncollected Writings.* Boston: Beacon Press, 2000.

NBl   Kurt Johnson and Steve Coates, *Nabokov's Blues: The Scientific Odyssey of a Literary Genius.* Cambridge, MA: Zoland Books, 1999.

NWL   *Dear Bunny, Dear Volodya: The Nabokov-Wilson Letters, 1940–1971.* Rev. ed. Edited by Simon Karlinsky. Berkeley: University of California Press, 2001.

PF   *Pale Fire.* New York: Vintage International, 1989.

SelL   *Selected Letters, 1940–1977.* Edited by Dmitri Nabokov and Matthew J. Bruccoli. New York: Harcourt Brace, 1989.

SO   *Strong Opinions.* New York: Vintage International, 1990.

SM   *Speak, Memory: An Autobiography Revisited.* New York: G. P. Putnam's Sons, 1966.

VNA   Vladimir Nabokov Papers, Berg Collection, New York Public Library

VNAY   Brian Boyd, *Vladimir Nabokov: The American Years.* Princeton, NJ: Princeton University Press, 1991.

VNRY   Brian Boyd, *Vladimir Nabokov: The Russian Years.* Princeton, NJ: Princeton University Press, 1990.

Ahuja, Nitin. "Nabokov's Case against Natural Selection." *Tract* (blog), Spring 2006. http://www.hcs.harvard.edu/tract/nabokov.html.

Aikhenval'd, Iulii. *Siluety russkikh pisatelei* [Silhouettes of Russian writers]. 5th ed. Berlin: Slovo, 1928.

Aksakov, S. T. *Babochki* [Butterflies]. Moscow: Detskaya literatura, 1938.

Alexander, Victoria N. *The Biologist's Mistress: Rethinking Self-Organization in Art, Literature, and Nature.* Litchfield Park, AZ: Emergent Publications, 2011.

———. "Nabokov, Teleology, and Insect Mimicry." *Nabokov Studies* 7 (2002–3): 177–213.

———. "Neutral Evolution and Aesthetics: Vladimir Nabokov and Insect Mimicry." Working Papers Series 01-10-057. Santa Fe, NM: Santa Fe Institute, 2001.

———. "The Poetics of Purpose." *Biosemiotics* 2 (2009): 77–100.

Alexander, Victoria N., and Stanley Salthe. "Monstrous Fate: The Problem of Authorship and Evolution by Natural Selection." *Annals of Scholarship* 19 (2010):45–66.

Alexandrov, Vladimir. "A Note on Nabokov's Anti-Darwinism; Or, Why Apes Feed on Butterflies in *The Gift.*" In Elizabeth Cheresh Allen and Gary Saul Morson, eds., *Freedom and Responsibility in Russian Literature: Essays in Honor of Robert Louis Jackson,* 239–44. Evanston, IL: Northwestern University Press, 1995.

Allen, Thomas J. *The Butterflies of West Virginia*

and Their Caterpillars. Pittsburgh: University of Pittsburgh Press, 1997.

Amis, Martin. "*Lolita* Reconsidered." *Atlantic Monthly,* September 1992, 109–21.

Babikov, Andrei. "'*Dar*' za chertoi stranitsy." *Zvezda* 2015.4, http://magazines.russ.ru/zvezda/2015/04/7bab-pr.html.

Bailey, James S. "Center, N.Y., Entomologically Considered." *Canadian Entomologist* 9 (1877): 115–19.

Barabtarlo, Gennady. "Nabokov's Trinity (On the Movement of Nabokov's Themes)." In Julian W. Connolly, ed., *Nabokov and His Fiction: New Perspectives,* 109–38. Cambridge: Cambridge University Press, 1999.

Barnes, Hugh. *The Stolen Prince: Gannibal, Adopted Son of Peter the Great, Great-Grandfather of Alexander Pushkin, and Europe's First Black Intellectual.* New York: HarperCollins, 2006.

Barnes, Wm., and J. McDunnough. *Check List of the Lepidoptera of Boreal America.* Decatur, IL: Herald Press, 1917.

Barthes, Roland. "The Death of the Author." Orig. publ. 1967. Reprinted in *Contributions in Philosophy* 83 (2001): 3–8.

Bates, H. W. "Contributions to an Insect Fauna of the Amazon Valley. *Lepidoptera: Heliconidae.*" *Transactions of the Linnean Society* 23 (1862): 495–566.

Bateson, William. *Materials for the Study of Variation, Treated with Special Regard to Discontinuity in the Origin of Species.* London: Macmillan, 1894.

Belyi, Andrei. "Sravnitel'naia morfologiia ritma russkikh lirikov v iambicheskom dimetre" [Comparative morphology of the rhythm of Russian lyric poets in iambic dimeter]. In *Simvolizm* [Symbolism], 331–95. Orig. publ. 1910. Reprint, Munich: Fink.

Berenbaum, May. "Blue Book Value." *Science* 290 (2000): 57–58.

Berger, Alan, and Dorion Sagan. "A Brief History of the Future: The Metaphysics of Ruination." *Cabinet: A Quarterly of Art and Culture* 21 (2006): 40–43.

Blackwell, Stephen H. "A New or Little-Known Subtext in *Lolita.*" *Nabokovian* 60 (2008): 51–55.

———. "Notes on a Famous First Line ('Light of My Life')." *Nabokovian* 64 (2010): 39–44.

———. *The Quill and the Scalpel: Nabokov's Art and*

the Worlds of Science. Columbus: Ohio State University Press, 2009.

Boswell, Emma. *How Do You Solve a Problem Like Lolita?* BBC TV film. Directed and produced by Emma Boswell. Broadcast premiere December 13, 2009. Posted by Few Docs, October 14, 2012, https://www.youtube.com/watch?v=TnvvBL6set4.

Boyd, Brian. "*Lolita:* What We Know and What We Don't." *Cycnos* 24 (2007):215–28.

———. "Nabokov, Literature, Lepidoptera." In *NB,* 1–31.

———. *Nabokov's "Ada": The Place of Consciousness.* 2nd ed. Rochester, MN: Cybereditions, 2002.

———. *Nabokov's "Pale Fire": The Magic of Artistic Discovery.* Princeton, NJ: Princeton University Press, 1999.

———. "Pinning Down Krolik." *Nabokovian* 48 (2002): 23–27.

———. *Stalking Nabokov: Selected Essays.* New York: Columbia University Press, 2011.

Boyle, Robert H. "An Absence of Wood Nymphs." Orig. publ. 1959. Reprinted in *At the Top of Their Game,* 123–33. New York: Nick Lyons Books, 1983. Excerpted in *NB,* 528–37.

Brooks, Karl L. *A Catskill Flora and Economic Botany: IV (Part 2).* Albany: New York State Museum, 1984.

Brown, F. Martin. "Measurements and Lepidoptera." *Lepidopterists' News* 4 (1950): 4–5.

———. "The Types of the Pierid Butterflies Named by William Henry Edwards." *Transactions of the American Entomological Society* 99 (1973): 97–101.

Brown, F. Martin, Donald Eff, and Bernard Rotger. *Colorado Butterflies.* Denver, CO: Denver Museum of Natural History, 1957.

Butler, Diana. 1960. "Lolita Lepidoptera." In *New World Writing,* 58–84. Philadelphia: J. B. Lippincott, 1960.

Butler, Kirstin. "Nabokov's Legacy: Bequeathing Butterfly Theory." Brain Pickings (blog), 2011. http://www.brainpickings.org/index.php/2011/07/01/nabokov-butterflies/.

Calhoun, John V. "The Extraordinary Story of an Artistic and Scientific Masterpiece: *The Butterflies of North America* by William Henry Edwards, 1868–1897." *Journal of the Lepidopterists' Society* 67 (2013): 73–110.

Cech, Rick, and Guy Tudor. *Butterflies of the East Coast: An Observer's Guide.* Princeton, NJ: Princeton University Press, 2005.

Chamberlain, Nicola L., et al. "Polymorphic Butterfly Reveals the Missing Link in Ecological Speciation." *Science* 326 (2009): 847–50.

Coates, Steve. "Nabokov's Work, on Butterflies, Stands the Test of Time." *New York Times,* May 27, 1997.

Conniff, Richard. "Vlad the Impaler: An Examination of Nabokov as Lepidopterist." *New York Times Book Review,* February 20, 2000.

Cook, L. M., et al. "Selective Bird Predation on the Peppered Moth: The Last Experiment of Michael Majerus." *Biology Letters* 8 (2012): 609–12.

Cracraft, Joel. "Phylogenetic Models and Classification." *Systematic Zoology* 23 (1974): 71–90.

Darwin, Charles. *On the Origin of Species by Means of Natural Selection.* London: John Murray, 1859.

Dasmahapatra, K. K., et al. "Butterfly Genome Reveals Promiscuous Exchange of Mimicry Adaptations among Species." *Nature* 487 (2012): 94–98.

Dean, Bashford. "A Case of Mimicry Outmimicked? Concerning Kallima Butterflies in Museums." *Science* 16 (1902): 832–23.

Desmortiers, Patricia. *Vladimir Nabokov.* Film in series *Un siècle d'écrivains.* Directed by Patricia Desmortiers. 45 mins. Paris: Cinétévé, 1997.

Diaconis, Persi, and Frederick Mosteller. 1989. "Methods for Studying Coincidences." *Journal of the American Statistical Association* 84 (1989): 853–61.

Dirig, Robert. "Definitive Destination, McLean Bogs Preserve: Finger Lakes Region, New York." *American Butterflies* 10, no. 4 (2002): 4–16.

———. "Nabokov's Rainbow." *American Butterflies* 9, no. 3 (1999): 4–10.

———. "Theme in Blue: Vladimir Nabokov's Endangered Butterfly." In Gavriel Shapiro, ed., *Nabokov at Cornell,* 205–18. Ithaca, NY: Cornell University Press, 2003.

Dirig, Robert, and Akito Y. Kawahara. "Moonbeams in the Forest: *Speyeria diana* at a West Virginia Enclave." *Southern Lepidopterists' News* 35, no. 4 (2013): 170–78.

Dobzhansky, Theodosius. "Catastrophism versus Evolution." *Science* 92 (1940): 356–58.

———. *Genetics and the Origin of Species.* New York: Columbia University Press, 1937.

Dobzhansky, T., and C. Epling. *Contributions to the Genetics, Taxonomy, and Ecology of* Drosophila pseudoobscura *and Its Relatives.* Publication no. 554. Washington, DC: Carnegie Institution, 1944.

Dolinin, Alexander. "The Signs and Symbols in Nabokov's 'Signs and Symbols.'" Zemblarchive (blog), http://www.libraries.psu.edu/nabokov/dolinin.htm. Reprinted in Yuri Leving, ed., *Anatomy of a Short Story,* 257–59. New York: Continuum, 2012.

dos Passos, Cyril F. 1949. "A Visit to the Home of the Late William Henry Edwards at Coalburg, West Virginia." *Lepidopterists' News* 3, no. 6 (1949): 61–62.

Dufour, Léon. "Anatomie générale des Diptères." *Annales des Sciences Naturelles* 1 (1844): 244–64.

Eberhard, William G. *Female Control: Sexual Selection by Cryptic Female Choice.* Princeton, NJ: Princeton University Press, 1996.

———. *Sexual Selection and Animal Genitalia.* Cambridge, MA: Harvard University Press, 1985.

Edwards, William Henry. *Butterflies of North America.* 2 vols. Boston: Houghton, Osgood, 1868–84.

———. "Description of New Species of Diurnal Lepidoptera Found within the United States." *Transactions of the American Entomological Society* 3 (1870): 13–14.

———. "Description of the Female of *Argynnis diana.*" *Proceedings of the Entomological Society of Philadelphia* 3 (1864): 431–34.

———. "On *Pieris bryoniae* Ochsenheimer, and Its Derivative Forms in Europe and America." *Papilio* 1, no. 6 (1881): 95–99.

Eimer, T. *Orthogenesis der Schmetterlinge* [Orthogenesis of butterflies]. Leipzig: Wilhelm Engelmann, 1897.

Eldredge, Niles, and Joel Cracraft. *Phylogenetic Patterns and the Evolutionary Process: Method and Theory in Comparative Biology.* New York: Columbia University Press, 1985.

Favareau, Donald, ed. *Essential Readings in Biosemiotics: Anthology and Commentary.* New York: Springer, 2010.

Fet, Victor. "Zoological Label as Literary Form." *Nabokovian* 60 (2008): 18–26.

Field, Andrew. *Nabokov: His Life in Part.* New York: Viking, 1977.

Flower, Timothy F. "The Scientific Art of Nabokov's 'Pale Fire.'" *Criticism* 17 (1975): 223–33.

Forister, M. L., and C. F. Scholl. 2012. "Use of an Exotic Host Plant Affects Mate Choice in an Insect Herbivore." *American Naturalist* 179 (2012): 805–10.

Forister, M. L., et al. "After Sixty Years, an Answer to the Question: What Is the Karner Blue Butterfly?" *Biology Letters* 7 (2010): 399–402.

Foucault, Michel. "What Is an Author?" Orig. publ. 1969. Reprinted in *Contributions in Philosophy* 83 (2001): 9–22.

Franclemont, John G. "Remembering Nabokov." In George Gibian and Stephen Jan Parker, eds., *The Achievements of Vladimir Nabokov: Essays, Reminiscences, and Stories from the Cornell Nabokov Festival*, 227–28. Ithaca, NY: Center for International Studies, Committee on Soviet Studies, 1984.

Glassberg, Jeffrey. "Photograph of Early Hairstreak." *American Butterflies* 17, no. 2 (2009): 40–41.

Goldschmidt, Richard. "Einige Materialien zur Theorie der abgestimmten Reactionsgeschwindigkeiten" [Some material on the theory of concerted reactions speeds]. *Archiv für mikroskopische Anatomie und Entwicklungsgeschichte* 98 (1923): 292–313.

———. *The Material Basis of Evolution*. New Haven: Yale University Press, 1940.

———. "Mimetic Polymorphism, a Controversial Chapter of Darwinism." *Quarterly Review of Biology* 20 (1945): 147–64.

———. "Mimetic Polymorphism, a Controversial Chapter of Darwinism (Concluded)." *Quarterly Review of Biology* 20 (1945): 205–30.

———. *Physiologische Theorie der Vererbung* [Physiological theory of heredity]. Berlin: Julius Springer, 1927.

———. "Untersuchungen zur Entwicklungsphysiologie des Flügelmusters der Schmetterlinge" [Studies on developmental physiology of the wing pattern of the butterfly]. *Archiv für Entwicklungsmechanik der Organismen* 47 (1920): 1–24.

Gompert, Z., et al. "Genomic Regions with a History of Divergent Selection Affect Fitness of Hybrids between Two Butterfly Species." *Evolution* 66 (2012): 2167–81.

———. "Geographically Multifarious Phenotypic Divergence during the Speciation Process." *Ecology and Evolution* 3 (2013): 595–613.

———. "Homoploid Hybrid Speciation in an Extreme Habitat." *Science* 314 (2006): 1923–25.

———. "Secondary Contact between *Lycaeides idas* and *L. melissa* in the Rocky Mountains: Extensive Introgression and a Patchy Hybrid Zone." *Molecular Ecology* 19 (2010): 3171–32.

Gould, Stephen Jay. *I Have Landed: The End of a Beginning in Natural History*. New York: W. W. Norton, 2002.

———. *The Structure of Evolutionary Theory*. Cambridge, MA: Belknap Press of Harvard University Press, 2002.

Grant, Bruce S. "Industrial Melanism." *eLS* (2012). DOI: 10.1002/9780470015902.a0001788.pub3.

Grayson, Jane. *Vladimir Nabokov*. New York: Penguin, 2001.

Grice, Grayson. Review of *NBl*. *BookForum* Winter 1999.

Groueff, Stephane. *Crown of Thorns: The Reign of King Boris III of Bulgaria, 1918–1943*. Lanham, MD: Madison Books, 1998.

A Guide to the Theodore L. Mead Collection. Winter Park, FL: Archives and Special Collections, Rollins College, 2006.

Higgins, L. G. "The Correct Name for What Has Been Called *Lycaeides argyrognomon* in North America." *Journal of the Lepidopterists' Society* 39 (1985): 145–46.

Hoebeke, E. Richard, Richard B. Root, and James K. Liebherr. "John George Franclemont, April 15, 1912–May 26, 2004." Memorial Statements, Cornell University Faculty, 2003–4. Ithaca, NY: Office of the Dean of Faculty, Cornell University, 2004.

Holland, W. J. 1903. *The Moth Book: A Popular Guide to a Knowledge of the Moths of North America*. New York: Doubleday, Page, 1903.

Hull, David L. *Darwin and His Critics: The Reception of Darwin's Theory of Evolution by the Scientific Community*. Chicago: University of Chicago Press, 1983.

———. *Science as a Process: An Evolutionary Account of the Social and Conceptual Development of Science*. Chicago: University of Chicago Press, 1990.

Hull, David L., and Michael Ruse. *The Cambridge Companion to the Philosophy of Biology*. Cambridge: Cambridge University Press, 2007.

Hustak, Carla, and Natasha Myers. "Involutionary Momentum: Affective Ecologies and the Sciences of Plant/Insect Encounters." *differences* 23 (2012): 74–118.

Huxley, Julian. *Evolution: The Modern Synthesis*. London: George Allen and Unwin, 1942.

Jeon, Kwang W. "Amoeba and x-Bacteria: Symbiont Acquisition and Possible Species Change." In Lynn Margulis and René Fester, eds., *Symbiosis as a Source of Evolutionary Innovation*, 118–31. Cambridge, MA: MIT Press, 1991.

Johnson, Kurt. "Lepidoptera, Evolutionary Science and Nabokov's Harvard Years—More Light and Context." Paper presented at the annual meeting of the American Literature Association, Cambridge, MA, May 25, 2001.

Joron, Mathieu, et al. "Chromosomal Rearrangements Maintain a Polymorphic Supergene Controlling Butterfly Mimicry." *Nature* 477 (2011): 203–6.

Karges, Joann. *Nabokov's Lepidoptera*. Ann Arbor, MI: Ardis, 1985.

Khakhina, Liya Nikolaevna. *Concepts of Symbiogenesis: A Historical and Critical Study of the Research of Russian Botanists*. Edited by Robert Coalson, Lynn Margulis, and Mark McMenamin. New Haven: Yale University Press, 1992.

Klinghoffer, David. 2008. "Vladimir Nabokov, 'Furious' Darwin Doubter." *Evolution News and Views* (blog), July 17, 2008, http://www.evolutionnews.org/2008/07/vladimir_nabokov_furious_darwi008971.html.

Klots, Alexander B. *A Field Guide to the Butterflies of North America, East of the Great Plains*. Cambridge, MA: Houghton Mifflin, 1951.

Korshunov, Yu. P. "O cheshuekrylykh so slov Vladimira Nabokova" [Vladimir Nabokov as lepidopterologist]. http://jugan.narod.ru/nabokov.htm.

Kozo-Polyansky, Boris M. *Symbiogenesis: A New Principle of Evolution*. Orig. publ. 1924. Annotated ed. Cambridge, MA: Harvard University Press, 2010.

Kuznetsov, N. Ya. *Lepidoptera*. Vol. 1: *Fauna of Russia and Adjacent Countries*. Orig. publ. 1915. Translated by A. Mercado. Jerusalem: Israel Program for Scientific Translations, 1967.

———. "O stremlenii k nazyvaniiu kak odnom iz techenii v entomologicheskoi literature (*Namengeberei* nemtsev)" [On the passion for naming as one of the trends in entomological literature (*Namengeberei* of the Germans)]. *Revue Russe d'Entomologie* 12 (1912): 256–76.

Leving, Yurii. *Vokzal, garazh, angar: Vladimir Nabokov i poetika russkogo urbanizma* [Train station, garage, hangar: Vladimir Nabokov and the poetics of Russian urbanism]. Saint Petersburg: Izd-vo Ivana Limbakha, 2004.

Lintner, Joseph Albert. 1873. "Calendar of Butterflies for the Year 1869." *Annual Report of the New York State Museum, 23*, 182. Albany: New York State Museum, 1873.

Lucas, L. K., J. A. Fordyce, and C. C. Nice. "Patterns of Genitalic Morphology around Suture Zones in North American *Lycaeides* (Lepidoptera: Lycaenidae): Implications for Taxonomy and Historical Biogeography." *Annals of the Entomological Society of America* 101 (2008): 172–80.

Mallet, J. "A Species Definition for the Modern Synthesis." *Trends in Ecology and Evolution* 10 (1995): 294–299.

Margulis, Lynn, and Dorion Sagan. *Acquiring Genomes: A Theory of the Origin of Species*. New York: Basic Books, 2003.

Mayr, Ernst. *Systematics and the Origin of Species*. New York: Columbia University Press, 1942.

McDunnough, J. "Check List of the Lepidoptera of Canada and the United States of America. Part I: Macrolepidoptera." *Memoirs of the Southern California Academy of Sciences* 1 (1938): 1–275.

———. "New North American *Eupithecias* I (Lepidoptera, Geometridae)." *Canadian Entomologist* 77, no. 9 (1945): 168–76.

———. "Revision of the North American Species of the Genus *Eupithecia* (Lepidoptera, Geometridae)." *Bulletin of the American Museum of Natural History* 93 (1949): 533–728, pls. 26–32.

Meinhardt, Hans. *Models of Biological Pattern Formation*. New York: Academic Press, 1982.

Nabokov, Vladimir. *Ada, or Ardor: A Family Chronicle*. New York: Vintage International, 1990.

———. "Butterfly Collecting in Wyoming." *Lepidopterists' News* 7 (1953): 49–52. In *NB*, 489–94.

———. *Conclusive Evidence*. New York: Harper and Brothers, 1951.

———. *Despair*. New York: Vintage International, 1989.

———. *Drugie berega* [Other shores]. New York: Chekhov House, 1954.

———. *The Enchanter*. Translated by Dmitri Nabokov. New York: Putnam's, 1986.

———. "Father's Butterflies." [Second Addendum to *The Gift*.] Orig. publ. 1939. In *NB*, 198–234.

———. "The Female of *Lycaeides argyrognomon sublivens*." *Lepidopterists' News* 6 (1952): 35–36.

———. *The Gift.* New York: Vintage International, 1991.

———. *Laughter in the Dark.* Norfolk, CT: New Directions, 1960.

———. "Lines Written in Oregon." *New Yorker,* August 19, 1953. In *NB,* 499–500.

———. *Look at the Harlequins!* New York: McGraw-Hill, 1974.

———. "*Lysandra cormion,* a New European Butterfly." *Journal of the New York Entomological Society* 49 (1941): 265–67.

———. "The Nearctic Forms of *Lycaeides* Hüb. (Lycaenidae, Lepidoptera)." *Psyche* 50 (1943): 87–99.

———. "The Nearctic Members of *Lycaeides* Hübner (Lycaenidae, Lepidoptera)." *Bulletin of the Museum of Comparative Zoology, Harvard College* 101 (1949): 479–541.

———. *Nikolai Gogol.* New York: New Directions, 1961.

———. "Note on *Catocala fraxini.*" *Lepidopterists' News* 1 (1947): 34.

———. "Notes on Neotropical *Plebejinae* (Lycaenidae, Lepidoptera)." *Psyche* 52 (1945): 1–61.

———. "Notes on the Morphology of the Genus *Lycaeides* (Lycaenidae, Lepidoptera)." *Psyche* 51 (1944): 104–38, http://psyche.entelub.org/pdf/51/51-104.pdf.

———. "Novelist as Lepidopterist." *New York Times Magazine,* July 27, 1975, 46.

———. "On a Book Entitled *Lolita.*" *Anchor Review,* June 1957. In *Lolita,* 311–17.

———. "On Discovering a Butterfly." *New Yorker,* May 15, 1943, 26.

———. *Pnin.* New York: Vintage International, 1991.

———. "Rebel's Blue, Bryony White." Review of L. C. Higgins and N. D. Riley, *Field Guide to the Butterflies of Britain and Europe. Times Educational Supplement,* October 1970.

———. "Remarks on F. Martin Brown's 'Measurements and Lepidoptera.'" *Lepidopterists' News* 4 (1950): 75–76.

———. *Sobranie sochinenii russkogo perioda v piati tomakh: Stoletie so dnia rozhdeniia, 1899–1999* [Collected works of the Russian period in five volumes: Centennial, 1899–1999]. 5 vols. Edited by N. I. Artemenko-Tolstaia and A. A. Dolinin. Saint Petersburg: Symposium, 2000.

———. "Some New or Little Known Nearctic *Neonympha* (Lepidoptera: Satyridae)." *Psyche* 49 (1942): 61–80.

———. "The Swift." Recording. New Rochelle, NY: Spoken Arts, 1983.

———. "The Vane Sisters." Orig. publ. 1959. In Dmitri Nabokov, ed. and trans., *The Stories of Vladimir Nabokov.* New York: Alfred A. Knopf, 1995.

———. "Yesterday's Caterpillar." *New York Times Book Review,* June 3, 1951.

Negenblya, I. Ye. 2011. "Pervyi menedzher aviatsii v Yakutii" [The first aviation manager in Yakutia]. *Yakutskii arkhiv* 2 (2011): 47–54.

Nelson, Gareth, and Norman Platnick. *Systematics and Biogeography: Cladistics and Vicariance.* New York: Columbia University Press, 1981.

Nepomnyashchy, Catharine Theimer, Nicole Svobodny, and Ludmilla A. Trigos, eds. *Under the Sky of My Africa: Alexander Pushkin and Blackness.* Evanston, IL: Northwestern University Press, 2006.

Nice, C. C., et al. "The History and Geography of Diversification within the Butterfly Genus *Lycaeides* in North America." *Molecular Ecology* 14 (2005): 1741–54.

Nichols, B. *Butterflies and Skippers of the Great Smoky Mountains: A Checklist.* Gatlinburg, TN: Great Smoky Mountains Association, 2010.

Nijhout, H. F. *The Development and Evolution of Butterfly Wing Patterns.* Washington, DC: Smithsonian Institution Press, 1991.

———. "Elements of Butterfly Wing Patterns." *Journal of Experimental Zoology Part B: Molecular and Developmental Evolution* 291 (2001): 213–25.

———. "Pattern and Process." In H. F. Nijhout, Lynn Nadel, and Daniel L. Stein, eds., *Pattern Formation in the Physical and Biological Sciences,* 269–98. Reading, MA: Addison-Wesley, 1997.

Norman, Will, and Duncan White. "Transitional Nabokov: Conference Overview." *Nabokov Online Journal* 2 (2008). http://www.nabokovonline.com/uploads/2/3/7/7/23779748/v2_15_conference.pdf.

Pelham, Jonathan P. "A Catalogue of the Butterflies of the United States and Canada: With a Complete Bibliography of the Descriptive and Systematic Literature." *Journal of Research on the Lepidoptera* 40 (2010): i–xiv, 1–658.

———. *A Catalogue of the Butterflies of the United States and Canada.* Online resource: http://butterfliesofamerica.com/US-Can-Cat.htm [revised 22 June 2014].

Popper, Karl. *Conjectures and Refutations: The Growth of Scientific Knowledge.* London: Routledge and Kegan Paul, 1973.

Porter, Adam H., and Arthur M. Shapiro. "Lock-and-Key Hypothesis: Lack of Mechanical Isolation in a Butterfly (Lepidoptera: Pieridae) Hybrid Zone." *Annals of the Entomological Society of America* 83 (1990): 107–14.

Poulton, E. B. 1913. "Mimicry and the Inheritance of Small Variations." *Bedrock: A Quarterly Review of Scientific Thought* 2 (1913): 295–312.

Punnett, Reginald Crundall. *Mimicry in Butterflies.* Cambridge: Cambridge University Press, 1915.

Pushkin, A. S. *Eugene Onegin.* Translation and commentary by Vladimir Nabokov. Rev. ed. 4 vols. Princeton, NJ: Princeton University Press, 1975.

Pyle, Robert Michael. "Between Climb and Cloud: Nabokov among the Lepidopterists." In *NB,* 32–76.

———. "Lookee Here!" In Robert Michael Pyle, *The Tangled Bank: Writings from "Orion,"* 50–57. Corvallis: Oregon State University Press, 2012.

———. *Sky Time in Gray's River: Living for Keeps in a Forgotten Place.* Boston: Houghton Mifflin Harcourt, 2007.

———. *Walking the High Ridge: Life as Field Trip.* Minneapolis: Milkweed Editions, 2000.

Ramey, James. "Parasitism and *Pale Fire's* Camouflage: The King-Bot, the Crown Jewels and the Man in the Brown Macintosh." *Comparative Literature Studies* 41 (2004): 185–213.

Reid, R. G. B. *Biological Emergences: Evolution by Natural Experiment.* Cambridge: MIT Press, 2007.

Riley, Norman D. *A Field Guide to the Butterflies of the West Indies.* London: Collins, 1975.

Ritland, D., and L. Brower. 1991. "The Viceroy Butterfly Is Not a Batesian Mimic." *Nature* 350 (1991): 497–98.

Rittner, Don, ed. *Pine Bush—Albany's Last Frontier.* Albany, NY: Pine Bush Historic Preservation Project, 1976.

Rowe, W. W. *Nabokov and Others: Patterns in Russian Literature.* Ann Arbor, MI: Ardis, 1979.

Ryan, Frank P. *The Mystery of Metamorphosis: A Scientific Detective Story.* White River Junction, VT: Chelsea Green, 2011.

Sapp, Jan. *The New Foundations of Evolution: On the Tree of Life.* New York: Oxford University Press, 2009.

Sartori, M., ed. *Les Papillons de Nabokov* [The butterflies of Nabokov]. Lausanne: Musée cantonal de Zoologie, Lausanne, 1993.

Schuh, Randall T., and Andrew V. Z. Brower. 2009. *Biological Systematics: Principles and Applications.* 2nd ed. Ithaca, NY: Cornell University Press, 2009.

Schwanwitsch, B. N. "Evolution of the Wing-Pattern in the Lycaenid Lepidoptera." *Proceedings of the Zoological Society of London* 119 (1949): 189–263.

———. "On the Groundplan of the Wing-Pattern in Nymphalids and Certain Other Families of Rhopalocerous Lepidoptera." *Proceedings of the Zoological Society of London,* ser. B, 34 (1924): 509–28.

Scudder, Samuel Hubbard. *The Butterflies of the Eastern United States and Canada, with Special Reference to New England.* 3 vols. Cambridge, MA: Published by the author, 1889.

Senderovich, Savely, and Yelena Shvarts. "The Juice of Three Oranges: An Exploration in Nabokov's Language and World." *Nabokov Studies* 6 (2000–2001): 75–124.

Shapiro, Arthur M. "Butterflies and Skippers of New York State." *Search* 4, no. 3 (1974): 21.

———. 1970. "Butterfly Mysteries at McLean." *Cornell Plantations* 26, no. 1 (1970): 3–6.

Shapiro, Gavriel. *The Sublime Artist's Studio: Nabokov and Painting.* Evanston, IL: Northwestern University Press, 2009.

Shapiro, James A. *Evolution: A View from the 21st Century.* Upper Saddle River, NJ: FT Press Science, 2011.

Simpson, George G. *The Meaning of Evolution.* New Haven: Yale University Press, 1949.

Smith, David Spencer, Lee D. Miller, and Jacqueline Y. Miller. *The Butterflies of the West Indies and South Florida.* Oxford: Oxford University Press, 1994.

Snow, C. P. *The Two Cultures.* Cambridge: Cambridge University Press, 1959.

Snyder, Gary. "What You Should Know to Be a Poet." In Gary Snyder, *Regarding Wave,* 40. New York: New Directions, 1970.

Sturtevant, A. H. "*Drosophila pseudoobscura.*" Review of *Contributions to the Genetics, Taxonomy, and Ecology of Drosophila pseudoobscura and Its Relatives* by Th. Dobzhansky and Carl Epling. *Ecology* 25 (1944): 476–77.

Süffert, F. "Zur vergleichende Analyse der Schmetterlingszeichnung." *Biologisches Zentralblatt* 47 (1927): 385–413.

Swanson, Henry F. *20 Years of Butterfly Revelations.* Winter Park, FL: Presbyterian Women, First Presbyterian Church, 1998.

Talavera, Gerard, et al. 2012. "Establishing Criteria for Higher-Level Classification Using Molecular Data: The Systematics of *Polyommatus* Blue Butterflies (Lepidoptera, Lycaenidae)." *Cladistics* 29:166–92.

Tammi, Pekka. *Russian Subtexts in Nabokov's Fiction: Four Essays.* Tampere, Finland: Tampere University Press, 1999.

——. "The St. Petersburg Text and Its Nabokovian Texture." *Cycnos* 10 (1993): 123–33. http://revel.unice.fr/cycnos/index.html?id=1311.

Taylor, Robert. "Nabokov Exhibition at Harvard Shows Off His Other Passion: Butterflies." *Boston Globe,* January 29, 1988.

Todd, Rebecca M., et al. "Affect-Biased Attention as Emotion Regulation." *Trends in Cognitive Sciences* 16 (2012): 365–72.

Turner, J. R. G. "Mimicry: The Palatability Spectrum and Its Consequences." In R. I. Vane-Wright and P. R. Ackery, eds., *The Biology of Butterflies,* 141–61. London: Academic Press, 1984.

Vascular Plants of the Great Smoky Mountains: A Checklist. Gatlinburg, TN: Great Smoky Mountains Association, 2008.

Vila, R., et al. "Phylogeny and Palaeoecology of *Polyommatus* Blue Butterflies Show Beringia Was a Climate-Regulated Gateway to the New World." *Proceedings of the Royal Society B: Biological Sciences* 278 (2011): 2737–44.

West Virginia Atlas and Gazetteer. Yarmouth, ME: DeLorme, 1997.

White, Peter, et al. *Wildflowers of the Smokies.* Gatlinburg, TN: Great Smoky Mountains Association, 1996.

Whitehead, Alfred North. *Alfred North Whitehead: An Anthology.* Selected by F. S. C. Northrop and Mason W. Gross. New York: MacMillan, 1953.

——. *Essays in Science and Philosophy.* New York: Philosophical Library, 1948.

——. *Science and the Modern World.* Orig. publ. 1926. Reprint, New York: New American Library, 1962.

Wimsatt, W. K., and Monroe Beardsley. "The Intentional Fallacy." *Sewanee Review* 54 (1946): 468–88.

Zachos, J., et al. "Trends, Rhythms, and Aberrations in Global Climate 65 Ma to Present." *Science* 292 (2001): 686–93.

Zaleski, Philip. "Nabokov's Blue Period." *Harvard Magazine,* July–August 1986, 36–38.

Zimmer, Carl. "Nabokov 2.0: Expanded Story, plus Reactions." *The Loom* (blog), January 31, 2011. http://blogs.discovermagazine.com/loom/2011/01/31/nabokov-2-0-expanded-story-plus-reactions/#.VRMhg_nF9x4.

——. "Nonfiction: Nabokov Theory on Butterfly Evolution Is Vindicated." *New York Times,* January 25, 2011.

——. "Vladimir Nabokov, Evolutionary Biologist Extraordinaire." The Loom (blog), January 25, 2011. http://blogs.discovermagazine.com/loom/2011/01/25/vladimir-nabokov-evolutionarybiologist-extraordinaire/.

Zimmer, Dieter E. *A Guide to Nabokov's Butterflies and Moths.* Hamburg: Privately printed, 2001. Revised online edition, 2012, http://www.d-e-zimmer.de/.

——. "A Chronology of *Lolita*." 2012. http://www.dezimmer.net/LolitaUSA/LoChrono.htm.

——. *Nabokov reist im Traum in das Innere Asiens* [Nabokov travels in a dream to the interior of Asia]. Reinbek: Rowohlt, 2006.

——. "Nabokov's Whereabouts." 2008. http://www.dezimmer.net/HTML/whereabouts.htm.

Zimmer, Dieter E., and Sabine Hartmann. "The Amazing Music of Truth: Nabokov's Sources for Godunov's Central Asian Travels in *The Gift*." *Nabokov Studies* 7 (2002–3): 33–74.

Contributors

Victoria N. Alexander, Director, Dactyl Foundation and Public Scholar for the New York Council for the Humanities, is a novelist (*Smoking Hopes, Naked Singularity, Trixie,* and *Locus Amœnus*) and a philosopher of science, whose *The Biologist's Mistress: Rethinking Self-Organization in Art, Literature and Nature* and numerous articles on biosemiotics and evolution are all inspired by Nabokov's writings on chance.

Dubi Benyamini is President of the Israeli Lepidopterists Society and a world authority of the biology of the Blues (butterfly family Lycaenidae) of the southern South American Andes and Patagonia. He has discovered more than thirty species new to science. He is the author of two books and more than two hundred scientific papers on butterflies.

Andrew Berry is Research Associate, Museum of Comparative Zoology, and Lecturer, Department of Organismic and Evolutionary Biology, Harvard University. He is a historian of science specializing in evolutionary biology, and his books include *Infinite Tropics: An Alfred Russel Wallace Anthology* and, with James D. Watson, *DNA: The Secret of Life.*

Stephen H. Blackwell, Professor of Russian at the University of Tennessee, has written many articles and two books on Nabokov, including *The Quill and the Scalpel: Nabokov's Art and the Worlds of Science.* He is currently working on a study of trees in Nabokov, and on the relation of aesthetics to epistemology in literature. He is a past president of the International Vladimir Nabokov Society.

Brian Boyd, University Distinguished Professor (English), University of Auckland, has writ-ten many books and hundreds of articles on Nabokov and has edited many volumes of his works, including *Nabokov's Butterflies.* He has also written on authors from Homer to Art Spiegelman, on literature, art, evolution, and cognition, and on the arts and sciences.

Thomas Dai received a BA in biology from Harvard University in 2014. His senior thesis in Naomi Pierce's laboratory was on the androconial organs of lycaenid butterflies in the tribe Eumaeini, and he prepared the data and design to map the geographic distribution of specimens collected by Vladimir Nabokov in the Museum of Comparative Zoology, Harvard University.

Robert Dirig has studied Toothwort Whites since 1966. He researched Nabokov's butterfly specimens at the Cornell University Insect Collection and has worked as a curator in the Cornell herbaria. His many publications on Lepidoptera, plants, and lichens (including twenty featuring Nabokov and the Karner Blue) frequently blend science, history, and art.

Rodney Eastwood is Curator of the ETH Zürich Entomological Collection at the Institute for Agricultural Sciences in Zurich, Switzerland. He conducted research on lycaenid butterflies as a Fulbright Fellow in Naomi Pierce's laboratory and oversaw the digitization of the butterfly collection, including Nabokov's specimens, while working for Collections Operations at the Museum of Comparative Zoology, Harvard University.

Victor Fet (Marshall University) is a poet and a zoologist who has written on Nabokov since 2003. He has translated from the Russian *Symbiogenesis: A New Principle of Evolution* (co-edited with Lynn Margulis) and *The Dawn of Human*

Genetics, has published four books of poetry in Russian, and is the co-author of *Catalog of the Scorpions of the World.*

James Fordyce is an evolutionary ecologist at the University of Tennessee. He has worked on topics ranging from community assemblages of tropical butterflies to chemically mediated trophic interactions while maintaining a keen interest in the Blues of western North America. His recent work aims at understanding how ecological factors facilitate evolutionary divergence in butterflies.

Matt Forister is an ecologist and Associate Professor at the University of Nevada, Reno. He has studied and published papers on lizards and rats and bacteria but prefers above all little blue butterflies on the tops of big blue mountains. He is particularly interested in trophic interactions and the evolution diet breadth.

Kurt Johnson is coauthor of *Nabokov's Blues* (1999) and more than two hundred articles on Lepidoptera. He is currently associated with the Florida State Collection of Arthropods, McGuire Center for Lepidoptera and Biodiversity, University of Florida, Gainesville, and Forum 21 Institute at the United Nations. He also specializes in comparative culture and religion and is coauthor of the influential *The Coming Interspiritual Age* (2013) and *Ethics, Values and the New UN Development Agenda* (2015).

Lauren Lucas is a lecturer and research scientist in the Department of Biology at Utah State University. She has written more than a dozen publications on hybridization and morphological evolution in *Lycaeides* butterflies and conservation genetics of animals endemic to freshwater springs. She is also interested in developing best practices in biology pedagogy.

James Mallet is Distinguished Lecturer at Harvard University and Emeritus Professor at University College London. He is the author of more than two hundred publications on evolution, particularly on warning color, mimicry, and speciation in butterflies, with some on the history of evo-

lution. In 2009 he received the Darwin-Wallace Medal from the Linnean Society of London.

Chris Nice is an evolutionary ecologist and Professor of Biology at Texas State University. His publications include papers on the ecology, evolution, genetics, and behavior of butterflies and other organisms. He and collaborators have recently used genomic data to investigate the consequences of hybridization in butterflies.

Naomi E. Pierce is Curator of Lepidoptera, Museum of Comparative Zoology, and Hessel Professor of Biology, Department of Organismic and Evolutionary Biology, Harvard University. She studies behavioral ecology and evolution, focusing on species interactions, life history evolution, and systematics of Lepidoptera. Her interest in the evolution of lycaenid butterflies led her to test Nabokov's hypotheses about the origin of Neotropical *Polyommatus* Blues.

Robert Michael Pyle is a butterfly biogeographer and professional writer. His eighteen books include *Evolution of the Genus* Iris: *Poems, Mariposa Road, Chasing Monarchs, Wintergreen,* and several standard butterfly works. He co-edited *Nabokov's Butterflies* and has investigated Nabokov in Colorado. A Guggenheim Fellow, Pyle founded the Xerces Society for Invertebrate Conservation.

Dorion Sagan's stories and essays have appeared in *After Hours, Cabinet, Wired, Natural History,* and the *Smithsonian,* among others. His coauthored books include *What Is Life?, Microcosmos, Death and Sex, Dazzle Gradually,* and *Into the Cool.* His interests include the interrelationships among art, philosophy, culture, science, and the history of science.

Roger Vila is CSIC Scientist at Institute of Evolutionary Biology (CSIC-UPF) in Barcelona, Spain, where he leads the Butterfly Diversity and Evolution Lab. He has been a passionate lepidopterist since childhood, obtained a PhD in biochemistry, and tested Nabokov's hypotheses on butterflies as a postdoctoral researcher at Harvard University.

Illustration Credits

With the exception of the inscription butterflies drawn for Véra, all drawings by Nabokov are reproduced by permission of the Vladimir Nabokov Estate and of the Henry A. and Albert W. Berg Collection of English and American Literature, The New York Public Library, Astor, Lenox, and Tilden Foundations. Inscription drawings are reproduced with the kind permission of Glenn Horowitz Bookseller and the Vladimir Nabokov Estate. Plates to the essays are the property of the authors.

Abbreviations for Berg Collection locations:

BA: Butterflies of America [seven boxes]

BMisc: Butterflies. Miscellaneous notes [seven boxes]

SWP: Butterflies. Scrapbook, containing diagrams and notes on the evolution of wing patterns

Note: Due to the storage practice for these materials at the Berg Collection, card locations are approximate, and not guaranteed to be permanent. Positions "a" and "b" are on the front or *recto* of the card holders; "c" and "d" on the *verso.*

Black and White Figures

Figure 1 [*cormion, coridon*] BA1-55d

Figure 2 [*cormion, coridon*] BA1-56c

Figure 3 [*cormion, coridon*] BA1-59d

Figure 4 [*menalcas*] BA1-11c

Figure 5 [*iphigenides*] BA1-11d

Figure 6 [*admetus*] BA1-18d

Figure 7 [*agestis*] BA1-39c

Figure 8 [*agestis*] BA1-125d

Figure 9 [fifty-four specimens] BA1-40a

Figure 10 [fifty-four specimens, continued] BA1-42c

Figure 11 [*cramera*] BA1-42d

Figure 12 [*semiargus*] BA1-47d

Figure 13 [*vogelii*] BA1-51d

Figure 14 [*Iolana*] BA1-76c

Figure 15 [*helena*] BA1-119d

Figure 16 [*hyrcana*] BA1-123d

Figure 17 [*lycormas*] BA1-73c

Figure 18 [*persephatta*] BA1-49b

Figure 19 [*icarus*] BA1-60d

Figure 20 [dorsal genital terminus comparative, Old World Blues] BA5-110c

Figure 21 [*argyrognomon, idas*] BMisc5-11d

Figure 22 [*argyrognomon, idas*] BMisc5-15c

Figure 23 [*piasus*] BA7-92c

Figure 24 [*cleotas*] BA7-130d

Figure 25 [furci of *Plebejus, Lycaeides*] BA1-26d

Figure 26 [*Turanana*] BA1-68d

Figure 27 [*Glaucopsyche* wing pattern] BA1-96c

Figure 28 [*oro*] BA1-70d

Figure 29 [*laetifica*] BA1-74d

Figure 30 [*shasta*] BA1-139c

Figure 31 [*shasta*] BA1-140a

Figure 32 [*icarioides*] BA1-140d

Figure 33 [*neurona*] BA1-144c

Figure 34 [*rita*] BA1-84d

Figure 35 [*idas/longinus* hybrid] BMisc3-49b

Figure 36 [*melissa*] BMisc3-50c

Figure 37 [*anna*] BMisc3-54a

Figure 38 [unlabeled] BMisc3-92a

Figure 39 [*melissa*] BMisc3-99a

Figure 40 [*samuelis*] BMisc2-28a

Figure 41 [*samuelis*] BA5-13d

Figure 42 [*melissa*] BA7-61d

Figure 43 [*melissa*] BMisc5-5d

Figure 44 [*Cyclargus*] BA1-96c

Figure 45 [*Cyclargus*] BA7-40d

Figure 46 [*thomasi, erembis*] BA1-117c

Figure 47 [*bethune-bakeri*] BA7-41c
Figure 48 [*dominica*] BA7-45d
Figure 49 [*martha*] BA2-1c
Figure 50 [*isola*] BA2-1d
Figure 51 [*isola*] BA7-34c
Figure 52 [*isola*] BA7-34d
Figure 53 [*huntingtoni*] BA1-149b
Figure 54 [*huntingtoni*] BA2-2a
Figure 55 [*huntingtoni*] BA2-2b
Figure 56 [*Hemiargus*, redefinition] BA1-149a
Figure 57 [*Hemiargus*, redefinition] BA7-13c
Figure 58 [*Hemiargus*, redefinition] BA7-13d
Figure 59 [*Hemiargus*, redefinition] BA7-17c
Figure 60 [*hanno*, neotype] BA7-22a
Figure 61 [*Hemiargus*, redefinition] BA7-27c
Figure 62 [*hanno*] BA7-104a
Figure 63 [*inconspicua*] BMisc5-7b
Figure 64 [*inconspicua*] BMisc5-9c
Figure 65 [*chilensis*] BA7-33d
Figure 66 [*chilensis*] BA7-31d
Figure 67 [*titicaca*] BA7-89c
Figure 68 [*speciosa*] BA7-89d
Figure 69 [*speciosa*] BA7-91d
Figure 70 [*speciosa*] BA7-91c
Figure 71 [*maniola*] BA7-9b

Figure 72 [*dorothea*] BA6-10a
Figure 73 [*barbouri*] BA7-102b
Figure 74 [*faga*] BA7-103b
Figure 75 [*koa*] BA7-114c
Figure 76 [*koa*] BA7-125d
Figure 77 [*babhru*] BA7-115d
Figure 78 [*bornoi*] BA7-120d
Figure 79 [locality map] BA2-folder 2
Figure 80 [phylogeny] BA1-59c
Figure 81 [*Rassenkreis* circle of races] BA2-folder 1
Figure 82 [geographic and phylogenetic relationships] BA3-15d
Figure 83 [geographic and phylogenetic relationships] BA3-15c
Figure 84 [transformation series, *Lycaeides*] Scrapbook—Genitalia
Figure 85 [*collina*] BA7-51c
Figure 86 [*candidus*] BA7-61a
Figure 87 [magic triangles] BMisc3-111a
Figure 88 [comparative wing maculation] BA2-42d
Figure 89 [comparative wing maculation] SWP
Figure 90 [wing maculation schematic] SWP
Figure 91 [wing maculation schematic] SWP
Figure 92 [measurement chart] BA1-139c

Color Plates

Color Plate 1 [*cormion, coridon*] BA1-47a
Color Plate 2 [*pheretes*] BA1-7a
Color Plate 3 [*posthumus*] BA1-8c
Color Plate 4 [*phyllis*] BA1-10c
Color Plate 5 [*iphigenides*] BA1-12c
Color Plate 6 [*iphidamon*] BA1-13c
Color Plate 7 [*hoppferi*] BA1-14a
Color Plate 8 [*glaucias*] BA1-16b
Color Plate 9 [*nivescens*] BA1-14b
Color Plate 10 [*bellargus*] BA1-17d
Color Plate 11 [*admetus*] BA1-20c
Color Plate 12 [*argus*] BA1-31c
Color Plate 13 [*argus*] BA1-31c
Color Plate 14 [*chinensis*] BA1-38c
Color Plate 15 [*chinensis*] BA1-130a
Color Plate 16 [*miris*] BA1-122d
Color Plate 17 [*argyrognomon, insularis*] BA2-115c
Color Plate 18 [*ismenias* subspecies] BMisc4-4b
Color Plate 19 [*thius*] BA7-63a
Color Plate 20 [*Lucia*] BA7-66c
Color Plate 21 [*Scolitantides*] BA7-94c
Color Plate 22 [*divina*] BA7-95d
Color Plate 23 [*Brephidium*] BA7-100b

Color Plate 24 [*Agriades*] BA7-110d
Color Plate 25 [*Agrodiaetus*] BA7-111a
Color Plate 26 [*felicis*] BA7-111d
Color Plate 27 [*sonorensis*] BA1-88c
Color Plate 28 [*ferniensis*] BMisc3-56a
Color Plate 29 [*idas, argyrognomon*] BMisc3-76a
Color Plate 30 [*anna*] BMisc5-6d
Color Plate 31 [*samuelis*] SWP
Color Plate 32 [*thomasi*] BA7-40c
Color Plate 33 [*martha*] BA1-148d
Color Plate 34 [*gyas*] BA7-17d
Color Plate 35 [*chilensis*] BA7-84b
Color Plate 36 [*titicaca*] BA7-133d
Color Plate 37 [magic triangles] BMisc3-75d
Color Plate 38 [various wings, butterflies, and schematics] SWP
Color Plate 39 [various wings, butterflies, and schematics] SWP
Color Plate 40 [various wings, butterflies, and schematics] SWP
Color Plate 41 [various wings, butterflies, and schematics] SWP
Color Plate 42 [wing scales, *samuelis*] SWP

Color Plate 43 [*cleobis*] SWP

Color Plate 44 [*boeticus*] SWP

Color Plate 45 [maculation comparative] SWP

Color Plate 46 [wing cell comparative] SWP

Color Plate 47 [wing sector, *Lycaeides*] SWP

Color Plate 48 [wing sector, *samuelis*] SWP

Color Plate 49 [wing sectors, *melissa*] SWP

Color Plate 50 [wing sector comparative] SWP

Color Plate 51 [maculation study, *icarioides*] SWP

Color Plate 52 [wing cross section] SWP

Color Plate 53 [wing scale comparative] SWP

Color Plate 54 [wing scale comparative] SWP

Color Plate 55 [*argyrognomon, ismenias*] SWP

Color Plate 56 [maculation comparative] SWP

Color Plate 57 ["*Vanessa verae*"] *Véra's Butterflies* 24

Color Plate 58 ["*Vanessa incognita*"] *Véra's Butterflies* 12

Color Plate 59 ["*Colias lolita* Nab."] *Véra's Butterflies* 5

Color Plate 60 ["*Paradisia radugaleta*"] *Véra's Butterflies* 44

Color Plate 61 ["*Maculinea aurora* Nab." (male)] *Véra's Butterflies* 10

Color Plate 62 ["*Polygonia thaisoides* Nab."] *Véra's Butterflies* 269

Color Plates to Essays

Color Plate E1 Photographs by Dubi Benyamini

Color Plate E2 Photograph by Lauren K. Lucas

Color Plate E3 Photograph by Lauren K. Lucas using a Leica microscope and camera

Color Plate E4 Illustration created by Zachariah Gompert with the software environment R

Color Plate E5 Photograph by James Mallet

Color Plate E6 Photograph by Gabriel A. Miller; inset, photograph by Rodney Eastwood

Color Plate E7 Photographs by Gabriel A. Miller

Color Plate E8 Photographs by Gabriel A. Miller; map prepared by Scott Walker, Digital Cartography Department, Pusey Library, Harvard University

Color Plate E9 Photographs by Robert Dirig

Color Plate E10 Photographs by Robert Dirig, courtesy of the Cornell University Insect Collection

Color Plate E11 Photographs by Robert Dirig

Color Plate E12 Photographs by Robert Dirig

Color Plate E13 Photographs by Robert Dirig, except specimen photographs, which are courtesy of the Carnegie Museum of Natural History, Pittsburgh

Index

Page numbers in *italics* refer to illustrations.